Supercritical Fluid Chromatography with Packed Columns

CHROMATOGRAPHIC SCIENCE SERIES

A Series of Monographs

Editor: JACK CAZES
Cherry Hill, New Jersey

ADDITIONAL VOLUMES IN PREPARATION

Supercritical Fluid Chromatography with Packed Columns

Techniques and Applications

edited by

Klaus Anton
Novartis Pharma AG
Basel, Switzerland

Claire Berger
Ciba Specialty Chemicals, Inc.
Basel, Switzerland

Marcel Dekker, Inc. New York · Basel · Hong Kong

Library of Congress Cataloging-in-Publication Data

Supercritical fluid chromatography with packed columns : techniques and
applications / edited by Klaus Anton, Claire Berger
    p. cm. – (Chromatographic science ; v. 75)
    Includes bibliographical references and index.
    ISBN 0-8247-0013-9 (hardcover : alk. paper)
    1. Supercritical fluid chromatography. I. Anton, Klaus. II. Berger, Claire.
III. Series.
QD79.C45S85 1997
543'.0896–dc21

                                        97-34426
                                              CIP

Cover figure: chromatograms of samples from an epoxy resin manufacturing process. I = injections; a_t = separation time; b_t = recovery time. Chromatographic conditions: column, 100 mm × 2 mm (Spherisorb CN, 3 μm): mobile phase, CO_2 (2.5 mL/min); modifier, methanol (0.275 mL/min); temperature, 80°C; pressure program, 125–250 bar in 2 min; detector, UV at 205 nm. From A. Giorgetti, N. Periclès, H. M. Widmer, K. Anton, and P. Dätwyler, *J. Chromatogr. Sci.*, *27*: 318 (1989). Reproduced from the *J. Chromatogr. Sci.* by permission of Preston Publications, A Division of Preston Industries, Inc.

The publisher offers discounts on this book when ordered in bulk quantities. For more information, write to Special Sales/Professional Marketing at the address below.

This book is printed on acid-free paper.

MARCEL DEKKER, INC.
270 Madison Avenue, New York, New York 10016
http://www.dekker.com

Current printing (last digit):
10 9 8 7 6 5 4 3 2 1

PRINTED IN THE UNITED STATES OF AMERICA

Preface

During the last five years, packed column supercritical fluid chromatography (pSFC) has achieved maturity. It can now be considered a valuable normal phase type selectivity complementary chromatographic method in the major fields of the industry. Since its introduction in 1962 by Klesper, Corwin, and Turner, SFC has experienced several ups and downs in its development.

Until the early 1980s, progress in the field, as reflected in the scientific literature, was limited to fewer than five reports annually. The development of the technique was clearly limited by instrument and experimental difficulties due to the high temperatures and pressures needed to maintain the mobile phase in a supercritical state.

However Klesper, Corwin, and Turner, while demonstrating the feasibility of using supercritical fluids as chromatographic mobile phase, developed the first historical pSFC application, the separation of nickel porphyrin isomers. They continued to develop pSFC by applying the technique to the analysis of polystyrene oligomers. Subsequent work by Giddings and coworkers and Sie and Rijnders were also carried out on packed columns. Progress in pSFC prior to 1978 has been extensively reviewed by Klesper.

During the 1980s, pSFC was put on hold, whereas capillary SFC (cSFC) experienced an explosive growth mainly due to the novel combination of supercritical mobile phase with open tubular fused silica capillary column technology introduced by Novotny, Lee, and coworkers in 1981. The growth was also due to the introduction of reliable commercial instrumentation for capillary SFC in 1986.

SFC on packed columns underwent a renaissance in interest at the end of the 1980s, when the limitations of cSFC became obvious and important progress in pressure gradient techniques of mixed mobile phases was achieved for pSFC.

Taking advantage of the numerous ameliorations of high-pressure chromatographic instrumentation that were primarily achieved in high-performance liquid chromatography (HPLC), pSFC was developed on a new basis in the early 1990s. Sophisticated commercial instruments allowing independent flow control under pressure gradient came out in 1992, boosting the development of applications in every major field of the industry.

Therefore, it seemed timely to collect in book form a variety of representative new developments and applications covering the main fields of the industry. Fifteen chapters, written by experts in the field, summarize the important practical aspects from an analytical to industrial preparative scale of pSFC:

Solubility is a key parameter in pSFC. The reader will find in the first chapter a comprehensive method to determine solubility data prior to developing a new application.

The second chapter discusses some basic principles that have been decisive for the development of today's high-performance instrumentation.

The third chapter reflects the state of the art regarding instrumentation combined with various applications in different fields.

SFC on packed columns was first developed with UV-visible detection and was therefore claimed as being unsuitable for molecules without chromophore. This problem can now be circumvented by using evaporative light-scattering detection as demonstrated in the fourth chapter.

The fifth chapter deals with packed capillary columns, as an innovative way to expand the frontiers of pSFC toward cSFC. As in cSFC, high efficiencies are achieved and universal flame ionization detection (FID) can be used.

The numerous ways to adjust selectivity are characteristic for pSFC. The authors of the sixth chapter show how an intelligent coupling of different stationary phases can be used to resolve complex mixtures containing compounds covering a large polarity range.

Another innovative way to expand the application field of pSFC is to use subcritical mobile phases. This method discussed in the seventh chapter is particularly attractive for high molecular weight thermolabile compounds.

SFC on packed columns has been shown to have a huge potential for separation of drugs and in particular chiral molecules. The reader will find in the eighth and ninth chapters comprehensive reviews on chiral packed Subcritical (SubFC) and SFC.

A special focus on applications of pSFC in the pharmaceutical industry will be given in the tenth chapter. SFC on packed columns can easily replace HPLC in a lot of applications. The technique has been included in the Swiss pharmacopoeia in 1994 and is therefore officially recognized as an appropriate method for the analysis of drug substances and products.

The potential of pSFC in the development of polymer additives is shown in the eleventh chapter where real world examples as well as an extended review of pSFC separations, including a table with references for more than 50 commonly used polymer additives, are presented.

SFC on packed columns was applied for separation of polymers and oligomers since the very beginning. pSFC still appears as a powerful method in the analysis of polymers and oligomers. The twelfth chapter describes the use of pSFC as a fractionation method.

Supercritical fluids possess excellent characteristics for application in environmental analysis. They are used both in sample preparation and in chromatographic analysis as illustrated in the thirteenth chapter.

The benefit of pSFC for the resolution of complex mixtures can also be used at a process scale as described in the two final chapters. The authors emphasize the economic and ecological parameters in the breakthrough of industrial preparative scale chromatography.

Finally, a detailed compound and subject index lists all the important keywords in text, including figures, and figure captions.

We believe that the significance of pSFC for industrial separations will grow in the next years under the strong pressure of regulatory policies regarding environmental issues (in particular the use of solvents) and as a consequence of the ongoing rationalization efforts in every field of the industry. Therefore it is our goal in presenting the information in a collected form to help readers in their practical daily work by giving an updated, praxis-oriented overview of recent developments and applications of pSFC.

We are grateful to all contributors to this volume, without whose continuous effort this work could not have been completed. We address special thanks to our industrial colleagues, who agreed to disclose applications, which in other companies are considered proprietary. We also thank Dr. E. Francotte of Novartis Pharma AG, for reviewing all parts dealing with chiral separations.

Klaus Anton
Claire Berger

Contents

Contributors

Klaus Anton Chemical and Analytical R&D, Novartis Pharma AG, Basel, Switzerland

Monique Bach Chemical and Analytical R&D, Novartis Pharma AG, Basel, Switzerland

Danielle Barth Laboratoire de Thermodynamique des Séparations, Ecole Nationale Supérieure des Industries Chimiques, Nancy, France

Claire Berger Additives Division, Ciba Specialty Chemicals, Inc., Basel, Switzerland

Terry A. Berger Berger Instruments, Newark, Delaware

Gerd Brunner Thermische Verfahrenstechnik, Technical University of Hamburg-Harburg, Hamburg, Germany

Keith Coleman Anachem Ltd., Luton, Bedfordshire, United Kingdom

Frank David Research Institute for Chromatography, Kortrijk, Belgium

Olle Gyllenhaal Analytical Chemistry, Pharmaceutical R&D, Astra Hässle AB, Mölndal, Sweden

Monika Johannsen High Pressure Center Kaiseraugst, F. Hoffmann–La Roche Ltd., Basel, Switzerland

Pascal Jusforgues G.C. and S.F.C. Preparative Chromatography Division, Prochrom R&D, Champigneulles, France

Anders Karlsson Analytical Chemistry, Pharmaceutical R&D, Astra Hässle AB, Mölndal, Sweden

Agata Kot Department of Analytical Chemistry, Technical University of Gdańsk, Gdańsk, Poland

Milton L. Lee Department of Chemistry and Biochemistry, Brigham Young University, Provo, Utah

Peter Lembke Research and Development, K.D.-Pharma GmbH, Bexbach, Germany

Eric Lesellier L.E.T.I.A.M., University Institute of Technology, Orsay, France

Andrei Medvedovici Department of Analytical Chemistry, University of Bucharest, Bucharest, Romania

Nico Periclès Ciba Specialty Chemicals, Inc., Kaisten, Switzerland

William H. Pirkle School of Chemical Sciences, University of Illinois, Urbana, Illinois

Pat Sandra Department of Organic Chemistry, University of Ghent, Ghent, Belgium

Mohamed Shaimi Laboratoire de Thermodynamique des Séparations, Ecole Nationale Supérieure des Industries Chimiques, Nancy, France

Yufeng Shen Department of Chemistry and Biochemistry, Brigham Young University, Provo, Utah

Roger M. Smith Department of Chemistry, Loughborough University, Loughborough, Leicestershire, United Kingdom

J. Thompson B. Strode III Department of Chemistry, Virginia Tech, Blacksburg, Virginia

Larry T. Taylor Department of Chemistry, Virginia Tech, Blacksburg, Virginia

Alain Tchapla L.E.T.I.A.M., University Institute of Technology, Orsay, France

Koichi Ute Department of Chemistry, Osaka University, Osaka, Japan

Francis Vérillon Department of Marketing, Gilson S.A., Villiers le Bel, France

Jörgen Vessman Analytical Chemistry, Pharmaceutical R&D, Astra Hässle AB, Mölndal, Sweden

Christian Wolf School of Chemical Sciences, University of Illinois, Urbana, Illinois

1

Determination of Solubility Data with On-Line Packed Column Supercritical Fluid Chromatography

Monika Johannsen

F. Hoffmann–La Roche Ltd., Basel, Switzerland

Gerd Brunner

Technical University of Hamburg-Harburg, Hamburg, Germany

I. INTRODUCTION

During the last 30 years, interest in analytical and technical use of supercritical fluids has increased significantly. There are now numerous important applications in the fields of extraction (SFE), chromatography (SFC), and reactions (SFR) with supercritical fluids [1,2].

Extensive investigations on fundamental aspects, such as phase equilibria, are an essential part of the development of new applications, especially for preparative separations. The characteristic phase equilibria is a key element in the optimization for technical scale processes, which have a great impact on the economics. Separation costs may compose up to 50% of the total investment and running costs of a production process.

There is currently a need for rapid and simple methods for determining the solubilities of materials in neat and mixed (sub-/supercritical) fluids. Information on solubility simplifies the choice of conditions for SFE, SFR, and the optimization of analytical and preparative packed column SFC (pSFC) (see also Chapters 3, 14 and 15).

The determination of the phase behavior of low-volatile substances in fluids under high pressure is initially carried out experimentally. The experimental determination of phase equilibria, especially at extreme highs of pressure or temperature, is quite costly. Therefore, in general, it is desirable to reduce the number of experiments and to calculate the phase equilibria for different sets of conditions from selected experimental data. A review on calculations of phase equilibria using equations of state was recently presented [3].

For calculation of phase equilibrium, some experimental data have to be measured in order to prove the accuracy of the calculation. Even today, for complicated mixtures calculations are difficult and often do not correspond sufficiently to experimental data.

Equilibrium solubilities in neat or mixed (sub-/supercritical) fluids are experimentally determined by dynamic or by static equilibrium methods with off-line or on-line analysis [4–8]. Two extensive reviews about systems for which high-pressure solubility data have been published in recent years are given in Refs. 8 and 9. In any case, different methods for rough estimation of solubilities from chromatographic data are known [10,11]. The use of retention measurements in SFC to estimate solubilities has been described [12,13].

In a static method measurement, both phases remain in the equilibrium cell until such time as equilibrium is attained (see also Chapter 7). Samples of the equilibrated phase mixtures are then taken and analyzed for composition. In a dynamic method measurement, only the condensed phase remains in the cell, whereas the gaseous phase flows through it. Conditions of equilibrium must be carefully observed during these experiments. This problem can be solved by using the recirculation method, in which the gaseous flow recirculates through the cell so that equilibrium concentrations are easier to achieve [9,11].

On-line sampling minimizes problems occurring during sampling and sample transfer, and the results are directly available. A very small sample volume guarantees minimum disturbance of the equilibrium of the system during sampling. On-line sampling has been coupled with different chromatographic methods. Most frequently, the mobile phase used in SFC is supercritical carbon dioxide. Therefore it is attractive to couple a system for determination of solubilities in supercritical carbon dioxide with packed column SFC (pSFC) in order to make on-line measurements possible without any solvent change or pressure reduction. In this application it is less attractive to use capillary column SFC (cSFC) because the retention times are generally longer and modifier addition to the mobile phase cannot be used with the flame ionization detector (FID).

An experimental method will be shown to allow fast and accurate determination of equilibrium solubilities of compounds in neat or mixed (sub-/supercritical) fluids.

II. METHODS FOR MEASUREMENT OF SOLUBILITIES IN SUPERCRITICAL FLUIDS

In this section, an apparatus for the measurement of equilibrium solubilities with on-line coupling to a pSFC system is described. Due to different systems, especially different mixed (sub-/supercritical) fluids and different pressures, slightly different configurations of the apparatus are needed. The configurations used for the measurement of solubilities in pure carbon dioxide and in carbon dioxide/methanol mixtures at pressures in the range of 200–350 bar are described in detail in the following sections. Measurements of solubilities at pressures lower than 200 bar are discussed in a separate section.

Solubilities of natural compounds from two different product groups are presented in the application sections. Xanthines were investigated because of their special role in the decaffeination of coffee and tea. Fat-soluble vitamins occur naturally in vegetable oils and are mainly produced synthetically today. However, their isolation from natural sources, e.g., by SFE, is of increasing interest.

A. Measurement of Solubilities in Pure Carbon Dioxide at Pressures in the Range of 200–350 Bar

1. Experimental

(a) Apparatus. The apparatus used for measurement of equilibrium solubilities in pure carbon dioxide over a pressure range of 200–350 bar and temperatures between 25°C and 85°C and direct coupling with a supercritical fluid chromatographic system is shown in Fig. 1 [14–16]. A syringe pump serves to pressurize the equilibrium cell and the chromatographic system with liquid carbon dioxide. The pressure is controlled by a pressure gauge with an accuracy of ±1 bar. The cylinder (volume 260 mL) of the pump is cooled. After the syringe pump, the carbon dioxide flow is split into two parts, one serving the chromatographic system and the other the solubility apparatus. Both parts are protected by check valves and can be isolated by turning off the high-pressure valves. Temperature is maintained and controlled with a heating cabinet. By this method, temperature is regulated with an accuracy of ±1°C. Methanol used as modifier can be added with a high-performance liquid chromatographic (HPLC) pump for chromatographic elution. Mixing is achieved in a piece of hollow stainless steel tubing (1/8 in.), which also serves as a preheating coil for the mobile phase. The equilibrium cell consists of an empty stainless steel HPLC column (6 or 8 mL inner volume) equipped with frits at both ends. Samples are taken through a Rheodyne six-port valve (equipped with a 50- or 100-μL sample loop) or a Rheodyne four-port valve (equipped with an internal 1-μL loop). A syringe adapter is connected for calibration. Samples are separated on an HPLC packed column with silica gel material, which is suitable in most cases. Eluted solutes are detected with ultraviolet-visible (UV-vis) absorbance detection at a suitable wavelength (e.g., xanthines at 272 nm, tocopherols at 295 nm, etc.) and chromatograms are recorded on a computer.

For the determination of pure component solubilities, the chromatographic conditions are very simple. In most cases it is possible to use a silica gel analytical column. A chromatographic column is necessary for analysis even if there are only one-component systems being considered. This is a consequence of the

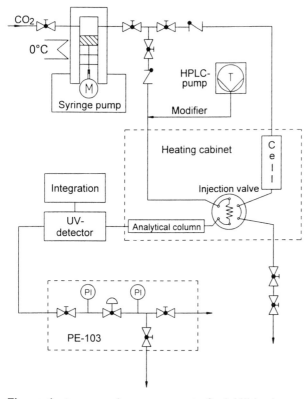

Figure 1 Apparatus for measurement of solubilities in pure carbon dioxide at a pressure in the range of 200–350 bar and temperatures between 25°C and 85°C [14–16].

high sensitivity of the detector, which is necessary for determining the low amounts of components in the samples. Every disturbance in the chromatographic flow, e.g., during injection, influences the baseline of the chromatogram. With a chromatographic column, the "injection pulse" and solvent are is separated from the substance peak.

A two-stage expansion of the fluid after detection is controlled by an expansion module (PE-103, NWA, Lörrach, Germany). In the first stage the pressure is reduced to 10 bar. Modifiers and dissolved compounds precipitate during expansion and are separated from the gas in a gravitational separator. The liquid formed by expansion is removed after certain time intervals. The gas flows to the second expansion, a mechanically operated gas flow controller that allows a constant flow of gas through a metering valve, independent of the pressure on the high-pressure side.

(b) Procedure. The equilibrium cell is loaded with the solid of interest and some stainless steel balls are added for sample mixing. Glasswool is used for retaining the solid in the equilibrium vessel. After installing the cell in the cabinet, the system is pressurized up to the desired conditions. The time for establishing equilibrium is set to 20 min, which has been proven sufficient in preliminary experiments. In the meantime, the chromatographic system is adjusted. Samples are loaded with the Rheodyne valve. The sample loop is flushed for 30 s at a low flow rate which is adjusted by a fine regulation valve. Afterward the sample is injected into the analytical column by switching the valve.

(c) Quantitative Evaluation. Solubilities are determined from elution peak areas in comparison with external calibration curves. These curves are obtained by injecting different standard solutions of known concentrations through the injection valve. The solubility in [g/mL] determined from calibration curves can be converted to solubility in [g/kg], respectively, mole fraction solubility with respect to the density of the neat fluid at corresponding conditions. The change in density of the neat fluid induced by the presence of the solute is neglected. Density of, for example, pure carbon dioxide is calculated from the Bender equation [17,18].

2. Applications

This section presents the results of the solubility measurements of natural compounds in supercritical carbon dioxide at temperatures of 40, 60, and 80°C and over a pressure range of 200–350 bar. The influence of pressure, density, and temperature is discussed. The solubility values presented are mean values of at least five samplings. Standard deviations of the measurements are within 5%.

(a) Influence of Density on the Solubility of Vitamin K_1 in Supercritical Carbon Dioxide. The experimental data of the solubility of vitamin K_1 in supercritical carbon dioxide at a temperature of 40, 60, and 80°C and a pressure in the range of 200–350 bar are presented in Fig. 2 as a function of the carbon dioxide density in a semilogarithmic scale. The solubility of vitamin K_1 measured under these conditions is in the range of 3.0–35 g/kg [14,19]. The lowest solubility was found at a temperature of 80°C and a pressure of 200 bar (density about 600 kg/m^3) and the highest one at a temperature of 40°C and 60°C and a pressure of 350 bar (density about 860 kg/m^3).

Two effects are observed: At constant temperature solubility increases with increasing density. This is based on increasing solvent power of carbon dioxide at higher density. The nearly linear relationship between the density of carbon dioxide and the logarithm of the solubility given by the correlation after Mitra and Wilson [20] was verified for vitamin K_1 in the region of pressure and temperature investigated. And at constant density (e.g., at 750 kg/m^3), a rise of

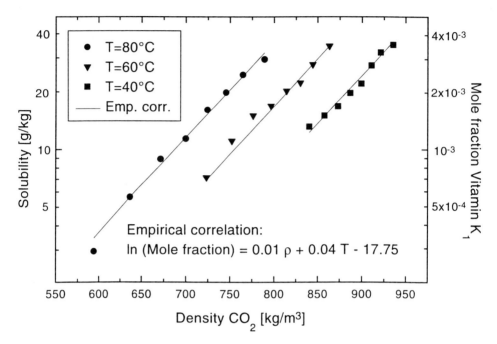

Figure 2 Solubility of vitamin K_1 in supercritical carbon dioxide as a function of density at different temperatures [14].

temperature results in a solubility increase. This is due to the increasing vapor pressure of the solid.

(b) Influence of Pressure on the Solubility of β-Carotene in Supercritical Carbon Dioxide. In Fig. 3, the solubility of β-carotene in supercritical carbon dioxide at 40, 60, and 80°C is shown as a function of pressure. The solubility of β-carotene under these conditions ranges from 1×10^{-3} g/kg ($T = 40°C$ and $P = 200$ bar) to 4×10^{-2} g/kg ($T = 80°C$ and $P = 320$ bar), [14,19,21–24]. The solubility of β-carotene was found to be three orders of magnitude lower than that of the other fat-soluble vitamins [14,19].

The 80°C isotherm (by a polynomial fit) is higher than the 60°C isotherm; the 40°C isotherm is the lowest. For all pressures investigated, solubility increases with increasing temperature. The two contrary effects are decisive: First, the solvent power of carbon dioxide decreases because of a decrease of density; second, solubility increases due to the increasing vapor pressure of β-carotene. In this case, the effect of the increasing vapor pressure predominates over the effect of the decreasing solvent power when temperature is increased.

(c) Influence of Temperature on the Solubility of Theobromine in Supercritical Carbon Dioxide. These two contrasting effects are also decisive concerning the

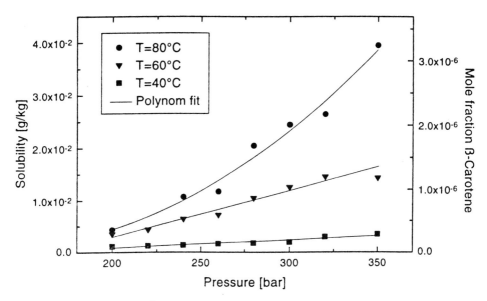

Figure 3 Solubility of β-carotene in supercritical carbon dioxide as a function of pressure at different temperatures [14].

influence of temperature on solubility. In Fig. 4, results of experimental measurements of the solubility of theobromine in supercritical carbon dioxide at a pressure in the range of 200–350 bar are shown as a function of temperature in a semilogarithmic scale. The solubility of theobromine is between 3.6 mg/kg ($T = 313$ K, $P = 210$ bar) and 19 mg/kg ($T = 353$ K, $P = 340$ bar) [14,15,25].

At every pressure in the range investigated the solubility increases with increasing temperature, although the carbon dioxide density decreases at the same time. Obviously, here the increase of vapor pressure of theobromine exerts a greater influence on solubility than the decrease of density of carbon dioxide. This behavior is similar to that seen for β-carotene in the previous section but contrasting to that found for vitamin K_1.

B. Measurement of Solubilities in Carbon Dioxide/Methanol Mixtures

The addition of modifiers such as methanol to supercritical carbon dioxide can significantly increase solubilities of polar organic solutes and selectivity of the separation technique [26–28].

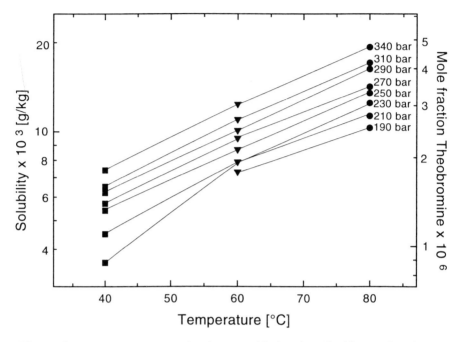

Figure 4 Solubility of theobromine in supercritical carbon dioxide as a function of temperature at different pressures [14].

All work with binary mobile phases assumes complete miscibility of the primary and the modifier fluid at the concentration of interest. Therefore, a knowledge of vapor–liquid equilibrium in the system is imperative. A large number of systems, e.g., carbon dioxide/methanol, have been reported in the literature, e.g. [29].

1. Experimental

(a) Mixing Device and Solubility Apparatus. Mixtures of accurately known compositions have to be prepared for solubility measurements in mixed (sub-/supercritical) fluids. In order to prepare homogeneous mixtures, the pumping system can be filled with premixed fluids. The system described before can be extended for that purpose. A simple method for the preparation of mixed mobile phases (especially gas and liquid mixtures) for SFC similar to the one shown by Fields [30,31] is described. The fluid mixing device used in this work, consisting of a vacuum pump, two stainless steel sample cylinders (60 and 300 mL, rated to 120 bar), two 3-way ball valves (one with a syringe adapter), and four quick-connects, is shown in Fig. 5 [14]. The packing material of the valve has to be carefully selected. Polytetrafluorethylene (PTFE) was found to be the most suitable material.

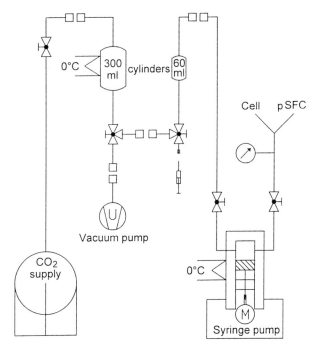

Figure 5 Mixing device for preparing mixed (sub-/supercritical) fluids of accurately known compositions [14].

The equilibrium cell and the chromatographic part of the apparatus used for measurement of solubilities in mixed (sub-/supercritical) fluids are almost unchanged compared to the apparatus for solubility measurement in neat fluids described in Sec. II.A.1 (Fig. 1). The HPLC pump used previously to add modifier in the chromatographic system is not needed any longer because modifier is already introduced in the cylinder of the syringe pump.

(b) Procedure. To prepare a mixed (sub-/supercritical) fluid, the mixing device, syringe pump, and all relevant lines are evacuated with a vacuum pump. The sample cylinders are then disassembled by means of quick-connects and weighed. After weighing on a balance to an accuracy of ±0.01 g, the two cylinders are installed and an appropriate amount of carbon dioxide is condensed into the large cooled cylinder within 30 min. The desired amount of modifier is injected into the small cylinder with a syringe through the syringe adapter. After that, both cylinders are weighed again. The mass of modifier added is then determined by difference. After reassembling the cylinders, methanol is sucked slowly into the pump cylinder by withdrawing the cylinder piston to the bottom. After purging the lines with carbon dioxide, the carbon dioxide is condensed

into the pump cylinder again within 30 min. An additional weighing of the sample cylinders provides actual mass of carbon dioxide transferred to the pump. After closing the syringe pump valve, the pump is pressurized up to the desired conditions and equilibrated for about 12 h overnight.

Afterward the procedure of solubility measurement in mixed (sub-/supercritical) fluids is analogous to the measurements in neat fluids (see Sec. II.A.1(b)).

(c) Quantitative Evaluation. Quantitation of solubilities in mixed (sub-/supercritical) fluids is made analogously to that in neat fluids (see Sec. II.A.1(c)). The concentration of the sample in [g/mL mixed fluid] is determined from peak areas with an external calibration curve. This value can be converted to the solubility in [g/kg mixed fluid] with respect to the density of the mixed fluid.

However, density of the mixed (sub-/supercritical) fluids is needed for calculation of the solubility values. Unfortunately, the densities of the binary mixtures are largely unknown. Because measured values are rarely published, the accuracy of predictions using modified equations of state is uncertain. Alternatively, density values of carbon dioxide/methanol mixtures have been calculated [14,32] by extrapolation of experimental data [33,34].

2. Applications

In the previous application section, it was shown how solubility is influenced by temperature, pressure, and density of the fluid. Furthermore, solubility depends on the composition of the fluid, on the modifier, and on the mixture proportions. Here, results of the solubility measurements of xanthines in supercritical carbon dioxide/methanol mixtures are shown.

(a) Influence of Fluid Composition on the Solubility of Xanthines in Carbon Dioxide/Methanol Mixtures. The influence of adding methanol to supercritical carbon dioxide on the solubilities of xanthines at 220 bar and 40°C is shown in Fig. 6 [14]. The conditions under which two phases are formed from carbon dioxide/methanol have been investigated, e.g. [29]. However, at 220 bar and 40°C, a single phase exists.

Solubility of caffeine in carbon dioxide/methanol mixtures varies between 1.5 g/kg carbon dioxide without methanol and 16 g/kg (with 11 mass% methanol). The solubility of theophylline ranges from 40 mg/kg in pure carbon dioxide up to 2.8 g/kg (with 15 mass% methanol). The solubility of theobromine varies from 4 mg/kg without modifier to 100 mg/kg (with 21.5 mass% methanol). Consequently, the solubilities of all xanthines were enhanced by methanol addition in the fluid [14,32]. For example, with about 10 mass% of methanol in the mixed fluid, the solubility of every xanthine increases by one order of magnitude. Although the chemical structures of the xanthines are very similar, their solubilities in supercritical carbon dioxide mixtures vary substan-

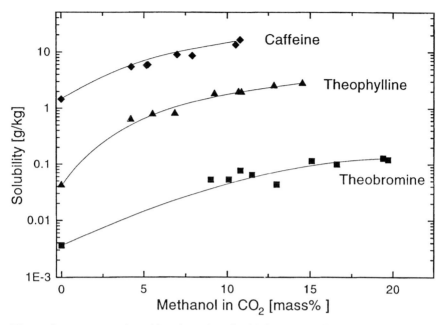

Figure 6 Solubility of xanthines in carbon dioxide/methanol mixtures at $T = 40°C$ and $P = 220$ bar [14].

tially as in pure carbon dioxide. The solubility of caffeine (1,3,7-trimethylxanthine) in pure supercritical carbon dioxide turned to be about one order of magnitude higher than that of theophylline (1,3-dimethylxanthine) and about two orders of magnitude higher than that of theobromine (3,7-dimethylxanthine) [14,15]. The addition of methanol to the carbon dioxide shows no influence on the relative difference in solubilities between the xanthines [14,32].

C. Measurement of Solubilities at a Pressure Lower than 200 bar

1. Experimental

(a) Apparatus and Procedure. For the measurement of solubilities at a pressure in the range of 50–200 bar, a second unit is needed to deliver carbon dioxide because the chromatographic unit and the equilibrium cell no longer run at the same pressure. Additional commercially available pressure modules are used in the chromatographic part of the apparatus. The apparatus used for measurement of solubilities at a pressure lower than 200 bar is shown in Fig. 7 [14].

The syringe pump here only serves to pressurize the equilibrium cell with carbon dioxide. The carbon dioxide flow in the chromatographic system is

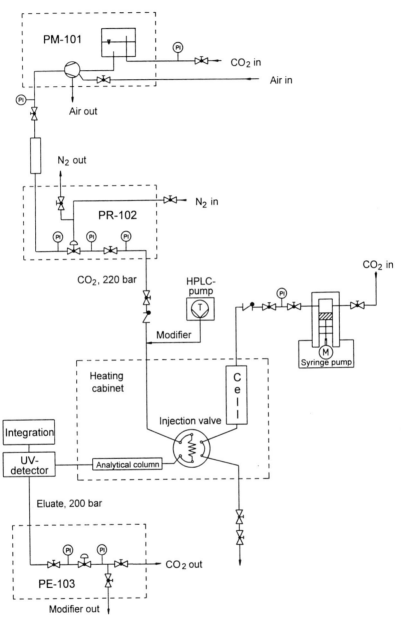

Figure 7 Apparatus for measurement of solubilities in pure carbon dioxide at a pressure in the range of 50–200 bar between 25°C and 85°C [14].

controlled by the modules PM-101 and PR-102 (NWA, Lörrach, Germany). In the pressure enhancement module (PM-101) the mobile phase is liquefied in a cooled reservoir (P_{max} = 200 bar). The liquefied gas is pressurized up to the desired initial pressure with a pneumatically driven pump. The liquefied gas flow is expanded in a controlled pressure reducer to operating pressure using the pressure regulation module (PR-102). This expansion simultaneously acts as pulsation dampener. Methanol can be added as a modifier with an HPLC pump to help chromatographic elution, parallel to the apparatus for measurements at higher pressure. The procedure and quantitative evaluation are carried out as described before (Sec. II.A.1).

2. Applications

Here results of the solubility measurements of caffeine and α-tocopherol at a pressure in the range of 80–350 bar are shown.

(a) Solubility of Caffeine in Carbon Dioxide. The solubility of caffeine in supercritical carbon dioxide has been studied extensively, so that these data can be used to check the method described here. In Fig. 8, experimental solubility data of caffeine in carbon dioxide at a pressure in the range of 80–350 bar and a

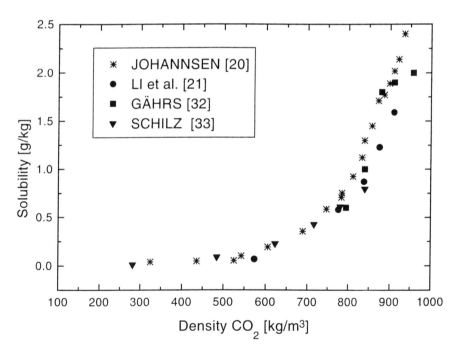

Figure 8 Solubility of caffeine in carbon dioxide at T = 40°C as a function of density [14].

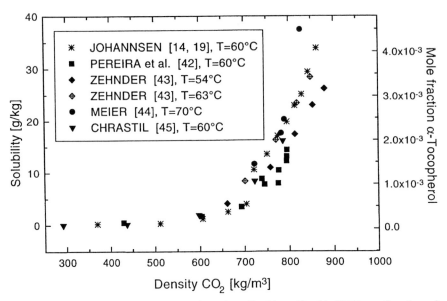

Figure 9 Solubility of α-tocopherol in carbon dioxide at T = 54–70°C as a function of density [14].

temperature of 40°C are presented as a function of density in comparison which data from the literature. The solubility of caffeine in carbon dioxide under these conditions is in the range of 0.04–2.4 g/kg [14,15,25,35–37]. Despite different standard materials, different purities of carbon dioxide, and different methods, all of the datasets are in close agreement.

(b) Solubility of α-Tocopherol in Carbon Dioxide. The preparative separation of tocopherols from natural sources by SFE and SFC is of increasing interest [38–40]. Therefore, the solubility of α-tocopherol in carbon dioxide was investigated at a pressure in the range of 110–350 bar and a temperature of 333 K [14,19,41]. The solubility of α-tocopherol in carbon dioxide under these conditions is in the range of 0.2–34 g/kg. In Fig. 9, the experimental results are presented as a function of density in comparison with data from literature [42–45]. All data correspond. As described in Sec. A.2(a), solubility increases with increasing density at constant temperature.

III. CONCLUSIONS

A system for fast and accurate experimental determination of equilibrium solubilities in neat and mixed (sub-/supercritical) fluids was described. Determination of solubilities has been performed by using an analytical method with direct coupling of an equilibrium cell to a pSFC system with UV detection. An apparatus like this can easily be built from most commercial pSFC equipment. Along with solubility data, the user will simultaneously receive information on the retention behavior of the compound. It is then easy to compare the solubility and retention behavior of a substance.

Solubilities are measurable in the range of 10^{-4} g/kg up to 10^2 g/kg, which covers six orders of magnitude. The range for possible solubility measurement is fixed by selection of a suitable sample loop. In pure carbon dioxide 5–10 experimental solubility values can be determined at different conditions on a daily basis. The measurement of solubilities in mixed (sub-/supercritical) fluids is more time consuming.

The limits for using this apparatus are given by the limits of the chromatographic system. The solutes have to be detectable, that means, for example, showing UV activity. Consequently, the detection limits the determination of solubilities. In cases of low UV absorption other detectors, such as light-scattering detectors, can be alternatively used.

It is possible to determine the solubility in other neat fluids, diverse mixed (sub-/supercritical) fluids, or at lower temperature by simple expansion or rebuilding of the apparatus. For measurements in neat fluids other than carbon dioxide a configuration like the one for the measurement of solubility at lower pressure (Sec. II.C.1) can be used. In that case, the syringe pump is filled with a different fluid. The mixing device shown in Sec. II.B.1 can be used for other mixed (sub-/supercritical) fluids. For solubility measurements at lower temperatures, the equilibrium cell and the chromatographic column have to be controlled separately. Moreover, it is possible to determine solubilities of solute mixtures, if the chromatographic system is suitable to separate out each substance for quantitative determination.

REFERENCES

1. G. Brunner, *Gas Extraction*, Steinkopff Verlag, Darmstadt, Germany (1994).
2. C. A. Eckert, B. L. Knutson, and P. G. Debenedetti, *Nature*, *383*:313 (1996).
3 R. Dohrn, *Berechnung von Phasengleichgewichten*, Friedr. Vieweg and Sohn, Braunschweig, Germany, 1994.
4. G. M. Schneider, *Phase Equilibria of Liquid and Gaseous Mixtures. Experimental Thermodynamics*, Vol. 25, *Experimental Thermodynamics of Nonreacting Fluids*, Chapter 16 (Part 2), IUPAC, B. LeNeindre and B. Vodar (eds.), Butterworth, London, 1975.
5. C. L. Young, *Experimental Methods for Studying Phase Behaviour of Mixtures at High Temperatures and Pressures. Chemical Thermodynamics*, Vol. 2, (M. L. McGlashan, ed.), Chemical Society, London, UK (1978).
6. U. K. Deiters, G. M. Schneider, *Fluid Phase Equilibria*, *29*:145 (1986).

7. K. N. Marsh, *Fluid Phase Equilibria, 52*:385 (1989).
8. R. Dohrn, G. Brunner, *Fluid Phase Equilibria, 108*:213 (1995).
9. R. E. Fornari, P. Alessi, and I. Kikic, *Fluid Phase Equilibria, 57*:1 (1990).
10. D. J. Miller and S. B. Hawthorne, *Anal. Chem., 67*:273 (1995).
11. L. Frederiksen, K. Anton, P. van Hoogevest, H. R. Keller, and H. Leuenberger, accepted for publication in *J. Pharm. Sci.*
12. K. D. Bartle, A. A. Clifford, and S. A. Jafar, *J. Chem. Eng. Data, 35*:355 (1990).
13. J. Yang and P. R. Griffiths, *Anal. Chem., 68*:2353 (1996).
14. M. Johannsen, Ph. D. thesis, Hamburg-Harburg, Germany, 1995.
15. M. Johannsen and G. Brunner, *Fluid Phase Equilibria, 95*:215 (1994).
16. M. Johannsen and G. Brunner, *Am. Lab., 26*:85 (1994).
17. E. Bender, "Equation of State Exactly Representing the Phase Behaviour of Pure Substances," Proceedings of Fifth Symposium on Thermophysical Properties. Am. Soc. Mech. Engrs., New York, pp. 227–235 (1970).
18. U. Sievers, *Die thermodynamischen Eigenschaften von Kohlendioxid*, Fortschritt-Berichte VDI-Z, Reihe 6 No. 155, VDI Verlag, Düsseldorf, Germany, 1984.
19. M. Johannsen and G. Brunner, *J. Chem. Eng. Data, 42*:106 (1997).
20. S. Mitra and N. K. Wilson, *J. Chromatogr. Sci., 29*:305 (1991).
21. M. L. Cygnarowicz, R. J. Maxwell, and W. D. Seider, *Fluid Phase Equilibria, 59*:57 (1990).
22. A. J. Jay, D. C. Steytler, *J. Supercrit. Fluids, 5*:274 (1992).
23. K. Sakaki, *J. Chem. Eng. Data, 37*:249 (1992).
24. M. Skerget, Z. Knez, and M. Habulin, *Fluid Phase Equilibria, 109*:131 (1995).
25. S. Li, G. S. Varadarajan, and S. Hartland, *Fluid Phase Equilibria, 68*:263 (1991).
26. R. K. Gilpin, S. S. Yang, and G. Werner, *J. Chromatogr. Sci., 26*:388 (1988).
27. H. G. Janssen, P. J. Schoenmakers, and C. A. Cramers, *HRC, 12*:645 (1989).
28. H. G. Janssen, P. J. Schoenmakers, and C. A. Cramers, *J. Chromatogr., 552*:527 (1991).
29. E. Brunner, W. Hültenschmidt, and G. Schlichthärle, *J. Chem. Thermodynam., 19*:273 (1987).
30. S. M. Fields, Ph.D. thesis, Brigham, UK, 1987.
31. D. E. Raynie, S. M. Fields, M. M. Djordevic, K. E. Markides, and M. L. Lee, *HRC, 12*:51 (1989).
32. M. Johannsen and G. Brunner, *J. Chem. Eng. Data., 40*:431 (1995).
33. T. A. Berger and J. F. Deye, *Anal. Chem., 62*:1181 (1990).
34. T. A. Berger, *HRC, 14*:312 (1991).
35. A. Birtigh, K. Liu, M. Johannsen, and G. Brunner, *Sep. Sci. Technol., 30*:3265 (1995).
36. H. J. Gährs, *Ber. Bunsenges. Phys. Chem., 88*:894 (1984).
37. W. Schilz, Ph.D. thesis, Saarbrücken, Germany, 1978.
38. M. Saito, Y. Yamauchi, K. Inomata, and W. Kottkamp, *J. Chromatogr. Sci, 27*:79 (1989).
39. M. Saito and Y. Yamauchi, *J. Chromatogr., 505*:257 (1990).
40. J. W. King, F. Favati, and S. L. Taylor, *Sep. Sci. Technol., 31*:1843 (1996).
41. A. Birtigh, M. Johannsen, G. Brunner, and N. Nair, *J. Supercrit. Fluids, 8*:46 (1995).

42. P. J. Pereira, M. Gonçalves, B. Coto, E. Gomes de Azevedo, and M. Nunes da Ponte, *Fluid Phase Equilibria*, *91*:133 (1993).
43. B. H. Zehnder, Ph.D. thesis, Zurich, Switzerland, 1992.
44. U. Meier, Ph.D., thesis, Zurich, Switzerland, 1992.
45. J. Chrastil, *J. Phys. Chem.*, *86*:3016 (1982).

2
Chemical Basis for the Instrumentation Used in Packed Column Supercritical Fluid Chromatography

Terry A. Berger

Berger Instruments, Newark, Delaware

I. INTRODUCTION

In recent years there has been a marked resurgence of interest in packed column supercritical fluid chromatography (pSFC). However, during the 1980s most SFC instrumentation sold was designed to be used with capillary columns. Many users have subsequently attempted to reuse their capillary hardware with packed columns. These attempts are often driven by managers unwilling to purchase additional hardware when an "SFC" instrument is in the lab. Unfortunately, there are dramatic differences between the instrumentation required for use of packed or capillary columns in SFC. Attempting to use inappropriate capillary instrumentation with packed columns is usually difficult at best and often results in complete failure.

Packed column SFC resembles high-performance liquid chromatography (HPLC). Binary or tertiary mobile phases are common. The composition of the mobile phase is almost always more important than its density in controlling retention. Packed columns are best operated with pumps in a flow control mode. Flow rates are typically several ml/min (cm^3/min). Pressure is controlled by a back pressure regulator mounted downstream from the column, but pressure programming is a secondary control method. Temperatures are generally close to the critical temperature. Standard ultra-violet (UV) detectors with a high-pressure flow cell are used.

Capillary column supercritical fluid chromatography (cSFC) resembles gas chromatography (GC) at high pressures, except that pressure (density) programming replaces temperature programming. Modestly elevated temperatures (i.e., 100°C) are common. Binary fluids and composition programming are seldom used. Because it is controlled using fixed restrictors, flow varies with pressure, temperature, and composition. The total flow rate is typically measured in μL/min (mm³/min). Solutes are quantified with a flame ionization detector (FID) after the mobile phase is expanded to atmospheric pressure.

Superficially, differences in commercial hardware for packed and capillary operation appear to be driven primarily by practical problems, such as the difficulty of accurately pumping very low flow rates, or the unavailability of very low dead volume back pressure regulators. If that were true, advances in engineering could solve such problems.

Most comparisons of the two techniques have concentrated on the superficial similarities between them (i.e., they are both called SFC). Such comparisons typically use the same solutes and phases. Which technique "wins" such a comparison depends on the choice of solutes and phases, and on the biases of the user. Comparing capillary and packed columns using the same conditions inherently and unfairly favors one type over the other. Such comparisons are driven by attempts to find similarities.

The substantial differences between the two column types have not been adequately appreciated. In this chapter, both the practical and the chemical basis for pSFC instrumentation is explored and clarified. The effect of physical parameters on real separations is demonstrated.

II.   WHY SFC?

A. Speed and Efficiency

Packed column chromatographers are interested in carbon dioxide as a mobile phase because even at high densities its viscosity is more like that of a gas than a liquid [1–3]. Solutes in dense carbon dioxide also retain diffusion coefficients 10 times larger than those in normal liquids [4–7].

The lower viscosity and higher diffusivity in such fluids, compared to normal liquids, allows faster separations and higher efficiency per time than in HPLC. As an example, a 2.2-m-long column packed with 5-μm particles generated 220,000 plates in < 9 min, at 450 plates/s and with a pressure drop of 160 bar [8]. Both composition and pressure could be rapidly programmed, greatly increasing the apparent efficiency. In HPLC, the best published example [9] used a packed capillary column to generate 225,000 plates but required a hold-up time of 33 min, with a pressure drop of 360 bar. The long hold-up time precludes effective programming.

B. Separation of Thermally Labile Molecules

With supercritical fluids, solvation energy can replace thermal energy to overcome intermolecular interactions. Many solutes decompose before they can be volatilized for GC analysis. Chromatographers have been attracted to SFC because it offers the possibility of GC-like separations but at lower temperatures.

C. Selectivity

SFC is usually a normal phase technique without most of the problems usually associated with normal phase HPLC. Composition, pressure, temperature, and stationary phase identity all contribute to provide a very wide range of selectivity adjustment. Equilibration is very fast, usually requiring flow turnover of little more than three to five column void volumes. Composition, pressure, and temperature can all be programmed with rapid recovery to initial conditions between analyses. Traces of water do not cause the variations in retention often encountered in normal phase HPLC.

Normal phase separations offer a truly complementary technique to reversed phase HPLC for analyses requiring two independent methods. Chromatographers are presently faced with using two relatively similar reversed phase HPLC methods for lack of an alternative.

For low-polarity solutes, like triglycerides or carotenes, packed column SFC with nonpolar stationary phases and high concentrations of polar modifiers, such as methanol and methanol/acetonitrile mixtures, can be a reversed phase technique, as shown in Chapter 7. At high modifier concentrations, further increases in modifier concentration cause an increase in retention.

III. INCENTIVES FOR USING NEAT CARBON DIOXIDE

A. Safety

Neat carbon dioxide is relatively safe, because it is a product of human respiration. It is nonflammable. It is also relatively inexpensive. Compared to many of the organic liquids used as solvents in HPLC, carbon dioxide in environmentally friendly. The carbon dioxide released into the environment by a packed column instrument pumping at 2 mL/min for 24 h is equivalent to burning 1.3 L of gasoline. A chromatographer driving the average American car to work 4.5 miles (~7 km) each way produces as much carbon dioxide. Because the carbon dioxide is often distilled out of the air, its use in SFC constitutes a form of recycling.

B. A Universal, Easy-to-Use Detector

Another important consideration in using pure carbon dioxide as the mobile phase is its compatibility with the FID. One of the biggest problems with HPLC is the lack of an easy-to-use, stable, sensitive, universal detector with a uniform response factor, like the FID. If HPLC-like separations could be performed with pure carbon dioxide, the FID would allow detection of molecules without chromophores and normalize the response of all molecules.

C. Absence of Alternative Fluids with Greater Polarity

Pure carbon dioxide is not very polar, making it a mobile phase most appropriate for the elution of low to moderately polar solutes. Considerable effort has gone into finding a more polar fluid that retains the desired fluid characteristics. It is debatable as to whether a fluid dramatically more polar than carbon dioxide will be found that also has a low critical temperature. Several non-ozone-depleting fluorocarbons (F-23 and F134a) have much higher dipole moments and hold some promise.

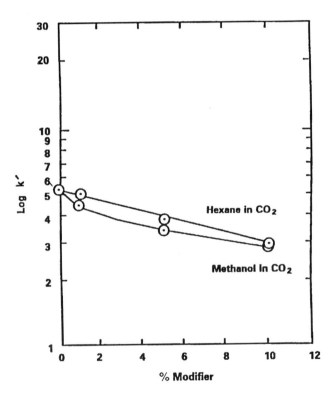

Figure 1 The retention of chrysene [11] on an HPLC column packed with octadecyl-coated silica particles as a function of modifier concentration. Note that the addition of neither hexane or methanol significantly changes the retention of the solute. Column: 2×100 mm, 5μm Hypersil ODS. Other conditions: 2 mL/min, 200 bar, 50°C, 254 nm, 1 μL of 100 μg/mL in hexane injected.

Some pure supercritical fluids have much higher solvent strength than carbon dioxide. Unfortunately, they also have much more daunting critical parameters. Supercritical water is an extreme solvent, but its critical parameters are >374°C and >218 bar. Very high critical temperatures are incompatible with thermally labile molecules.

D. Physical (vs. Chemical) Control

In the 1980s, it was widely believed that the Giddings elutrophic series [10], published in 1968, was correct. Using estimated interaction parameters, Giddings postulated that dense carbon dioxide should be as polar as isopropyl alcohol. If true, a very wide range of solvent strength would be accessible simply by compressing gaseous carbon dioxide to liquid-like densities. The control of retention through manipulation of a physical variable, like pressure, is appealing. Furthermore, if carbon dioxide was already as polar as an alcohol, the need for modifiers, like methanol, and the expense of a second pump were questioned. There is some experimental evidence that appears to support these expectations.

Polar modifiers appear to have little effect on the retention of light, low-polarity solutes on reversed phase (low polarity) HPLC columns. An example [11] is shown in Fig. 1, where neither methanol or hexane added to carbon dioxide changed the retention of chrysene on a C18 coated silica column. One might expect the methanol to have a greater effect on retention if it were substantially more polar than the carbon dioxide.

Such results were originally interpreted as affirmation of the Giddings elutrophic series (carbon dioxide seemed to be as polar as alcohols). Some workers went so far as to suggest that even the small changes in retention, observed in Fig. 1, were due to changes in density caused by the additions of the modifier, not a change in solvent strength.

IV. THE DISINCENTIVES FOR USING MODIFIERS

Many disincentives to the use of modifiers existed in the past. The need for several high-pressure pumps made SFC more expensive than HPLC because, unlike HPLC, low-pressure mixing valves cannot be used. The addition of modifiers to the main fluid was instrumentally difficult, and often irreproducible, particularly when using syringe pumps. Using syringe pumps and fixed restrictors, the effects of several variables were convoluted together and highly confusing. With modified fluids, the FID could not be used because organic modifiers produced a large baseline offset.

Since they were thought to have little effect on retention, it was assumed that modifiers were only added to improve peak shapes by covering "active sites" [12–15] on the packings. This implied that the peak shape problem arose from the "poor" quality of packing materials. Thus, modifier addition was expensive and inconvenient, and thought to only cover up a problem that should not have been there to start with. Most efforts with packed columns during the mid-1980s involved attempts to deactivate stationary phases to get better peak shapes with pure carbon dioxide. The general lack of success in these attempts led many to try cSFC where it was thought that deactivation of active sites was much less problematic.

V. CHEMICAL REASONS FOR THE REEMERGENCE OF PACKED COLUMNS

A. The Inadequate Polarity of Carbon Dioxide

Many of the fundamental assumptions about packed columns, supercritical fluids, and modifiers, prevalent up to the late 1980s, were incorrect. Systematic measurement of solvent strength [16,17] showed that carbon dioxide is an extremely nonpolar solvent. Recent evaluations [18] place it with fluorocarbons, less polar than short chain aliphatic hydrocarbons, even at high densities. The Giddings elutrophic series [10] was wrong.

The Giddings series is compared to an empirical solvent strength scale [19] obtained using a solvatochromic dye in Fig. 2. Such dyes changes color depending on the solvent strength of the solvent sheath surrounding it. The results from dyes confirmed that the inability to elute polar molecules using pure carbon dioxide arose from the inadequate solvent strength of carbon dioxide, not excessive interaction between the solutes and "active sites."

B. The Need for Appropriate Polarity Windows

The apparent irrelevance of polar modifiers, seen in Fig. 1, stemmed from an inadvertent use of inappropriate probes and columns. Carbon dioxide, hexane, and C18 stationary phases are all of similar low polarity, most appropriate for interacting with solutes of similar low polarity (the concept of "like dissolves like"). In nonpolar chromatographic systems (i.e., C18 and carbon dioxide), polar molecules (either solutes or modifiers) had little to interact with. Weak interactions often result in poor peak shapes.

The situation changes when a polar stationary phase is used. In Fig. 3, the retention of the same solute [20] used in Fig. 1 is much greater in pure carbon dioxide. The addition of methanol causes large changes in retention. In contrast, low-polarity hexane had only a small effect on the retention of the same solute. The only difference between Figs. 1 and 3 is the polarity of the stationary phase. In this example, the low-polarity carbon dioxide and the polar sulfonic acid column form a normal phase "polarity window" [21] appropriate for the separation of intermediate polarity solutes.

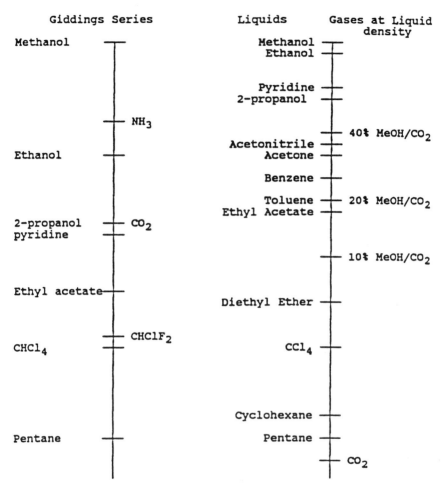

Figure 2 Comparison of Giddings elutrophic series from 1968 [10] (left), with a recent empirical solvent strength scale [19] based on the solvatochromic dye Nile red. Giddings placed carbon dioxide next to isopropanol in solvent strength, whereas the solvatochromic dye places it next to pentane.

When the chromatographic phases are much less polar than the solutes, there can be little interaction between the solutes and either phase. Solute solubility in the mobile phase and interactions with polar active sites then become concerns.

C. Increasing Solvent Strength with Modifiers

Modifiers increase the solvent strength of the mobile phase [16,17]. If the solvent strength is sufficiently increased, the relative interaction between solute

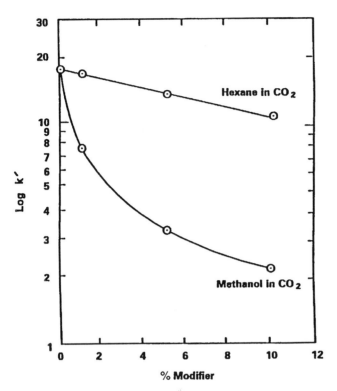

Figure 3 The retention of chrysene on a sulfonic acid ion exchange column [20], as a function of modifier concentration. Note that, compared to Fig. 1, retention is much greater. More importantly, methanol changes retention more dramatically than hexane. Same conditions as in Fig. 1.

and mobile phase approaches the level of interaction between the solute and stationary phase, resulting in elution of the solute.

Methanol is completely miscible with carbon dioxide. These two fluids make a solvent pair that covers a much wider range of solvent strength than any pair available in HPLC. This single pair covers half the range of solvent strength between hydrocarbons and water. (On Snyder's P' [22] scale, pentane and carbon dioxide are approximately 0, methanol is 5.1, and water is 10.4.)

1. Nonlinear Solvent Strength

Solvent strength is a nonlinear function of modifier concentration [16], as shown in Fig. 4. In fact, 2% methanol produced the solvent strength one would expect from 10% methanol, if the fluid solvent strengths were linearly additive. This nonlinear enhancement in solvent strength is caused by "clustering"

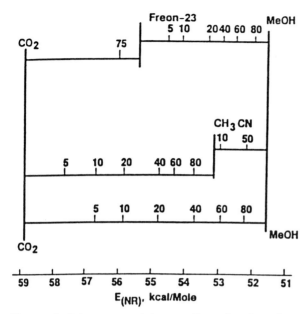

Figure 4 Solvent strength is a nonlinear function of composition [16]. The solvent strength of different mixtures is expressed in kcal/mole. The numbers beneath the vertical lines indicate the composition of the mixtures as % methanol in carbon dioxide.

together of the modifier molecules [23–27], creating enhanced local concentrations of modifier in carbon dioxide.

2. Nonlinear Retention

Plots of log k vs. modifier concentration are nonlinear, as shown in Fig. 5a [28]. This was initially interpreted to imply that SFC was "adsorption chromatography," and the nonlinearity followed the adsorption isotherm of the modifier. However, plots of log k vs. solvent strength (measured with solvatochromic dyes), which are independent of any surface phenomena, are also nonlinear. In fact, there are many instances [i.e., 17,29] where plots of (log) retention vs. solvent strength produce straight lines. An example is shown in Fig. 5b.

3. Composition Is More Important than Density

The enhanced solvent strength of modified fluids was not due to an increase in density, caused by the addition of the liquid phase modifier, as previously suggested by some workers, in the absence of density measurements. The density of methanol/carbon dioxide mixtures has been measured [29,30]. Retention changes [29], caused by variable composition at constant density and variable density at constant composition, showed that most of the enhancement

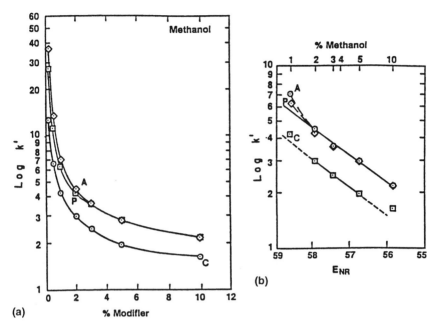

Figure 5 (a) Plots of log k vs. % modifier are nonlinear, implying the mechanism is similar to adsorption chromatography. (b) The same retention data is plotted against the measured mobile phase solvent strength and produces straight lines. This suggests the nonlinearity in Fig. 5a was due to nonlinear mobile phase solvent strength, not an adsorption mechanism [28].

was due to the change in composition, with changes in density (pressure) playing a secondary role. The relative effects of composition and density on the retention of one solute is presented in Fig. 6.

4.   Binary Fluids Have a Limited Density Range Available

Figure 7 is a form of phase diagram [30] for methanol in carbon dioxide mixtures, at one temperature. The enclosed, shaded area is a two-phase region. Single phases cannot exist within the two-phase region. When there are two phases in the column, separations are destroyed.

The diagonal lines at the top of Fig. 7 are constant density lines. Note that for most compositions densities below approximately 0.6 g/cm^3 *cannot* be made. They would occur in the two-phase region. Thus, less than half (>0.6 to <1 g/cm^3) the normal range of densities, used with pure carbon dioxide (0.01 to <1.0 g/cm^3), is available with binary mixtures of methanol in carbon dioxide at low temperatures.

In the past, some workers were preoccupied with the possibility of multiple phases at low pressures. This preoccupation undoubtedly arose from the then

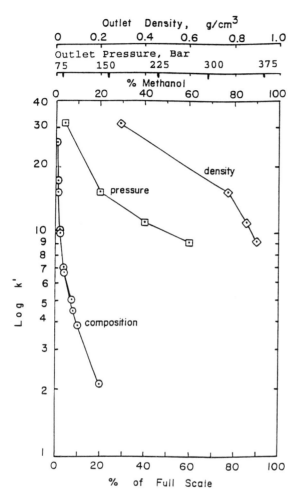

Figure 6 Relative effect of composition and density on retention. Composition has a larger impact on retention than density, using polar solutes and modified fluids [29]. Comparison of changing modifier concentration at constant density and changing density at constant modifier concentration.

dominance of capillary methods, where the primary control variable was the pressure (density) of a pure fluid. However, with binary fluids, the primary control variable is the concentration of modifier, not density. Two-phase regions are easily avoided by simply maintaining the column outlet pressure above the two-phase region at the temperature used. The instability of the fluids at low pressures and densities should be viewed as a minor inconvenience.

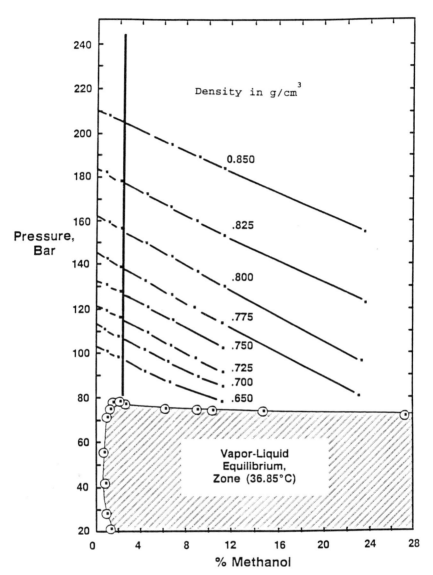

Figure 7 A phase diagram for carbon dioxide mixtures at 36.85°C. The shaded area represents a two-phase region where a single phase cannot exist. Outside the shaded area, a single phase always exists. The dark vertical line separates fluids defined as supercritical, on the left, from fluids defined as subcritical or liquids, on the right. The diagonal lines across the top are constant density lines with the density values indicated in g/cm³ . Note the continuity as fluids change from super- to subcritical [30].

D. Viscosity and Diffusion Coefficients

When small concentrations of modifier are added to carbon dioxide, the higher solute diffusion coefficients [31] and lower viscosity of pure carbon dioxide, compared to normal liquids, are maintained. Even 50% methanol in carbon dioxide creates only moderately higher pressure drops than 5% methanol in carbon dioxide.

E. Stationary Phase Effects

1. Surface Area

Retention is proportional to the surface area of the chromatographic stationary phase [32], as shown in Fig. 8 using column packings with different pore diam-

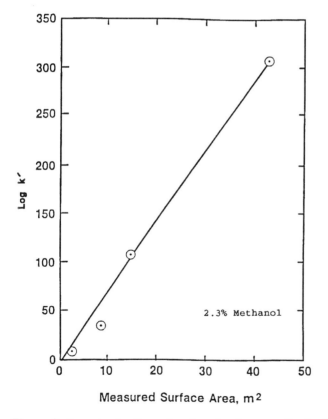

Figure 8 Retention is a linear function of surface area. By simply changing the surface area of silica particles, retention can be changed by at least 20 times [32].

eters. Modern totally porous packings can have <10 to >500 m^2 of surface area for void volumes on the order of 1 cm^3.

Packed columns have gigantic surface area to void volume ratios compared to capillaries. A 50-μm ID capillary has a surface area to void volume ratio of 0.08 m^2/cm^3. For columns without a bonded phase, retention on the packed column could be 6000 times greater than on the capillary column, due simply to greater surface area to void volume ratio of the packed column.

2. Film Thickness

Bonded stationary phases are generally much thicker on capillaries than on totally porous packing materials, which partially mitigates the differences in retentivity of the two column types. Retention is proportional to the ratio of the stationary phase volume (V_s) to the void volume (mobile phase volume) (V_{ms}). In chromatography, the inverse of this ratio is used to compare retentivity of columns and is called beta ($\beta = V_{ms}/V_s$). A 1-nm film on a 500 m^2/g packing (void volume approx. 1 cm^3) has a ratio of 0.5 ($\beta = 2$). A 50-μm id open tubular (capillary) column with a 0.1 μm film thickness has a ratio of 0.008 ($\beta = 125$). Thus, the packed column should still exhibit >60 times the retentivity of the capillary column.

3. Stationary Phase Polarity

If the mobile and stationary phases used with both packed and capillary columns are chemically the same, interactions with solutes will have the same *intensity*. However, the much more *extensive* relative surface area of packed columns results in much greater retention. To get the same retention on both packed and capillary columns, it should be clear to the reader that each would require a chemically different stationary phase on each of the two column types. When the same stationary phase has been used and compared in the past, the excess retentivity on packed columns has always been blamed on "active sites," not on the differences in phase ratios.

Superficially, the greater retention on packed columns suggests the use of stationary phases that are less polar than those used with capillaries. However, polar solutes need polar stationary phases.

Low-polarity stationary phases, like methylsilicone, widely used for nonpolar solutes in GC, yield very poor peak shapes and little retention of *polar* solutes in cSFC (and in GC) using neat carbon dioxide as the mobile phase. Similarly, octadecyl (C18) phases, common in reversed phase HPLC, produce inadequate retention and usually poor peak shapes with polar solutes in pSFC.

Use of low-polarity stationary phases and carbon dioxide results in little interaction between polar solutes and either chromatographic phase. Polar solutes will inherently interact strongly with even a few polar active sites, yielding a mixed retention mechanism. In the extreme, the bonded phase ceases to be the "stationary phase." Instead, the uncontrolled "active sites" determine retention.

Lee has approached the problem in a relatively unique manner, as shown in Chapter 5. He continues to develop highly deactivated but polar phases which he uses to separate polar solutes using only neat carbon dioxide. His success may be in part due to a decrease in the specific surface area caused by application of the deactivation and bonding of the stationary phase. He clearly elutes many compounds using neat carbon dioxide that previously had required modified fluids. The limits of such an approach remain poorly defined, but pure carbon dioxide is an ineffective solvent for very polar solutes.

4. Active Sites

Active sites have been blamed for virtually every problem encountered in pSFC. Most often, exposed silanols on the surface of silica are blamed for poor peak shapes or excessive retention. Packed columns are often thought to involve much greater problems with active sites than capillaries. Surprisingly, the relative density of silanols on packed and capillary columns is similar. Packings obviously have many more uncovered silanols but the number per unit area is similar on both column types.

While fused silica tubing is extremely pure and inert, the first step in making a capillary column is to totally hydrolyze the surface to silanols. The silanols are subsequently reacted with a deactivating agent. Stationary phases are then covalently bonded to the surface. The incomplete surface coverage by both deactivating agents and bonded phases is seldom discussed, but is often less than 75%, particularly in the narrow-bore tubing used in cSFC.

It was once thought that modified fluids changed retention more on packed columns than on capillaries [12–15]. It was also claimed that this was due to the deactivation of larger numbers of "active sites" on silica particles. The modifier was purported to change retention by covering those active sites. However, modifiers change retention on capillaries [33,34] about as much as they change retention on packed columns [35,36] using the same fluids. There is no evidence for any difference in the relative population of active sites on packed or capillary columns (sites/surface area).

5. Matching Solute and Stationary Phase Polarity

Most of the effort in developing columns for pSFC has involved attempts at deactivating of the stationary phase. The intent is to decrease the number of undesired *intense* side interaction with "active sites," so that neat carbon dioxide could be used as the mobile phase. However, it has been repeatedly shown that nonpolar stationary phases yield little retention and poor peak shapes for polar solutes.

A more effective approach is to increase the polarity of the stationary phase to increase the intensity of interaction with polar molecules. Interactions with active sites become irrelevant as the polarity of the stationary phase is increased.

A more polar stationary phase requires a more polar mobile phase. The most straightforward way to increase the polarity of the mobile phase is to add a polar modifier to carbon dioxide.

F. Carbon Dioxide Adsorption

In chromatography, operation at lower mobile phase densities produces faster separations, provided the stationary phase is not excessively thick and/or dense. Solute diffusion in a thick, dense film can be slow compared to its diffusion in a lower density mobile phase. If the chromatographic phases are improperly matched, the optimum linear velocity for diffusion in the mobile phase will be much faster than the optimum velocity for diffusion in the stationary phase. Since both phases cannot be operated at their own optimum simultaneously, efficiency is degraded.

Parcher [37,38] showed that carbon dioxide adsorbs extensively onto surfaces, as shown in Fig. 9. With neat carbon dioxide, this adsorption can create a very thick *effective* stationary phase, consisting of the phase purchased, plus a thick film of dense, adsorbed carbon dioxide. These adsorbed films are thickest where the temperature and the density are both relatively low. As seen from Fig. 9, adsorption at 40°C decreases from 23 μmol/m^2 at 80 bar, to 7 μmol/m^2 at 100 bar. It is generally acknowledged that silica has approximately 8 μmol/m^2 of surface sites. Thus, the adsorption is equivalent to between one and three monolayers of "condensed" (liquid-like) carbon dioxide. The numbers reported are actually surface excesses, so even more carbon dioxide is actually present on the surface.

Large changes in density and adsorption, associated with small changes in pressure at low temperatures, appear to correlate with changes in chromatographic efficiency [39]. In Fig. 10, plate height (inversely proportional to efficiency) is plotted vs. mobile phase velocity. The poor efficiency at 80 bar (upper curve) was obtained under the conditions producing the most extensive adsorption. The effective density of the adsorbed film of carbon dioxide probably approaches 1 g/cm^3. Under the same conditions, the mobile phase density is 0.28 g/cm^3. The poor efficiency at 80 bar, in Fig. 10, is consistent with a thick, dense stationary phase in contact with a low-density mobile phase. The dramatically improved performance at 100 bar (lower curve, Fig. 10) was obtained with a much thinner adsorbed film and a denser (0.62 g/cm^3) mobile phase.

1. Temperature

Adsorption decreases as the temperature is increased (see Fig. 9). At elevated temperatures, (i.e., >100°C), the adsorbed film is never thicker than a fraction of a monolayer, too thin to cause efficiency problems.

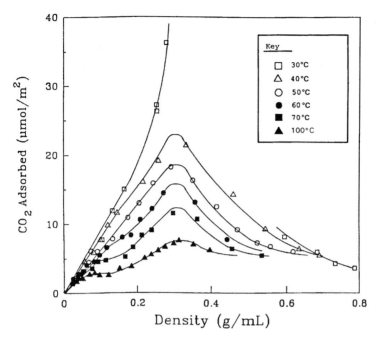

Figure 9 The adsorption of carbon dioxide onto a C18 stationary phase. Adsorption decreases as the density and temperature increases. From top to bottom: 30°, 40°, 50°, 60°, 70°, 100°C [37].

Capillaries are routinely used with low-density mobile phases. Similar to packed columns, capillary column peak shapes also degrade using both low densities and low temperatures. It is likely that carbon dioxide adsorption is the cause of this degradation of peak shape and the reason capillaries are most often used at >80°C. Elevated temperatures avoid mismatches between high diffusivity in the mobile phase and low diffusivity in a thick film of condensed mobile phase components, acting as part of the effective stationary phase.

G. Modifier Adsorption

Besides increasing the solvent strength of the mobile phase, modifiers can be extensively adsorbed (Fig. 11), increasing both the volume (extent) and polarity (intensity) of the stationary phase.

Surprisingly, most changes in adsorption occur between 0 and 1% or 2% modifier in the mobile phase. At higher concentrations, the adsorbed film does not change thickness or composition significantly [40] (at constant temperature

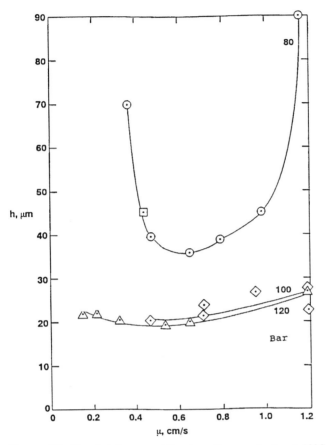

Figure 10 Plate height vs. mobile phase linear velocity at 40°C and 80, 100, and 120 bar outlet pressure, with minimal pressure drop across the column. The poor efficiency at 80 bar suggests the adsorbed film of carbon dioxide acts as a thick film stationary phase [39].

and pressure). Plots of log retention vs. solvent strength often show a nonlinearity below 1–2%, which seems to correlate with adsorption measurements.

With 2% methanol in the mobile phase, the adsorbed film [41] can be as much as 40% methanol, and the condensed film can be up to seven monolayers thick. This can be interpreted as a thick film stationary phase that is much more polar than the mobile phase (40% vs. 2% modifier). The addition of polar modifiers shifts the polarity of both chromatographic phases and allows the separation of much more polar solutes in a more polar "polarity window" [21].

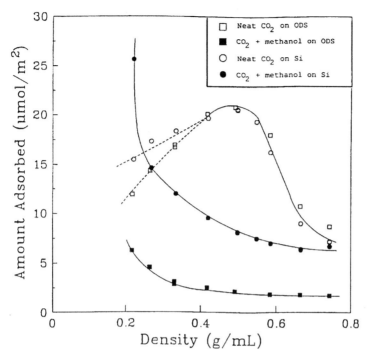

Figure 11 Simultaneous adsorption of carbon dioxide and methanol on silica and C18 (ODS) as a function of pressure, at 40°C. Open squares: Carbon dioxide on ODS. Shaded squares: methanol on ODS. Open circles: carbon dioxide on silica. Shaded circles: methanol on silica. Much more methanol adsorbs on the polar silica than on the nonpolar ODS. Approximately 8 μmol/m² corresponds to monolayer coverage [38].

Mixtures of methanol and carbon dioxide (Fig. 7) form a single phase only when the density of the mixture is above approximately 0.6 g/cm³. Since retention control is primarily accomplished by changing the mobile phase composition [29], there is little incentive to operate in pressure or density regions where there is a mismatch between the mobile and stationary phases. Low-temperature, high-density operation is common and seldom results in an efficiency loss.

Packed columns with *binary* fluids generally do not function well above 80–100°C, probably because adsorption is low at these temperatures. Packed columns with modified fluids operate best between 20°C and 80°C, where adsorption is extensive.

Temperature usually has more effect on selectivity than retention. This is likely to be associated with changes in both the thickness and composition of the adsorbed film of mobile phase accompanying temperature changes.

VI. INSTRUMENTATION FOR PACKED COLUMNS: INDEPENDENT CONTROL OF PRESSURE AND FLOW

A schematic diagram of a pSFC instrument is presented in Fig. 12. In pSFC, there are four instrumental (set from the keyboard) control variables: flow rate, composition, pressure, and temperature. The composition of the mobile phase is more important than its density in determining retention on packed columns (Fig. 6). Temperature is often the most effective parameter for changing selectivity. Pressure is a secondary but important control variable.

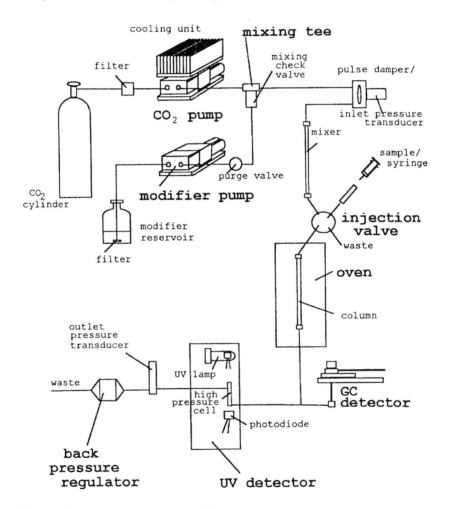

Figure 12 Schematic diagram of pSFC instrumentation.

Pumps are operated as flow sources, similar to HPLC. The same amount of fluid (i.e., grams per second) is delivered per unit time regardless of the system pressure. Constant flow allows the effects of flow, composition, and pressure to be clearly deconvoluted, without the confusion inherent using capillary instrumentation where a change in pressure also resulted in uncontrolled changes in flow, linear velocity, and efficiency.

Two (or more) high-pressure pumps are used to control the composition of the mobile phase. Mixing fluids together can require special considerations. Injection is performed using high-pressure switching valves common in HPLC. Temperature is more important in pSFC than in HPLC and impacts both retention and selectivity. The same columns used in HPLC are used in pSFC, but much longer columns are feasible.

The column outlet pressure is controlled by a back pressure regulator mounted downstream of both the column and any high-pressure detector. Both low-pressure GC and high-pressure HPLC detectors can be used. Specific details of each component are discussed in separate sections below.

A. Pumps

The most important instrumental advantage of packed column instrumentation, compared to capillary instruments, involves the direct, independent control of flow rate. The total flow in pSFC is typically 3–5 times greater than in HPLC, using the same column. Thus, a 4.6-mm-id packed column has an optimum flow rate near 3 mL/min (measured at the pump at a density near 0.95 g/cm^3).

High flow rates can be accurately controlled. Precise mixtures can be generated. However, there are subtle problems, not widely appreciated, that complicate the design of pumps used to deliver compressible fluids.

1. Pumping Compressible Fluids

With incompressible fluids, the first small movement of the piston increases the pressure to whatever pressure is required to cause flow. With ideal gases $PV = nRT$. Pressure and volume are inversely proportional to each other. Fluids like carbon dioxide are neither incompressible fluids nor ideal gases. Volume (and density) is not linearly proportional to pressure, particularly near the critical point. An exaggerated example helps make the point. At 40°C, increasing the pressure of carbon dioxide from 80 to 100 bar (+25%) increases the density from 0.28 to 0.62 g/mc^3 (+120%). Thus, to increase the pressure inside a 40°C pump by 25%, the piston of the pump would need to displace 55% of the volume in the cylinder.

While the piston is compressing the fluid to the new pressure, no flow emerges from the pump! Modern pumps compensate for the compressibility of the fluids. Carbon dioxide and most other fluids used in pSFC are much more compressible than the liquids used in HPLC. To pump these fluids, the pump requires an extended compressibility compensation range.

In inadequately compensated pumps, a significant proportion of the delivery stroke is wasted, compressing the fluid without any flow emerging. If the pumping system does not take into account the compression of the fluid, large errors in flow and composition can result.

There are numerous examples of such problems in the literature. HPLC pumps were used with a pump head chiller, but without increasing the compressibility compensation range. Typically, when the pressure was increased, progressively more of the stroke in the carbon dioxide pump went to compressing the fluid, not delivering flow. Because the modifier was an incompressible liquid, its flow did not drop. The total flow dropped, and the modifier fraction of the total increased. This created *unrecognized*, increasing composition, and decreasing flow gradients. The resulting chromatograms were attributed to pressure programs at a low "constant" modifier concentration and higher flow! Needless to say, such performance could not be reproduced on different hardware. Modifier concentrations as much as five times higher than suggested were needed to elute heavy, polar solutes. Modern pumps designed for pSFC have solved these problems.

In modern pumps, a compression stroke is differentiated from the delivery stroke. In the compression stroke, the primary purpose of the piston movement is to compress the fluid to system (column pressure, not to deliver flow. Modern pumps compress the fluid from the supply pressure to the system (column) pressure as rapidly as possible. The piston can then rapidly displace additional volume to make up for the lack of flow when the piston was refilling. After the fluid is compressed and some is injected to make up for lack of flow during the compression stroke, the pump switches to the delivery stroke. During the delivery stroke, the piston moves much more slowly to deliver the constant flow rate desired.

2. Compressibility Factor

Compressibility compensation requires calculation of the volume that needs to be displaced to raise the pressure in the pump to column pressure so the pump can deliver accurate flow. Some pumps use a compressibility factor "look-up table." These precalculated or premeasured tables are stored in the software or firmware used by the pump controller to set the speed and travel distance of the piston. Alternately, compressibility factors can also be calculated "on the fly," using an equation of state.

Calculated compressibility factors require accurate temperature and pressure sensing. The greater the deviation between measured and actual values of temperature and/or pressure, the larger the errors in flow and composition.

3. Pump Temperature

Liquids are easier to pump than gases or supercritical fluids. Most carbon dioxide pumps are chilled to well below the fluid's, critical temperature to ensure that it is pumped as a *liquid*. However, chilling the pump head does not completely compensate for changes in compressibility or density at different temperatures and pressures. Even at -20°C, the compressibility factor changes from 0.158 at 80 bar to 0.733 at 400 bar. The effect of temperature is less extreme. At 80 bar, the compressibility factor is 0.158 at -20°C, changing to 0.161 at 0°C (approx. 2% change).

At 0°C, the density of carbon dioxide changes by 12.3% between 80 and 400 bar (0.963–1.082 g/cm^3). At 100 bar, changing the temperature from −10 to +10°C changes the density by 8% (1.059–0.981 g/cm^3).

Clearly, simply making the pump head very cold does not eliminate changes in the delivery rate of the pump. Extreme cold is no substitute for adequate compressibility compensation and does not produce accurate flow control without variable compressibility compensation. Standard HPLC pumps without an extended compressibility factor range produce inaccurate flow and compositions.

The most common chillers involve an external recirculating bath filled with a fluid like ethylene glycol. The mobile phase is passed through a coil submerged in the bath, while some of the chilled ethylene glycol is pumped through a heat exchanger bolted to the pump head. Stainless steel is effectively a thermal insulator. Care must be taken with bolted on heat exchangers to ensure adequate heat transfer.

Pump heads can also be cooled using cryogenic cooling. A fluid like liquid carbon dioxide is expanded to atmospheric pressure inside the pump head. The expansion is accompanied by Joule-Thompson cooling. This form of control tends to produce small periodic variations in head temperature.

The most precise means of temperature control is the use of a thermoelectric (Peltier) cooler mounted directly on the pump head. With proper heat exchangers and precoolers the temperature of the fluid in the pump head can be accurately controlled.

Many "home-built" heat exchangers provide inadequate thermal transfer. The actual temperature in the pump head is often substantially higher than the temperature of the cooling bath. The electric motor heats one end of the pump while the heat exchanger cools the other. Stainless steel, the most common material in pump heads, is effectively a thermal insulator. Environmental conditions (such as relative humidity) can vary the actual heat transfer rate from the motor and heat exchanger day to day and instrument to instrument. If the compressibility compensation is based on bath temperature instead of the actual temperature in the pump head, inaccurate compensation will result. Long-term flow and composition precision will also be poor.

4. Compensating for Leaks

Chromatographers tend to forget that all real seals leak. Leaks get worse as equipment ages. Without compensation for leaks, pressure oscillations increase and detector baseline noise tends to degrade. Effectively, leaks are equivalent to an increase in the volume that needs to be compressed. Leaks make compressibility calculations inaccurate.

The seals around pistons typically allow 0–5 μL/min of mobile phase to escape from the system before they need to be replaced. Small leaks in check valves allow a small fraction of the flow to leak forward or backward, changing the amount of material in the pump cylinder available for compression or delivery.

One commercial pump starts with a theoretical optimum compressibility factor, calculated from an equation of state, then automatically seeks a superior empirical optimum that minimizes pressure and flow ripple. This pump compensates for leaks as well as inaccurate temperature and pressure readings. The result is less pressure ripple and lower detector noise.

5. Pumping Binary Fluids

Compared to HPLC, pSFC is often used with unusually low modifier concentrations. Figure 13 presents a typical example. Modifier concentrations as low as 2% produce only modest retention of some solutes, whereas concentrations above 10% result in little resolution between any of the solutes. Note that the selectivity does not change significantly with modifier concentration.

The most reproducible way to create modified fluids is to dynamically mix them using multiple high-pressure pumps, each in the flow control mode. Low-pressure mixing valves, now common in HPLC cannot be used to generate mixtures for pSFC since one of the fluids is supplied at elevated pressure in a gas cylinder. Two or more high-pressure pumps allow the composition to be dynamically varied to create composition gradients.

Because selectivity is not significantly altered by small changes in concentration, the purpose of system precision in generating binary fluids is control of retention time reproducibility.

(a) Premixed Fluids. Binary fluids can be purchased in cylinders from some gas vendors. In the past, they were often used by owners of single-channel syringe pump systems designed for capillary work, in an attempt to modify the their equipment for use with packed columns.

Premixed binary fluids change composition in the supply cylinder as they are used up [42,43]. The concentration of methanol in carbon dioxide changes by more than a factor of 2 during the lifetime of a cylinder. As a general rule of thumb, doubling modifier concentration cuts retention in half. Thus, pumping

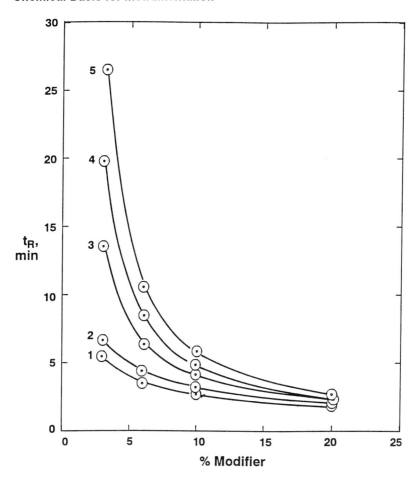

Figure 13 The effect of mobile phase composition on the retention of tricyclic antide-pressants [50]. Conditions: 3 mL/min MeOH (containing 0.5% isopropylamine) in carbon dioxide, 40°C, 200 bar. 4.6 × 250 mm, 5 μm dp Lichrospher cyanopropyl. Solutes: (1) amitriptyline, (2) impramine, (3) nortriptyline, (4) desipramine, (5) protriptyline. 220 nm, 5μL of 100 ppm each in methanol.

premixed binary fluids with single pumps produce significant drift in retention times. Use of such premixed fluids should be avoided.

(b) Padded Tanks. As an alternative to pump head chilling, helium has often been added to the head space of carbon dioxide cylinders to increase the supply pressure. This allowed the use of less capable pumps and made it less likely that

gas bubbles would form anywhere in the fluid delivery system (conditions always well away from the gas/liquid equilibrium line). However, for accurate, reproducible retention control, such padded tanks should also be avoided [43]. Helium dissolves slightly in carbon dioxide, changing its characteristics as a tank is used.

6.   Other Useful Pump Characteristics

Many modern HPLC pumps use multiple pistons. This allows one piston to fill while the other is delivering fluid. At the end of the stroke, the pistons change direction. A rapid compression stroke raises the pressure in the second piston to system pressure, whereupon it takes over fluid delivery. Such pumps are ideal for pSFC, provided the pump head is chilled and an extended compressibility range is provided for.

Older HPLC pumps used constant speed motors turning a cam that pushed a piston in and out of a cylinder to create a reciprocating action. The return stroke was used to refill the pump. The eccentric shape of the cam allowed for a rapid precompression of the fluid. However, each cam shape represents a single compressibility factor. To change the compressibility, a different cam or a variable-speed motor would be needed to change the speed of the cam. Changing cams is impractical. Because older pumps do not have variable speed motors, they typically make poor choices as pSFC pumps. The best modern pumps use stepper motors and digital drivers, which allow them to follow any compression or delivery stroke profile programmed by the user.

7.   Reciprocating vs. Syringe Pumps

Although syringe pumps make ideal pressure sources, they are a poor choice for flow control of a compressible fluid, when a different device controls pressure. When the pressure is increased, part of the displacement goes toward compressing the fluid, whereas part goes toward delivering flow. Each pressure requires a different speed to deliver a constant amount of fluid vs. time. Furthermore, when the pump is nearly full, the pump must move faster than when the pump is nearly empty.

The temperature of large syringes is far from uniform, particularly if there is rapid compression, causing adiabatic heating of the fluid. Changing temperatures causes flow instability.

If two syringes are used to create binary composition gradients (at constant outlet pressure), the head pressure of the column will change during the gradient because the viscosity of the fluid is a function of its composition. This causes further problems with compressibility compensation.

Multiple syringes with large volumes, when used to create composition gradients, can "beat" or "alias" against each other. This beating results in low-frequency variations in composition as each syringe partially responds to perturbations induced by the other pump. This was a major factor that lead to the replacement of syringe pumps in gradient elution HPLC and more recently caused the same problems with a commercial SFC instrument.

Syringe pumps, seldom have a delivery volume > 250 mL. At 2.5–3 mL/min, typical on a 4.6-mm-id packed column, syringe pumps need to be refilled at least every hour and 40 min. Refilling requires up to 20 min.

With a syringe pump and fixed restrictor, there is no convenient way to prevent the column from depressurizing when the pump is being refilled. Because mobile phase components adsorb onto stationary phases, intermittent depressurizations disrupt these adsorbed films. Furthermore, significant drops in pressure during pump refilling can cause binary mobile phases to separate into multiple phases. A significant portion of the syringe volume can be wasted in reestablishing the adsorbed films and mobile phase integrity after a pressure disruption.

Reciprocating pumps have unlimited delivery volume. They essentially never need to be refilled. Making binary mixtures is straightforward. They make it convenient to compensate for compressibility changes. Because individual strokes are small and rapid, variations in composition during a stroke can be averaged out with a mixer.

B. Mixing

The point at which the fluids are brought together is usually a tee in front of the injection port. During refill strokes, the flow from one pump momentarily stops. The fluid exiting the tee momentarily consists of 100% of the "other" fluid whose pump is not refilling. Short-term variations in the flow or composition are expressed as baseline noise in the detector output. Longer term variations also result in variations in retention time reproducibility.

Fluctuations in composition tend to be a greater problem in pSFC than in HPLC because one routinely uses lower concentrations of modifier. For example, using 2% methanol in carbon dioxide at 2.5 mL/min, requires a methanol flow rate of 50 μL/min. If the modifier pump stroke volume is 100 μL, the modifier pump would refill only once every 2 min. Deviations in flow usually occur during refill and recompression strokes. If the actual composition varies over a 2-min period, any mixer must be capable of completely mixing more than 2 min (i.e., >5 mL) of fluid flow. The mixer must hold >5 mL! Note that it is irrelevant as to whether the mixer is dynamic or static. A smaller dynamic mixer will not remove the perturbation.

A smaller modifier pump stroke simplifies the mixing problem. A 20-μL stroke volume, means that the pump refills five times more often than with a 100-μL stroke volume. The mixing volume required is one fifth as great (i.e., approx. 1 mL). A very large modifier (nearly incompressible fluid) pump stroke (i.e., a large syringe pump) also minimizes the problem, if it can be guaranteed that a refill stroke will not be required at any time during a chromatographic run.

With isocratic systems, the volume of the mixer is nearly irrelevant. However, mixers cause a delay between the start of a gradient and the time the gradient reaches the head of the column. Bigger mixers cause longer delays.

(a) Backflow. Another often overlooked issue is backflow. When each pump starts a refill stroke, no flow emerges for some finite time. With compressible fluids, the "other" fluid can backflow into the delivery line of the fluid whose pump is refilling, aggravating mixing problems because the flow through a mixing tee becomes 100% of the "other" fluid. One way to minimize such backflow is to put some form of shut-off valve at the tee. Some have used six-port high-pressure valves that rotate every time a pump must be refilled. More simply, a check valve can be installed in one or both lines. Valves minimize backflow but do not eliminate perturbations in the composition exiting the tee.

C. Injection

Standard HPLC injection valves are used to introduce samples. Because the fluids expand to fill the available volumn, partial loop injections in SFC are simple, although precision is poor compared to full-loop injections. Sample capacity is roughly proportional to surface area, so that standard packed columns can accommodate nearly 10,000 times more sample than a capillary column. At least 20 μL of some solvents can be directly injected onto a 4-mm-id packed column. Injection reproducibility can be $<<\pm1\%$ RSD.

Because SFC is normal phase, some difficulties can arise when larger injections or more polar sample solvents are used [44]. Less polar solutes require smaller sample volumes and less polar solvents. More polar solutes allow use of larger volumes of more polar solvents.

Only small amounts of water will dissolve in carbon dioxide. Nevertheless, 5 μL of water can be directly injected into a 4.6-mm column, using 8% methanol in carbon dioxide at 60°C as the mobile phase [45].

Very large water samples (at least 70 mL) can be introduced [46] by performing a form of solid phase extraction in the injection port. The loop of the injection valve is replaced with a cartridge precolumn 1–2 cm long. The water sample is pushed through the precolumn, trapping the solutes. The water is then blown off with a low-pressure gas and the valve thrown. Traces of water remaining appear to have little effect on retention time stability. A further description of a similar automated approach is provided in Chapters 3 and 13.

D. Temperature

In GC, a change of 20–30°C changes the retention by approximately a factor of 2. However, the retention of all solutes tends to change in a similar manner. In pSFC, temperature tends to cause modest changes in retention but sometimes dramatic changes in selectivity. Increasing temperature can cause an increase, a decrease, or no change in retention. Within one family of compounds, different members may exhibit different behaviors. The retention of antipsychotic drugs [47] is plotted as a function of temperature in Fig. 14. Crossing lines indicate peak reversals. The peaks numbered 4, 5, and 6 show complete overlap at 35°C, but at 65°C show $\alpha(= k_2/k_1) = 1.35$ between peaks 4 and 6. The retention of peak 6 increases with an increase in temperature. The retention of peak 5 decreases slightly, whereas the retention of peak 4 decreases more significantly. Other examples of large changes in selectivity with small changes in temperature include carbamate pesticides [48], stimulants [49], and tricyclic antidepressants [50].

On the other hand, temperature had almost no effect on the selectivity between five sulfonylurea herbicides [51], as shown in Fig. 15. All of the lines are essentially parallel, indicating little or no change in selectivity.

Large changes in selectivity with modest changes in temperature are probably due to changes in both the composition and thickness in the absorbed film of mobile phase components on the stationary phase. Increasing the temperature also makes the components of the mobile phase more miscible, decreasing the "clustering" phenomenon, and making solvent strength a more nearly linear function of composition.

1. Oven Characteristics

Because changes in selectivity due to temperature changes can be dramatic, the accuracy and precision of temperature control equipment is probably even more important in pSFC than in GC. The useful temperature range in pSFC tends to be smaller than in GC. Subambient temperatures are more generally useful in pSFC than in GC. However, silica-based packings can be dehydrated at high temperatures. With carbon dioxide–based fluids, packed columns are seldom useful above 120°C, although there have been reports using temperatures up to 200°C. The critical temperature of other fluids can be well above 250°C, but their use is generally avoided. Oven temperature should be controlled to $<\pm0.5°C$.

HPLC ovens usually control the temperature of a block of metal in contact with the column. There is usually poor thermal contact between the block and the column. Subsequently, the fluid is preheated to the block temperature by passing it through tubing embedded in the block. In pSFC, pressure drops in the column can cause localized adiabatic cooling. The inadequate thermal contact

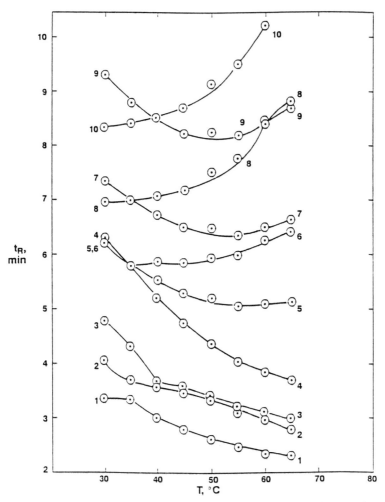

Figure 14 The effect of temperature on the selectivity of antipsychotic drugs [47]. Conditions: 3 mL/min of 6% MeOH (containing 0.5% isopropylamine) in carbon dioxide at 200 bar. 4.6 × 250 mm, 5 μm dp Lichrospher cyanopropyl. Solutes: (1) Triflupromazine, (2) carphenazine maleate, (3) methotrimeprazine, (4) promazine HCl, (5) molindone HCl, (6) perphenazine, (7) chloroprothinine, (8) desperpidine, (9) thiothixene, (10) reserpine. 5 μL of 100 ppm each injected.

between the column and the heated block can cause the temperature of the fluid to deviate dramatically (i.e., >–5°C) from the setpoint. Heated block ovens are a poor choice in pSFC. An air bath oven tends to provide better thermal contact but should still be used with a precolumn heat exchanger.

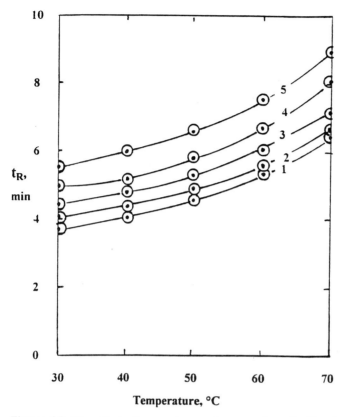

Figure 15 The effect of temperature on the separation of sulfonylurea herbicides [51]. Conditions: 2 mL/min of 10% MeOH in carbon dioxide at 200 bar outlet pressure. 4.6 × 250 mm, 5 m dp Lichrospher DIOL. Solutes: (1) benzsulfuron methyl, (2) chlorimuron ethyl, (3) chlorsulfuron, (4) metsulfuron methyl, (5) sulfmeturon methyl. 1 μL of approx. mg/mL each injected 220 nm.

2. The Irrelevance of Sub- vs. Supercritical Temperatures

For years it was often debated as to whether a specific separation was super- or subcritical. In most cases, the distinction is simply one of definition, with no real difference in chemical or physical characteristics. In a way, there is no such thing as SFC [52]. The characteristics that attract users to SFC are not unique to the supercritical condition. Diffusion coefficients, viscosity, density, and solvent strength show minimal change when the fluid changes from super- to subcritical, or vice versa. Fluids slightly above their critical temperature are almost identical to the same fluids slightly below their critical temperature. Note, for instance, the constant density lines across the top of Fig. 7. The solid vertical line divides sub- from supercritical conditions. There is no discontinuity when the fluids cross the line.

Because there is little penalty in speed or pressure drop, it is often desirable to try temperatures below the critical temperature of the mobile phase. Subcritical operation has been found to be particularly useful with chiral separations [53–57] where minimizing thermal agitation allows better shape fits.

E. Pressure Control

In the author's laboratory, pressure is used as a secondary control variable. Plots of retention vs. pressure in the literature show a range of effects. In Fig. 16a, pressure had almost no effect on the retention or selectivity of stimulants [49]. Several peaks merged at low pressures. Pressure had a slightly greater effect on the retention and selectivity of antipsychotic drugs [47], as shown in Fig. 16b.

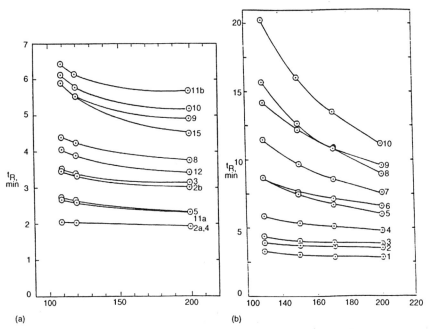

(a)   (b)

Figure 16 (a) The effect of pressure on the retention and selectivity of stimulants [49]. Conditions: 2 mL/min of 10% MeOH (containing 0.5% isopropylamine) in carbon dioxide, 40°C. (b) The effect of pressure on the retention and selectivity of same antipsychotics in Fig. 14. Conditions: 3 mL/min 6% MeOH (containing 0.5% isopropylamine) at 50°C. 5 μL of 110 ppm stimulants, 100 ppm antipsychotics injected.

Note the peak reversal. In yet another study, pressure had a relatively large effect on retention but almost no effect on selectivity [50].

The accuracy of pressure measurements is of secondary importance when using modified fluids and packed columns. Packed columns are usually operated at relatively high densities and low temperatures. Under such conditions, changes in density caused by changes in pressure are small.

1. Column Pressure Drops–Column Selection

The pressure drop is proportional to the viscosity of the fluid. Because both pure and binary supercritical and near-critical fluids have much lower viscosity (typically 1/10th) than normal liquids, pressure drops in pSFC are much lower than in HPLC. This lower pressure drop allows the use of smaller, more efficient packings; or very long columns; or higher flow rates. Several groups have used columns with 3-μm packings operated at up to 20 times their optimum flow rate for rapid screening.

With packed columns, the pressure drop is proportional to the square of the particle diameter. Switching from 5- to 3-μm particles increases pressure drop 2.8 times (25/9ths). The smaller particles are also both more efficient and have a higher optimum flow rate. Taking all effects into account, a 15-cm column packed with 3-μm particles produces the same efficiency in one third of the time but with 2.8 times higher pressure drop than a 25-cm column packed with 5-μm particles.

However, smaller particles occupy a larger fraction of the cross-sectional area of a tube, such as a column. Because they have smaller effective open cross-sectional area and require frits with smaller pores in their end fittings, they tend to generate even higher pressure drops and are easier to plug.

For years, the column pressure drop was thought to be a significant problem because changes in retention were so thoroughly associated with changes in pressure and density. Thus, a pressure drop means that local retention is different at every place along the column. However, with modified fluids, the retention gradient is relatively small and, even with pure fluids, causes few, if any, practical problems. Nevertheless, it is advisable to monitor pressure both before and after the column (two pressure transducers).

The position in the system with the worst solvent strength must be actively controlled. In the past, it was common to control the column inlet (i.e., the pump) pressure. Controlling the inlet pressure while generating uncontrolled pressure drops can create all sorts of potential problems. With increasing pressure drops, solute solubility near the outlet decreases, potentially leading to precipitation. At lower outlet pressures, binary fluids can become unstable, breaking down into two phases or causing excessive adsorption.

If the outlet pressure is controlled, an increase in the pressure drop means that the inlet pressure must rise. Solubility always improves. A single phase continues to exist and retention merely decreases slightly (due to slightly higher average density in the column).

Packed capillary columns continue to be used with column inlet pressure control since there are no adequate back pressure regulators with very low dead volumes that operate at the low flows required. A pressure transducer downstream of the column would excessively broaden peaks. Subsequently, the inlet pressure is controlled and the outlet pressure is unknown (see Chapter 5).

2. Back Pressure Regulators

The use of back pressure regulators creates one of the most important differences between capillary and packed column instrumentation. The downstream control of outlet pressure, combined with flow control pumps, allows independent control of pressure, composition, and flow (see also Chapter 3).

In the 1970s and 1980s, packed column systems used mechanical back pressure regulators. Some workers devised electromechanical drives for regulators to allow pressure programs. More recently, both chemical and instrument companies have developed solenoid activated [58,59], or piezoelectric [60–62] back pressure regulators that allow accurate control, easy programming, and keyboard settability. Since electronic pressure control has become commonplace on GCs, and as standalone accessories to GCs, pressure programming, using the low-pressure GC controllers with a pneumatic amplifier, offers an alternative low cost solution.

The expansion of the mobile phase inside the regulator results in dramatic Joule–Thompson cooling. Without heating, the device can intermittently clog with ice, causing flow and pressure fluctuations. Heating the body of the regulator need not cause thermal degradation of solutes. The surfaces in contact with the solutes is warmed by the heater but cooled by the expansion. Just enough heat should be added to avoid icing. There need not be any increase in the actual temperature of the surfaces. Furthermore, the sample residence time in the regulator is no more than a few milliseconds.

There are both outlet pressure transducers and electronic back pressure regulators that have internal volumes so small that peaks can pass through them without excessive band broadening. Peaks from standard bore columns can pass through such devices in enough volume to be collected in a relatively pure form. Furthermore, detectors, such as light scattering (ELSD) or mass spectrometers (MS), can be mounted downstream of the back pressure regulator without destroying the separation.

3. Fixed Restrictors

Fixed restrictors are passive devices, usually a pinhole or a frit, that restrict mass flow at a pressure dictated by a pump. Restrictors offer no easy way to change flow at constant pressure and temperature. Fixed restrictors slowly plug, changing flow uncontrollably. The degree of plugging is not obvious. Using fixed restrictors with binary fluids, changes in compositions cause changes in flow or changes in pressure drops or both [63,64]. These unpredictable changes make it difficult to deconvolute the effects of flow, composition, and pressure.

With packed columns, fixed restrictors are seldom used to control column flow (except with packed capillary columns). They are only used to introduce a small portion of the total flow into a GC like detector. The restrictor is usually mounted in a "tee" between the column outlet and the back pressure regulator. Most of the flow proceeds through the back pressure regulator, while a small fraction passes into the GC detector.

4. Back Pressure Regulators Are Not Variable Restrictors!

It has become common for back pressure regulators to be called variable restrictors. They are not. Restrictors are passive devices that restrict *mass flow* at a pressure dictated by the pumps. They are not true control devices. Control implies that flow could be varied on demand.

Back pressure regulators maintain the *pressure* of the system just upstream of their location, regardless of the flow rate or composition (or temperature). The system flow is dictated by a pump or pumps. Thus, restrictors and back pressure regulators belong to two very different control philosophies and *cannot* be interchanged.

5. Detectors

A discussion of detectors is too complex for a chapter like this. The reader should see any of several reviews (see Chapter 3 or Ref. 65). However, some general statements can be made.

In studying fuels and oils, the FID is widely used with packed columns. However, modified mobile phases are seldom compatible with the FID. For some solutes, other GC detectors are compatible with modified fluids and packed columns. For instance, the nitrogen phosphorus detector (NPD) or thermionic detector (TID), the flame photometric detector (FPD), the photo-ionization detector (PID), the chemiluminescence nitrogen detector, the electron capture detector (ECD), and the sulfur chemiluminescence detector are all compatible with at least small flow rates (i.e., 5 mL expanded) of moderately polar (i.e., up to 20% methanol in carbon dioxide) modified mobile phases.

Generally, modified fluids are used with the ultraviolet-visible (UV-vis) absorbance detector with the same limitations in pSFC that it exhibits in HPLC. The only modification required is the inclusion of a high-pressure flow cell. Solutes must contain a chromophore to be detected. However, carbon dioxide is extremely transparent to below 190 nm. It is easier to use short wavelengths in pSFC than in HPLC. Fluorescence detectors can also be used but also require a high-pressure flow cell.

One problem remaining in pSFC is the need to detect polar, labile molecules that do not contain a chromophore but require a modified fluid to elute. The evaporative light-scattering detector (ELSD) appears to be a possible solution as shown in Chapters 3 and 4.

The mass spectrometer also holds promise, but the development of interfaces for SFC-MS lags behind HPLC-MS or GC-MS. Some recent work suggests that SFC-MS is coming of age. An atmospheric pressure ionization, electrospray interface was reported [66] to give better detection limits than HPLC-MS using the same device.

VII. EXTENDING THE POLARITY RANGE: THE USE OF ADDITIVES

The single development that has allowed the extension of pSFC to much more polar molecules was not instrumental but involved changes to the mobile phase. Tertiary fluids [65–73] have allowed much more polar solutes to be eluted and separated. The third fluid component, called an *additive*, is usually much more polar than the modifier and by itself is often immisible with carbon dioxide. Typical additives include trifluoroacetic acid (TFA) and isopropylamine (IPAm). Tertiary fluids are usually prepared by making a solution of 0.1–1% additive in modifier. The modifier (plus additive) is then pumped together and mixed with carbon dioxide.

There are numerous examples in the literature showing no elution of solutes without additive, but sharp peaks with modest retention times when a small concentration of an additive is included in the same mobile phase. An example [47] is shown in Fig. 17.

Curiously, most effective additives belong to the same family as the solutes of interest [70,73]. Thus, stronger acids improve the peak shapes of weaker acids [70]. Similarly, stronger bases improve the peak shape of weaker bases [73]. Although there are a number of papers that talk about ion pairing, there are very few [68,69] where the mechanism actually appears to be ion pairing.

Steric hindrance is extremely important [73]. Quaternary and tertiary amines are much less effective additives than secondary amines, which are much less effective additives than primary amines for improving the elution of bases. This order is opposite to the order of increasing basicity. That is, the strongest bases are usually much less effective than weaker bases, if the stronger basic group is sterically hindered. The retention of bases also follows this inverse order. More hindered but stronger bases elute before less hindered, weaker bases.

In HPLC, additive effectiveness follows basicity. The most common basic additives in HPLC, such as trimethylamine, are relatively ineffective additives in pSFC.

The concentration of the additive can be important [69]. If the additive has a pK_a similar to that of the solutes, retention of the solutes can change as a func-

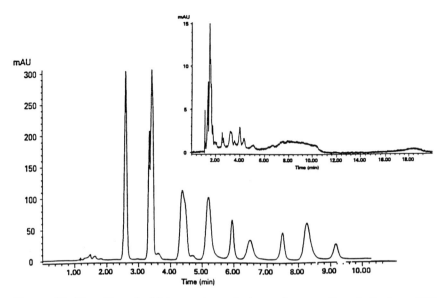

Figure 17 The effect of additive on the peak shape of same antipsychotic drugs [47]. Upper curve obtained using a mobile phase of methanol in carbon dioxide. Lower curve same with 0.5% isopropylamine added to the methanol. Solutes same as in Fig. 14. Conditions: 3 mL/min of 6% MeOH in CO_2, 50°C, 200 bar, 4.6 × 250 mm, 5 µm Lichrospher CN. 5 µL of 100 ppm each injected.

tion of the concentration of the additive. More typically, the additive chosen is much more polar than the solutes, suppressing their ionization.

Even from very dilute solutions, polar additives adsorb until they form a large fraction of a monolayer on the surface of the stationary phase [69]. If a very low concentration is used, it can take a long time for the monolayer to form. During the formation, retention time can drift monotonically.

The addition of additive does not appear to change the effect of the physical parameters. Changing the modifier concentration causes large changes in retention but little change in selectivity. Pressure only has modest effect on either retention or selectivity. Temperature tends to have only a minor effect on retention but a significant effect on selectivity. Programming conditions changes the nature of the adsorbed film. Negative peaks can result from adsorption or desorption of additive during composition, pressure, or temperature programs. The effects appear to be reproducible.

VIII. CONCLUSIONS

Packed column SFC is experiencing a renaissance, particularly for the analysis of small polar molecules like the drug substance in most pharmaceutical drug products. pSFC is as easy to use and understand as reversed phase HPLC but provides an alternate selectivity, higher speed, and the potential for higher efficiency. It also produces minimal waste and generally uses less expensive consumables.

Packed column SFC is dominated by the use of binary and tertiary mobile phases. To perform effective pSFC, the composition and the flow rate of the mobile phase must be precisely controlled. The adsorption of mobile phase components onto the stationary phase is largely overlooked but extremely important on understanding the retention process.

It should be clear from the analysis presented in this chapter that pSFC has little in common with cSFC except for the similarity in names. Attempts to convert capillary instrumentation for packed column work often results in complete failure. The two techniques attempt to exploit different aspects of the physical chemistry of super- and near-critical fluids.

REFERENCES

1. *Handbook of Chemistry and Physics*, 69th ed., CRC Press, Boca Raton, FL.
2. *Gas Encyclopedia*, Elsevier, Amsterdam, 1976.
3. J. H. Perry, *Chemical Engineers Handbook*, 4th ed., McGraw-Hill, New York, 1963.
4. Z. Balenovic, M. N. Meyers, and J. C. Giddings, *J. Chem. Phys.*, *52*:915 (1960).
5. I. Swaid and G. M. Schneider, *Ber. Bunsenges. Phys. Chem.*, *83*:969 (1979).
6. R. Feist and G. M. Schneider, *Sep. Sci. Tech.*, *17*:261 (1982).
7. H. H. Lauer, D. McManigill, and R. D. Bored, *Anal. Chem.*, *55*:1370 (1983).
8. T. A. Berger and W. H. Wilson, *Anal. Chem.*, *65*:1451 (1993).
9. K. E. Karlsson and M. Novotny, *Anal. Chem.*, *60*:1662 (1988).
10. J. C. Giddings, M. N. Meyers, L. M. McLaren, and R. A. Keller, *Science, 162*:67 (1968).
11. T. A. Berger, *Packed Column SFC*, RSC Monograph Series (R. M. Smith Ed.), Royal Society of Chemistry, Cambridge, UK, p. 57 (1995).
12. B. W. Wright and R. D. Smith, *J. Chromatogr.*, *355*:367 (1986).
13. A. L. Blilie and T. Greibrokk, *Anal. Chem.*, *57*:2239, (1985).
14. Ph. Morin, M. Caude, and R. Rosset, *J. Chromatogr.*, *407*:87 (1987).
15. S. M. Fields, K. E. Markides, and M. L. Lee, *J. Chromatogr.*, *406*:223 (1987).
16. J. F. Deye, T. A. Berger, and A. G. Anderson, *Anal. Chem.*, *62*:615 (1990).
17. T. A. Berger and J. F. Deye in *Supercritical Fluid Technology*, (F. Bright and M.E.P. McNally, eds.), ACS Symposium Series 488, American Chemical Society, Washington, DC, Chap. 11 (1992).
18. K. P. Johnston, K. L. Harrison, M. J. Clark, S. M. Howdle, M. P. Heitz, F. V. Bright, C. Carlier, and T. W. Randolph, *Science, 271*:624 (1996).
19. T. A. Berger, *Packed Column SFC*, RSC Monograph Series (R. M. Smith, ed.), Royal Society of Chemistry, Cambridge, UK, p. 56 (1995).
20. T. A. Berger, *Packed Column SFC*, RSC Monograph Series (R. M. Smith, ed.), Royal Society of Chemistry, Cambridge, UK, p. 58 (1995).

21. T. A Berger, *Packed Column SFC*, RSC Monograph Series (R. M. Smith, ed.), Royal Society of Chemistry, Cambridge, UK, p. 113 (1995).
22. L. R. Snyder, *J. Chromatogr.*, *92*:223 (1974).
23. J. Fugueras, *J. Am. Chem. Soc.*, *93*:3255 (1971)
24. S. Kim and K. P. Johnston, *AIChEJ.*, *33*:1603 (1987).
25. U. K. Deiters, *Fluid Phase Equilib.*, *8*:123 (1982).
26. W. B. Whiting and J. M. Prauznitz, *Fluid Phase Equilib.*, *9*:119 (1982).
27. C. A. Eckert and B. L. Knutson, *Fluid Phase Equilib. 83*:93 (1993).
28. T. A. Berger and J. F. Deye, *J. Chromatogr. Sci.*, *29*:280 (1991).
29. T. A. Berger and J. F Deye, *Anal. Chem.*, *62*:1181 (1990).
30. T. A. Berger, *J. High Resolut. Chromatogr.*, *14*:312 (1991).
31. S. V. Olesik and J. L. Woodruff, *Anal. Chem.*, *63*:670 (1991).
32. T. A. Berger and J. F. Deye, *J. Chromatogr.*, *594*:291 (1992).
33. C. R. Yonker, D. G. McMinn, B. W. Wright, and R. D. Smith, *J. Chromatogr.*, *396*:19 (1986).
34. C. R. Yonker and R. D. Smith, *J. Chromatogr.*, *361*:25 (1986).
35. A. L. Blilie and T. Greibrokk, *Anal. Chem.*, *57*:2239 (1985).
36. J. M. Levy, Ph.D. thesis, Case Western Reserve University, Cleveland, 1986.
37. J. R. Strubinger, H. Song, and J. F. Parcher, *Anal. Chem.*, *63*:98 (1991).
38. J. R. Strubinger, H. Song, and J. F. Parcher, *Anal. Chem.*, *63*:104 (1991).
39. T. A. Berger, *Chromatographia*, *37*:645 (1993).
40. C. H. Lochmuller and L. P. Mink, *J. Chromatogr.*, *471*:357 (1989).
41. T. A. Berger, *Packed Column SFC*, RSC Monograph Series (R. M. Smith, ed.), Royal Society of Chemistry, Cambridge, UK, 1995, Chap. 4.
42. F. K. Schweighardt and P. M. Mathias, *J. Chromatogr. Sci.*, *31*:207 (1993).
43. J. Via, L. T. Taylor, and F. K. Schweighardt, *Anal. Chem.*, *66*:1459 (1994).
44. R. M. Smith and D. A. Briggs, *J. Chromatogr. A*, *670*:161 (1994).
45. M. S. Klee and M. Z. Wang, *Hewlett Packard Appl. Note. No. 228–252* (Nov. 1993).
46. T. A. Berger, *Chromatographia*, *41*:471 (1995).
47. T. A. Berger and W. H. Wilson, *J. Pharm. Sci.*, *32*:281 (1994).
48. T. A. Berger, W. H. Wilson, and J. F. Deye, *J. Chromatogr. Sci.*, *32*:179 (1994).
49. T. A. Berger and W. H. Wilson, *J. Pharm. Sci.*, *34*:489 (1995).
50. T. A. Berger and W. H. Wilson, *J. Pharm. Sci.*, *32*:287 (1994).
51. T. A. Berger, *Chromatographia*, *41*:133 (1995).
52. T. A. Berger, *Packed Column SFC*, RSC Monograph Series (R. M. Smith, ed.), Royal Society of Chemistry, Cambridge, UK, p. 47 (1995).
53. N. Bargmann, A. Tambute, and M. Caude, *Analusis*, *20*:189 (1992).
54. F. Gasparrini, D. Misiti, and C. Villani, *J. High Resolut. Chromatogr.*, *13*:182 (1990).
55. F. Gasparrini, D. Misiti, and C. Villani, *TrAC*, *12*:137 (1993).
56. R. W. Stringham, K. G. Lynam, and C. C. Grasso, *Anal. Chem.*, *66*:1949 (1994).
57. K. Anton, J. Eppinger, L. Frederikson, E. Francotte, T. A. Berger, and W. H. Wilson, *J. Chromatogr.*, *666*:395 (1994).
58. J. Poole and M. Nickerson, U.S. patents 4,892,654 and 5,178,767. Jan., 1990; Jan. 12, 1993.
59. F. Verillon, D. Heems, B. Pichon, and J. C. Robert, "Supercritical Fluid Chromatograph with Independent Programming of Mobile Phase Pressure, Composition, and

Flow Rate," Proceedings of the Pittsburgh Conference, New Orleans, LA, p. 1124 (1992).
60. N. Pericles, U.S. patent no. 5,224,510, issued July 6, 1993.
61. A. Giorgetti, N. Periclès, H. M. Widmer, K. Anton, and P. Dätwyler, *J. Chromatogr. Sci.*, *27*:318 (1989).
62. K. Anton, N. Pericles, and H. M. Widmer, *J. High Resolut. Chromatogr.*, *12*:394 (1989).
63. T. A. Berger, *Anal. Chem.*, *61*:356 (1989).
64. T. A. Berger, *J. High Resolut. Chromatogr.*, *12*:96 (1989).
65. K. Anton, *Encyclopedia of Analytical Science*, (Alan Townshend, ed.), Acedemic Press, New York, p. 4856 (1995).
66. D. Pinkston, et al., paper presented at the 7th International Symposium on SFC/SFE, Indianapolis, IN, March 31–April 4, 1996.
67. M. Ashraf-Korassani, M. G. Fessahaie, L. T. Taylor, T. A. Berger, and J. F. Deye, *J. High Resolut. Chromatogr.*, *11*:352 (1988).
68. W. Steuer, M. Schindler, G. Schill, and F. Erni, *J. Chromatogr.*, *447*:287 (1988).
69. W. Steuer, J. Baumann, and F. Erni, *J. Chromatogr.*, *500*:469 (1990).
70. T. A. Berger and J. F. Deye, *J. Chromatogr.*, *547*:377 (1991).
71. T. A. Berger and J. F. Deye, *J. Chromatogr. Sci.*, *29*:26 (1991).
72. T. A. Berger and J. F. Deye, *J. Chromatogr. Sci.*, *29*:141 (1991).
73. T. A. Berger and J. F. Deye, *J. Chromatogr. Sci.*, *29*:310 (1991).

3

Analytical and Preparative Carbon Dioxide Chromatography with Automated Pressure Control and Different Detectors

Francis Vérillon

Gilson S.A., Villiers le Bel, France

Keith Coleman

Anachem Ltd., Luton, Bedfordshire, United Kingdom

I. INTRODUCTION

As a general technique, packed column supercritical fluid chromatography (pSFC) emerged in the late 1980s. Specifically designed commercial instruments, characterized by flow-independent automated pressure control, have been available since 1992. Modern pSFC is predominantly related to high-performance liquid chromatography (HPLC) by the choice of packed columns and ultraviolet-visible (UV-vis) absorbance detectors as well as by the implementation of binary composition gradients in the mobile phase. However, pSFC offers much more flexibility (pressure gradient, more systematic exploration of temperature, wider choice of detectors) and much higher resolution per time unit. These advantages should lead modern pSFC to replace, at least, many nonaqueous normal phase HPLC methods.

Carbon dioxide is the common basic component of supercritical fluid mobile phases. Nonflammable, nontoxic, relatively inexpensive, and without any disposal costs compared to organic combustible solvents, it can easily be brought above its critical pressure (74 bar) and temperature (32°C). In addition, it can be mixed with a diversity of polar solvents containing secondary additives (typically 0.1% amine and/or organic acid to neutralize stationary phase silanols with sample-like species, or even a few percents of water can be added to further extend the polarity), which results in higher critical parameters. Its miscibility with these modifiers is sufficient to offer a very wide composition range.

59

Generally, low amounts of modifiers (2–20%), elevated temperatures (50–150°C), and common stationary phases are used for complex samples. More modifier (20–40%), low temperature (<50°C), and specialized columns are used for simple mixtures of a few closely related compounds such as enantiomers. While the conditions for complex samples correspond to fluids defined as supercritical, generating high efficiencies from a kinetic optimization, the conditions for simple mixtures implement subcritical fluids enhancing selectivity through thermodynamic adjustments (subcritical fluid chromatography, subFC). No boundary exists between these gas-like and liquid-like operating conditions. Supercritical fluids (SFs) can be viewed as "representing the general fluid state, liquids and gases being its extreme forms," the only ones that exist at atmospheric pressure [1].

Because pSFC, in contrast with current HPLC and gas chromatography (GC), requires that the column outlet be pressurized, pSFC instruments must offer a flow-independent, automated pressure control at the column outlet. To generate a pressure gradient, the automated pressure control is more than a simple convenience: it is necessary for tracking maximum performance. Without this feature, positive pressure gradients can only be achieved by increasing the flow rate, which can spoil the resolution [2]. This chapter describes such an instrument— the Gilson Series SF3 system—using standard HPLC columns for analytical and preparative purposes. Several methods developed by different users of this instrument are presented in the three main areas of its industrial applications: pharmaceuticals where HPLC predominates, petrochemicals where GC predominates, and natural products where HPLC and GC are more equally in use. Each of these fields uses the UV-vis detector as well as mass-sensitive detectors to quantify not only compounds without chromophores, but also as many compounds as possible for quality control of raw materials and finished products.

II. INSTRUMENT

The Gilson Series SF3 system was developed with Ciba Geigy Ltd., which pioneered the work on its design in Basel, Switzerland [3–7]. It was released in 1992 as the first chromatograph for sub- and supercritical mobile phases with independent programming of the mobile phase pressure, composition, and flow rate [8–13]. Previously, pressure changes were obtained either by changing the flow rate through a fixed flow restrictor or, at constant flow rate, by manual adjustment of a pressure relief valve. The first solution is detrimental to efficiency and suffers from frequent plugging; the second cannot generate reproducible pressure gradients. The new instrument solved the problem of automating the pressure gradient without changing the mass flow rate of mobile phase through the column (see also Chapter 2).

A. System Components

The system is shown in Fig. 1 (analytical and preparative configuration) and a flow diagram of the basic analytical-only configuration is shown in Fig. 2. Carbon dioxide is delivered by pump A (308/10SC) equipped with a cooling jacket and a thermostatic kit, itself connected to a cryothermostat. Pump A can act as the system controller, communicating with three modules: pump B (306/5SC) for the modifier, the pressure regulator (821), and the temperature regulator (831). In this case, pump A is electrically linked with all the other modules: autoinjector (231 XL with 720 software for analytical purposes), pressurized cell UV-vis detector (119), and data processor. Carbon dioxide and modifier are blended at high pressure in a dynamic mixer (811). Centralized instrument programming and detector signal processing are available using the UniPoint system controller software.

The mobile phase pressure is measured at the column inlet and outlet, P$_1$ and P$_2$, respectively. P$_1$ is used to define high and low safety limits and to perform compressibility corrections for the flow rate accuracy of each pump. P$_2$ is the programmed pressure, measured between detection cell and pressure regulation valve. At the system outlet, a manual pressure relief valve (E) is installed inside

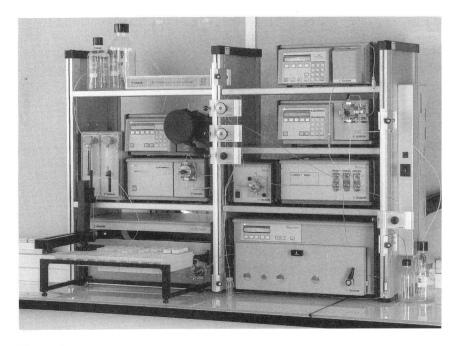

Figure 1 Gilson Series SF3 system for pSFC: analytical and preparative configuration with UV-vis detector.

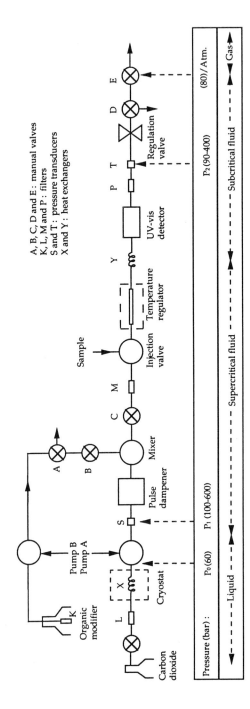

Figure 2 Flow diagram for the analytical configuration [9].

the column oven and adjusted above the critical value (typically 80 bar). Two models having different internal volumes are available for this valve E: modified Nupro SS-4R3A-EP, 1.1 mL (Willoughby, OH, USA) for analytical work with UV-vis detector, and Rheodyne 7037, 6 µL (now manufactured by Gilson under licence of Rheodyne, Cotati, CA) for preparative work and for coupling with nonpressurized cell detectors. Four manual valves A–D (SSI, State College, PA) permit isolation and priming/purging of each line. Column selection valves and a manual injection valve can also be mounted inside the column oven.

As compared with HPLC, requirements for the pumping system involve high-pressure mixing, a microflow rate for the modifier, a cooled and pressurized inlet line for carbon dioxide, and efficient compressibility compensation (see Chapter 2). The compressibility of carbon dioxide entering the pump head (1150 Mbar^{-1} at 7°C and 60 bar, 0.904 g/mL density) is almost 10 times higher than that of methanol (123 Mbar^{-1} at 22°C and atmospheric pressure, 0.791 g/mL density). These values are taken into account to deliver a constant flow rate irrespective of the downstream pressure [9]. This is a first requirement to obtain the same retention times during method transfer from the instrument of manufacturer A to the instrument of manufacturer B using the same column. A second requirement is the same temperature inside the carbon dioxide pump head, so that liquid carbon dioxide is pumped at the same density. The pumping density is 0.95 g/mL for the Hewlett-Packard system and 0.904 g/mL for the Gilson system, a small enough difference (5%) to generate similar chromatograms from direct method transfers between both instruments [14].

In the above conditions, 1 mL of pumped carbon dioxide corresponds to about 0.5 L of gas at the atmospheric outlets. Because one of these outlets is at the injection valve, a reasonable limit to safely expel the gas from the sample loop is 1 mL maximum for the loop capacity. Moreover, the injection must be performed in the total loop filling mode.

B. Pressure Control

A two-stage pressure control is implemented by two valves that perform two separate functions. First, the role of the pressure relief valve E (Fig. 2)—manual, spring-loaded, and located at the end of the line—is to maintain the upstream fluid at the critical state, while the downstream fluid becomes a gas or a mist. The gas expansion which takes place in this valve generates a temperature decrease proportional to the amount of carbon dioxide flowing through this valve (Joule–Thomson effect). However, because this valve is mounted in the column oven, it normally does not undergo the additional endothermic effect that would be associated with a change of state from liquid to gas. With carbon dioxide, this valve is generally set at 80 bar and there is no need to modify this adjustment even when the system is shut off. Full depressurization is obtained by opening the purge valve D (Fig. 2).

Second, the role of the pressure regulation valve—located upstream from the previous one, inside the pressure regulator (821)—is to pressurize the absorbance cell and the column outlet at various values (P_2) which are programmable as a gradient profile. This valve consists of a nozzle and an electromagnetically actuated rod that varies the nozzle obturation under the control of a pressure regulation software. Pressure is thus stable at better than 1% and different programmed values can be reached instantaneously, provided that the gradient slope (S_{max} depending on the flow rate F) is inferior to the specified maximum: S_{max} (bar/min) = 60F (mL/min).

As compared with a single-valve pressure control, this design should result in the pressure regulation valve working efficiently longer and with greater precision.

C. Working Range

The column temperature is adjustable from ambient plus 3°C up to 200°C. Three kinds of gradients can be implemented separately or simultaneously: pressure from 90 to 400 bar at the column outlet (up to 600 bar at the column inlet), volume composition from 0 to 100% modifier solvent in carbon dioxide (miscibility according to temperature and pressure), and total liquid flow rate from 0.5 to 5 mL/min (up to 20 mL/min by changing the pump heads). The carbon dioxide density available is thus from 0.1 to 1 g/mL (Fig. 3) [2]. Columns from 1 to 10 mm diameter can operate under optimum conditions.

Methanol and carbon dioxide are soluble in all proportions if pressure is higher and temperature lower than the following values: 80 bar and 40°C [15], 90 bar and 60°C [9], 160 bar and 110°C [16].

For trouble-free intensive use, it is recommended to restrict the available working range. With the analytical-only configuration, to eliminate the risk of solid carbon dioxide particles momentarily blocking the outlet line and creating spikes visible by the detector, the oven temperature must be higher than 50°C with unmodified carbon dioxide at 5 mL/min, and higher than 35°C with down to 1% modifier at 3 mL/min. With the analytical and preparative configuration, temperature and composition are not restricted because this configuration includes an extra pump for delivery of collection solvent (Sec. II.D), but pressure at the column outlet must be higher than 150 bar to avoid any risk of instability potentially due to the pressure relief valve (Rheodyne 7037).

Two other neat fluids were also used extensively as pSFC mobile phases with the Gilson SF3 system:

Supercritical ammonia (critical point: 113 bar, 132°C) for supercritical fluid synthesis (SFS), analytical and micropreparative pSFC on a carbon column, and supercritical fluid extraction (SFE) of ^{11}C-labeled compounds [17].

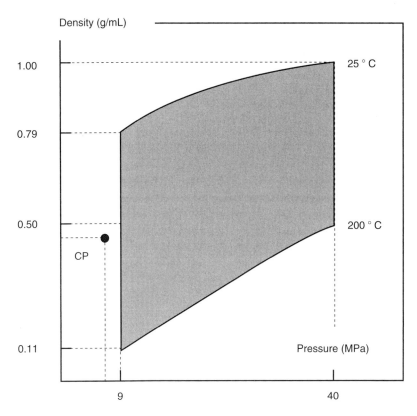

Figure 3 Range of column temperature, pressure at the column outlet, and density of carbon dioxide available with the Gilson SF3 system [2].

Supercritical and subcritical 1,1,1,2-tetrafluoroethane (HFC-134a) to obtain, for polar compounds on a silica column, different selectivities as compared with carbon dioxide and methanol-mixed carbon dioxide [18].

D. Analytical and Preparative Configuration

The most complete configuration, for both analytical and preparative work (Fig. 1) [19], includes a two-valve sampler (233 XL instead of 231 XL) for sample injection, fraction collection, and fraction reinjection, option for electrically cooled vials (832), and an extra pump (307/5SC) to add a collection solvent downstream from the UV-vis detector. As mentioned, the pressure relief valve E (Fig. 2) must be Rheodyne 7037 because of its negligible internal volume. For fraction collection, available criteria are elution time or signal amplitude (using the SFC-FC sampler software or the UniPoint system controller software).

Collection is usually performed in liquid phase, into 20-mL glass vials equipped with a prepierced cap and prefilled with a 2-mL solvent foot. The collection needle moves down to the vial bottom. Each vial receives several fractions of the same nature produced by repeated injections of the same sample. Purity and recovery are both determined from 20-µL reinjections into the same system. The addition of collection solvent (Fig. 4) is performed upstream from the pressure control valves at a flow rate of 5% to 40% of that of the mobile phase. This solvent keeps the solutes dissolved after depressurization of the mobile phase and generates two positive effects: recovery yields are increased up to full quantitative recovery and any risk of plugging potentially associated with high sample loading is eliminated.

Solventless injection of several samples in routine analysis is also automated by the 233 XL sampler. With this technique, peak splitting does not occur when the elution strength of the sample solvent is higher than that of the mobile phase, as illustrated in Sec. III.B.4 for the analysis of natural waxes [20]. Solventless injection requires two rotary six-port valves, a short cartridge to trap the sample solution, and a source of inert gas to remove the sample solvent. The three-step procedure is illustrated in Fig. 5. Solventless injection is also very useful for two other sample preparations: water removal from aqueous samples

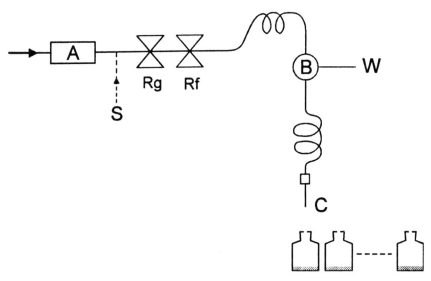

Figure 4 Fraction collection line. A, pressurized cell detector, B, fractionation valve; C, collection tip; S, collection solvent; W, waste; Rg, pressure regulation valve; Rf, pressure relief valve. Prepierced cap vials, possibly cooled. Internal volumes: AB, 140 µL; BC, 80 µL.

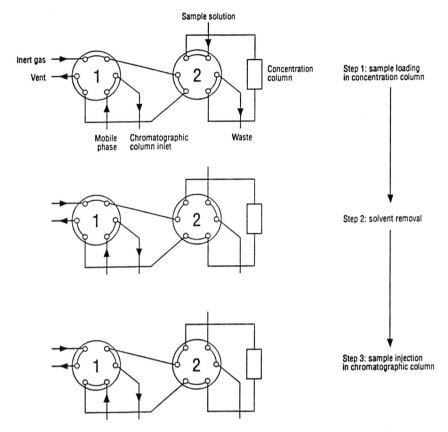

Figure 5 Solventless injection: flow diagram of the three-step procedure [20].

and trace enrichment of low-concentration analytes. This should permit an extended use of pSFC for biological and environmental analysis as shown in chap. 13.

Solventless collection of several fractions in solid phase could be automated using the Gilson ASPEC XL sampler. Fractions of adsorbed solutes, instead of dissolved solutes, would be collected into disposable SPE minicolumns and desorption obtained with less than 1 mL of solvent.

High-pressure collection has been described, into loop(s) of a rotary valve, and into special vessels immersed into liquid nitrogen. Carbon dioxide solidifies before the vessels are opened, then sublimates and leaves pure solutes [21].

E. Detectors

This section presents technical considerations and practical advice for the use of the three detectors most currently used with the Gilson SFC system. In addition to the UV-vis detector, the system is compatible with the flame ionization detector (FID) and the evaporative light-scattering detector (ELSD) without restricting their working range or their performance. Other couplings have been reported for the Gilson system: with the chemiluminescence nitrogen detector, also called thermal energy analyzer (TEA) for the analysis of nitrosamines [22], with the electrospray-ionization mass spectrometer (EI-MS) [23], and with the atmospheric-pressure chemical-ionization mass spectrometer (APCI-MS) for the analysis of cannabis [24]. An overview of SFC detectors was recently published [25].

1. Ultraviolet-Visible Absorbance Detector

A standard for analytical and preparative purposes, the UV-visible (UV-vis) detector is the most commonly used detector for pharmaceutically active substances, most of them having chromophores. The column effluent is detected below the critical temperature, at <30°C over the entire working range of the system (Fig. 2). Under these conditions, the baseline noise amplitude is the same as in HPLC (30 μAU/cm), which is also true for the limits of detection (LODs, typically 1 ppm in the sample) when the same column and the same detector equipped with the same flow cell are used for both pSFC and HPLC. In fact, the limits of detection (LODs) obtained with pSFC-UV-vis are generally similar to those of HPLC-UV-vis because shorter retentions in pSFC compensate for detrimental factors such as a possible smaller light path of pressurized flow cells, molecular extinction coefficients lower in carbon dioxide than those in hexane, etc. (see Fig. 21).

In contrast with the modifier gradient, the positive pressure gradient, specific to pSFC, has no delay time but generates a negative slope of the absorbance baseline. Figure 6 shows this slope on a high-speed chromatogram where 11 steroids were resolved in 5 min with superimposed pressure and modifier gradients. A more convenient horizontal baseline was obtained from the system software, which substracts all nonintegrated changes in the chromatogram. The amplitude of this slope depends on the cell geometry, the influence of both flow rate and wavelength being small [9]. It was 100 μAU/bar.cm (Fig. 6) with the Gilson 117 detector (2 mm light path); it has now decreased to 50 μAU/bar.cm with the Gilson 118 and 119 detectors (5 mm light path).

To detect as many compounds as possible (i.e., to be nonspecific), the UV-vis detector is increasingly used at low wavelengths, typically 210 nm, which is possible due to the UV transparency of carbon dioxide compared to nonaqueous HPLC mobile phase solvents. This is especially the case for the stability control of drug substances and drug products in the pharmaceutical industry (Sec. III.C.1). Under these conditions, the following perturbing phenomenon, known in HPLC both with high-pressure and low-pressure mixing systems, also exists in pSFC: when a small amount of modifier is automatically mixed with the primary solvent, composition pulsations often prevent the use of the maximum

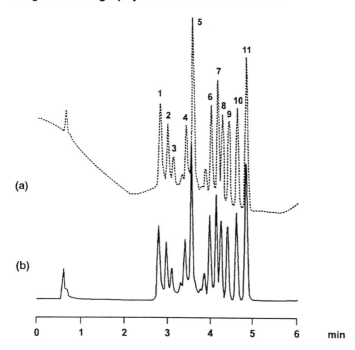

Figure 6 Analysis of steroids [2]. Detection: 210 nm and 0.1 AUFS (Gilson 117). (a): Without baseline correction. (b): With baseline correction. Injection: 5 µL, 0.02-0.4 µg/µL of each steroid in methanol. Column: Nucleosil diol-silica, 7 µm, 2 × 250 mm (Keystone). Mobile phase: carbon dioxide with 1–12% methanol in 5.5 min. Flow rate: 2.5 mL/min. Pressure (column outlet): 140–350 bar in 5.5 min. Temperature: 85°C. Peaks: 1, cholesterol; 2, progesterone; 3, androsterone; 4, testosterone; 5, estrone; 6, estradiol; 7, cortisone; 8, hydrocortisone; 9, prednisone; 10, prednisolone; 11, estriol.

sensitivity [5]. Figure 7 shows experimental results of pulsation amplitude vs. mixing volume. For 0.7 and 1.5 mL, the points correspond to mixing chambers of different capacity; 2.25 mL corresponds to a large tubing (0.8 × 1500 mm, i.e., 0.75 mL) installed between the 1.5 mL mixer and valve C; and 3 mL represents two 1.5-mL mixers in series. Qualitatively, this pertubing phenomenon is minimized, as in HPLC, by increasing mixing volume, amount of modifier, and wavelength, and, to a smaller extent, by decreasing pressure and temperature. High-quality carbon dioxide enhances the perturbation as seen in Fig. 7. Such problems do not exist using the mass detectors described hereafter.

2. Flame Ionization Detector

The most common GC detector, the flame ionization detector (FID), requires chromatographic effluent at a gas flow rate of 10–50 mL/min. This is only the case when capillary columns are used with supercritical fluids (capillary column

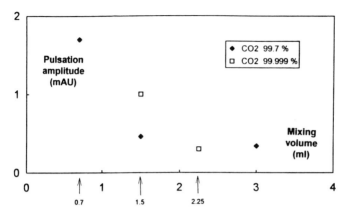

Figure 7 Pulsation amplitude vs. mixing volume. Detection: 210 nm (Gilson 117). Mobile phase: 5% methanol in carbon dioxide, 2.5 mL/min, 180 bar, and 40°C. See text.

supercritical fluid chromatography: cSFC). It burns this effluent in a hydrogen-air flame at about 2100°C. Its housing must be heated between 200°C and 300°C to prevent condensation of combustion water and of other nonvolatile substances. Its response is proportional to the number of oxidizable carbon atoms, producing ions and free electrons when burned in the hydrogen-air flame. It is a mass-specific, highly sensitive detector with very large linear dynamic range (10 pg to 0.1 mg, seven orders of magnitude). Its drawback is to exclude polar modifiers necessary for polar analytes (except water, formic acid, and formol, soluble at <1% in carbon dioxide).

Coupling an FID with the SF3 system has been successfully implemented by solving the flow split problem associated with the use of packed columns [26]. Figure 8 shows a flow diagram of this configuration. The FID is attached to the GC model 353 of GL Sciences (Tokyo, Japan). Two splits are designed to ensure proper operation of the 821 pressure regulation valve and of the FID, respectively. The pumped liquid flow rate F_1 is first split into F_2 and F_3 by the tee piece T_1, then F_3 is itself divided into F_4 and F_5 by the tee piece T_2.

Proper operation of the 821 pressure regulation valve requires a flow rate F_2 higher than 0.5 mL/min (Sec. II.C). The size of capillary A was selected for this purpose and the flow rate was checked by measuring the gas flow rate F_3 at the capillary outlet, after having disconnected T_2. Figure 9 is a graph of the results showing that F_3 increases (1) when the pressure at the column outlet increases from 100 to 400 bar, and (2) when the capillary temperature decreases from 100°C to 40°C. This means that carbon dioxide behaves like a gas inside capillary A (its viscosity increases with temperature). For F_1 of 2 mL/min of liquid, F_3 remains lower than 500 mL/min of gas (i.e., about 1 mL/min of liquid). Then

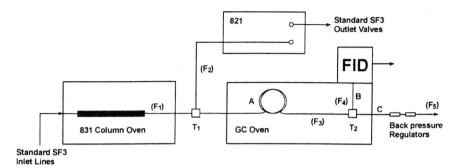

Figure 8 pSFC-FID flow diagram [26]. A: Polyimide-coated fused silica capillary, 75 μm × 10 m. B: Polyimide-coated fused silica capillary, 75 μm × 300 mm. C: PTFE tubing, 0.8 × 300 mm.

the complement F_2 fulfils the required condition to be higher than 0.5 mL/min of pumped carbon dioxide [26].

Proper operation of the FID requires, as already mentioned, a gas flow rate of 10–50 mL/min to this detector, with maximum sensitivity obtained for 30 mL/min. Two standard back pressure regulators commonly used in HPLC at

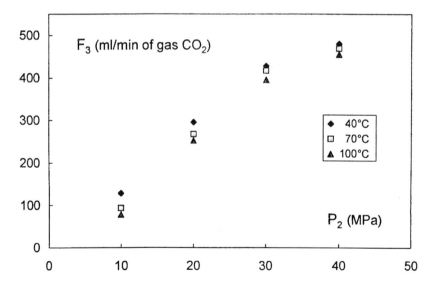

Figure 9 pSFC-FID split flow monitoring [26]. Flow rate of gas carbon dioxide (F_3) vs. column outlet pressure (P_2) for 2 mL/min of pumped carbon dioxide (F_1) and different temperatures of the split capillary (A). See Fig. 8 for symbols.

the outlet of the UV-vis detector answer this need. A 5-bar fixed back pressure regulator (Upchurch, Oak Harbour, WA) is in series with a 1- to 4-bar adjustable back pressure regulator (SSI, State College, PA). The second regulator was set so that, for F_1 of 2 mL/min and a pressure gradient from 90 to 400 bar at the column outlet, the gas flow rate F_4 in the FID varied only slightly around the optimum value, from 27.5 to 32.5 mL/min.

Finally, to avoid plugging it is advisable to inject low amounts of sample (10 ng) and to add a 0.1 mL/min outlet solvent, similarly to the collection solvent used for preparative work (Fig. 4). This can be done conveniently with the modifier pump if the system is dedicated to pSFC-FID.

3. Evaporative Light-Scattering Detector (ELSD)

An increasingly popular HPLC detector, the evaporative light-scattering detector (ELSD) overcomes the FID drawback of excluding common modifiers in pSFC. The model used, shown in Fig. 10, is Sedex 55 from Sedere (Alfortville, France). This detector requires chromatographic effluent at a liquid flow rate of 0.01–5 mL/min. At low temperature (maximum 80°C) it nebulizes, then evaporates, the mobile phase with an inert gas stream (air or N_2) to form small particles of nonvolatile analytes. It is compatible with polar modifiers, and its baseline remains horizontal during composition and/or pressure gradients. Its response is proportional to the particle size, which is approximately constant (1–10 μm); therefore its response is approximately the same for the same mass

Figure 10 Gilson SF3 system and Sedex 55 evaporative light-scattering detector.

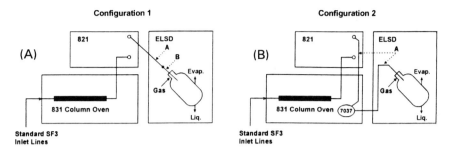

Figure 11 pSFC-ELSD flow diagrams. (A): PEEK tubing, 0.25 × 1.6 mm. (B): Fused silica capillary, 50 µm × 50 mm, inside a PEEK sleeve.

of each analyte. It is a mass-specific detector with medium sensitivity and medium linear dynamic range (logarithm of the signal vs. logarithm of the concentration is linear over two decades which are located between 10 ng and 10 µg). A higher performance (three decades) has recently been described for glucose and pure water elution from a C18 column [27]. However, the analytes must be nonvolatile (solids or semisolids), whereas the mobile phase must be fully volatile. A different model of ELSD is presented in Chapter 4.

For coupling with the SF3 system, there are two solutions (Fig. 11) which differ by the second stage of pressure control: either a fixed capillary restrictor supplied by the detector manufacturer (configuration 1 where pressure depends on flow rate and viscosity) or the 7037 valve (configuration 2 where pressure is independent of flow rate and viscosity). Both are equivalent in terms of peak broadening. However, to avoid formation of dry ice in the connection tubing to the nebulizor, configuration 2 requires either heating of this connection tubing, or at least 10% of modifier in the mobile phase. An interesting third solution (to be experimented) would consist of replacing the FID by the ELSD in the flow diagram of Fig. 8. This should generate both higher resolution (narrower peaks) and higher sensitivity (because a lower noise was already obtained with micro– flow rates [28]).

III. APPLICATIONS

The instrumental similarity of pSFC with HPLC makes method substitution very attractive and there are several reasons to try pSFC in R&D and routine analy- sis: long analysis times (>30 min), complex matrices, the need for normal phase type selectivities, the reduction of the number of chromatographic methods necessary to analyze a single finished product, the increasing number of samples, and the reduction of the consumption of organic combustible solvents.

As compared with HPLC, pSFC demonstrates significant separation advantages in addition to the economical and ecological benefits associated with reduced quantities of organic combustible solvents:

Analysis 3–10 times faster (optimum flow rate is about five times higher)
Higher resolution power (more efficiency per time unit)
Shorter method development (faster equilibrium)
Unique selectivities (specific conformation of solutes in carbon dioxide, more systematic exploration of temperature)
Lower running cost per sample, in spite of a higher flow rate (mobile phase is typically six times less expensive)
Improved safety and lower disposal cost (amount of organic solvent typically divided by one order of magnitude)
Analysis of new classes of compounds (polymers, alkanes, etc.)
Concentrated fractions for preparative and sample preparation purposes (spontaneous evaporation of the mobile phase)
Wider choice of detectors (FID and other GC detectors)
Easier coupling with mass spectrometer (MS), Fourier transform infrared (FTIR) spectrometer, and the quasi-universal ELSD

As compared with capillary GC, pSFC is penalized by lower efficiency per time unit and higher running cost per sample. Nevertheless pSFC still offers:

Shorter analysis time
Wider choice of stationary phases
Use of composition gradients that may compensate for a lower number of theoretical plates
Separation at lower temperatures (less aggressive toward column materials and sample components)
Preparative work (sample capacity is typically 500 times higher)
Higher sensitivity for compounds with chromopores (UV-vis detection is typically 10–100 times less concentration sensitive than FID, but with a 500 times higher sample mass, pSFC-UV-vis is typically 10 times more sensitive than GC-FID)

For these reasons, the three discussed chromatographic methods (HPLC, GC, pSFC) should generally be considered in the following preference order: first capillary GC, then pSFC, last HPLC. The next examples illustrate the above statements.

A. Petrochemical and Related Analyses

The UV-vis detector is used for oligomers, polymers additives, and resins; the FID for alkanes. Such analyses are impossible with HPLC. Generally performed with only a pressure gradient, the separations are rapid (no gradient delay) and require very little solvent.

For environmental testing, 16 polycyclic aromatic hydrocarbons were separated on a C18 column and detected by UV absorbance with the same detection limits as for HPLC [29]. The availability of a pressurized cell fluorescence detector should make pSFC competitive for this analysis. Fullerenes C60 and C70 were separated in 6 min on a tetrachlorophthalimidopropyl-bonded silica [30] (Fig. 12), whereas the same column took 30 min using hexane instead of carbon dioxide.

1. Oligomer and Polymer Additives

The Triton family of surfactants (octylphenol polyethylene glycol ethers manufactured by Rohm and Haas) gives classical examples of high-resolution power:

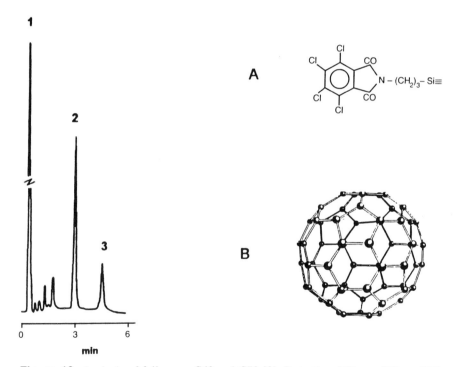

Figure 12 Analysis of fullerenes C60 and C70 [2]. Detection: 330 nm (Gilson 117). Injection: 5 μL (sample in toluene). Column: TCP-silica, 5 μm, 4.6 × 50 mm (Phase Separations). Mobile phase: carbon dioxide with methylene chloride, 2–7.5% in 6 min. Flow rate: 3 mL/min. Pressure (column outlet): gradient from 180 to 280 bar in 6 min. Temperature: 90°C. Peaks: 1, toluene; 2, fullerene C60; 3, fullerene C70. A: Chemical structure of tetrachlorophthalimidopropyl (TCP)–bonded silica. B: Chemical structure of fullerene C60 (network of pentagons and hexagons).

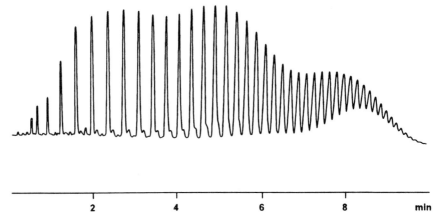

Figure 13 Analysis of a mixture of Triton X-114, X-165, and X-305 [9]. Detection: 210 nm (Gilson 117). Column: Spherisorb C18-silica, 3 μm, 2 × 100 mm (Stagroma). Mobile phase: carbon dioxide with 10% methanol. Flow rate: 2 mL/min. Pressure (column outlet): 125–350 bar in 10 min. Temperature: 140°C.

20 oligomers in 10 min for Triton X-165 [4], up to 40 oligomers in 10 min for a mixture of Triton X114, X-165, and X-305 [9] (Fig. 13). This has been achieved with C18-bonded silica, 10% methanol, 140°C, and pressure programming from 125 and 350 bar. A good column format is 3- or 5-μm particles in a 2-mm bore, permitting the use of a flow rate slightly higher than the optimum value (2 mL/min).

Polystyrenes have been analyzed under similar conditions but on polar stationary phases: silica for polystyrene 1050 [31], CN-bonded silica for polystyrene 580 [9]. In this case, the ELSD produced peaks that were more than 10 times higher than those of the UV-vis detector at 245 nm [9]. Analysis and fractionation of polymers are discussed in detail in Chapter 12.

Polymer additives, e.g., antioxidant additives (Metilox, Irgafos 168, and the Irganox 1076, 259, 1330, 1010, and 1098) were separated in 6 min by pSFC with two columns in series C8 and CN [4]. With HPLC, two methods were required. The analysis of these special additives is detailed in Chapter 11.

2. Resins

Ciba-Geigy Ltd, a major producer of epoxy resins, developed pSFC methods with pressure gradient on a CN column to considerably shorten analysis times as compared with reversed phase HPLC using composition gradient on a C18 column. Figure 14 shows both methods for the Badge resin [32]: the analysis took 10 min instead of 40. The elution order is reversed except for the oligomers (5 and 6), which are insensitive to the polarity change.

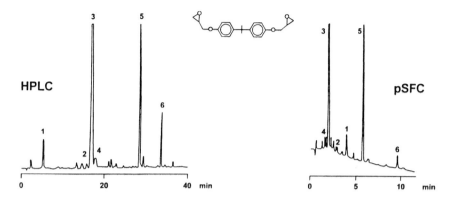

Figure 14 Analysis of epoxy resins: Ciba Geigy Ltd BADGE [32]. Detection: 280 nm. HPLC: Injection: 10 μL. Sample: 5 μg/μL in THF. Column: Nucleosil C18-silica, 5 μm, 4.6 × 125 mm. Mobile phase: water-acetonitrile-methanol 50:45:5, isocratic during 5 min, then composition gradient in 40 min to acetonitrile. Flow rate: 1 mL/min. Temperature: ambient. pSFC: Injection: 1 μL. Sample: 20 μg/μL in THF. Column: Spherisorb CN-silica, 3 μm, 4 × 125 mm. Mobile phase: carbon dioxide-methanol 89:11, with outlet pressure gradient, 2 min at 130 bar, then to 350 bar in 10 min, finally isobaric during 1 min. Flow rate: 2 mL/min. Temperature: 70°C. Peaks: 1, α-glycol; 2, α-chlorhydrine; 3, Badge; 4, optical isomer of Badge; 5, Badge dimer; 6, Badge trimer.

Similarly, Araldite GT 7004 was analyzed in 8 instead of 45 min [32]. The result was even better, 4 instead of 30 min, for the other epoxy resin shown in Fig. 15 [32].

Phenol-formaldehyde resins of British Petroleum were separated on a C18 column prior to further identification by mass spectrometry [33]. For this particularly complex mixture, the full flexibility of superimposed pressure, composition, and flow rate gradients was required.

3. Alkanes

Again time can be saved in the analysis of alkanes and paraffins contained in different types of fuels. A mixture of 22 paraffins, n-alkanes from C5 to C44 (Fig. 16), was resolved in 6 min with FID [26]. The C18 column of good format (3 μm, 2 × 100 mm) worked at a flow rate of 2 mL/min, close to optimum. High temperature (120°C) generated high efficiency: peak width at the basis was 2–8 s. Resolution was adjusted with a pressure gradient in the full range of 90–400 bar.

Polywax 500, a mixture of higher molecular weight alkanes from C14 to C70, gave a separation in 8 min using similar conditions, with higher temperatures for the column (150°C) and for the split capillary (250°C), but full elution was not checked. The mean limit of detection was approximately 50 pg. Rela-

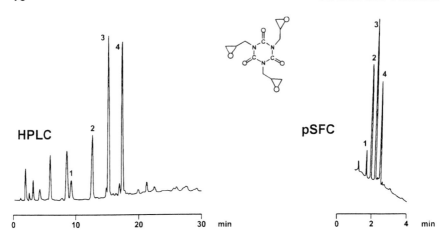

Figure 15 Analysis of epoxy resins: isocyanuric acid triglycidyl ether [32]. Detection: 205 nm. HPLC: Injection: 10 μL. Sample: 2 μg/μL in methanol. Column: Nucleosil C18-silica, 5 μm, 4.6 × 125 mm. Mobile phase: water-acetonitrile 90:10, isocratic during 5 min, then composition gradient in 7 min to water-acetonitrile 80:20, then in 8 min to water-acetonitrile 70:30, finally isocratic during 10 min. Flow rate: 1 mL/min. Temperature: 35°C. pSFC: Injection: 1 μL. Sample: 10 μg/μL in methanol. Column: Spherisorb CN-silica, 3 μm, 2 × 100 mm. Mobile phase: carbon dioxide-methanol, 93:7, with outlet pressure gradient, from 130 to 350 bar in 6 min, then isobaric during 1 min. Flow-rate: 1.5 mL/min. Temperature: 80°C. Peaks: 1, isocyanuric acid triglycidil ether; 2, monochlorhydrine; 3, dichlorhydrine; 4, trichlorhydrine.

tive standard deviation of retention times was lower than 1%, demonstrating the repeatability of pressure and flow rate. The relative standard deviation for peak areas was 2–6%. It would have been better if the sample, incompletely dissolved in CS_2 at ambient temperature, had been warmed before injection.

Quantitative analysis of complex petroleum fractions from n-C5 to n-C22, which covers the high value automotive fuel boiling range, was carried out by coupling the pSFC system with UV-vis detector, FID, and capillary GC-FID [34].

Compared with GC-FID and temperature gradient, pSFC-FID with pressure gradient can reduce the analysis time of nonpolar analytes in fossil fuels.

B. Quality Control of Food and Cosmetics

According to the sample complexity and the separation goal, resolution is obtained either with constant mobile phase, or with gradient of pressure only, or with pressure and composition, or even with pressure, composition, and flow rate. UV-vis detector and ELSD can be used alternatively or simultaneously.

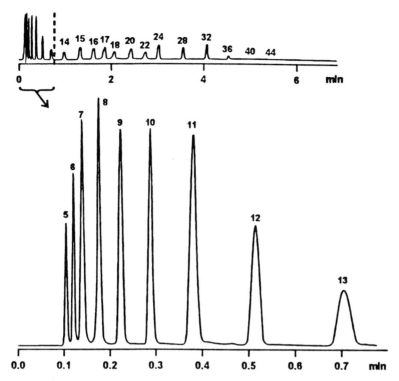

Figure 16 Analysis of *n*-alkanes from C5 to C44 [26]. Detection: FID (GL Sciences GC 353), 350°C. Injection: 0.5 μL (sample in CS$_2$). Column: Spherisorb ODS2-silica, 3 μm, 2 × 100 mm. Mobile phase: neat carbon dioxide. Flow rate: 2 mL/min. Pressure (column outlet): 90 bar during 1 min, up to 290 bar until 6 min, and up to 400 bar until 10 min. Temperature: column at 120°C, split capillary in GC oven at 150°C. Peaks: Numbers according to number of carbon atoms.

1. Lipids

Lipids, which are well-known thermosensitive compounds, are usually analyzed by GC after an esterification step. Direct analysis without saponification and subsequent esterification is now possible with pSFC. It was carried out according to the number of unsaturations for fatty acids with UV-vis detection (Fig. 17) [31], and according to the number of carbon atoms for triglycerides with ELSD (Fig. 23) [32]. A cyanosilica column (3 μm, 2 mm, in different lengths) was used in both cases with a gradient of pressure only.

Edible oils (peanut, olive, corn, sunflower, soybean, primrose, and rapeseed) were analyzed according to both criteria (number of unsaturations and carbon atoms) using the analytical and preparative system configuration described in Sec. II.D. Two complementary pSFC separations were performed [35]: first, on

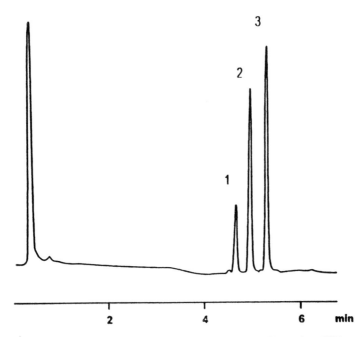

Figure 17 Analysis of unsaturated fatty acids [31]. Detection: 205 nm. Injection: 5 μL. Sample: 0.1 μg/μL in isopropanol. Column: CN-silica, 3 μm, 2 × 100 mm. Mobile phase: carbon dioxide with 1% isopropanol. Flow-rate: 2 mL/min. Pressure (column outlet): 125 to 350 bar in 12 min. Temperature: 80°C. Peaks: 1, triolein (18:1); 2, trilinolein (18:2); 3, trilinolenin (18:3).

a C18 column for separation and fraction collection by number of carbon atoms, then on a silver-loaded column for separation by number of double bonds. More recently, the sterolic fraction of these vegetable oils, which provides valuable information on the origin and quality of the product, was isolated in a fully automated way by pSFC of <8 min on aminopropyl-silica (10% methanol, 150 bar, 70°C) [36]. The collected fraction of free sterols was further analyzed by capillary GC-MS to quantify campesterol, stigmasterol, and sitosterol. This off-line technique gave good reproducibility and the sample preparation time was substantially reduced compared to classical wet chemical methods [36].

2. Essential Oils

Essential oils were analysed with UV-vis detector on polar stationary phases. A silica column at 125°C was used for the separation of 16 terpenes in 8 min and its application to the analysis of cinnamon oil (Fig. 18) [9]. Typical fingerprints of these complex samples are rapidly obtained by superimposed gradients of pressure, composition, and flow rate. For the analysis of saffron extracts, a

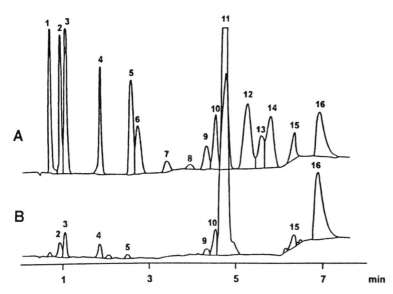

Figure 18 Analysis of an essential oil [9]. A: mixture of standards, B: cinnamon oil from Sri Lanka. Detection: 210 nm. Injection: 5 µL. Sample: 1–5 µg/µL in hexane for standards, 5 µg/µL in hexane for cinnamon oil. Column: Deltabond silica (Keystone), 5 µm, 4.6 × 150 mm. Mobile phase: carbon dioxide, isocratic during 0.5 min, then composition gradient with ethanol, to 0.5% in 0.5 min held during 4 min, and up to 7% in 1 min held during 2 min. Flow rate: 1.5 mL/min during 1 min, then gradient to 2 mL/min in 0.5 min held during 3.5 min, finally up to 4 mL/min in 1 min held during 2 min. Pressure (column outlet): 90 bar during 0.5 min and up to 150 bar in 2 min held during 5.5 min. Temperature: 125°C. Peaks: 1, α-pinene; 2, limonene; 3, p-cymene; 4, tr-caryophylene; 5, tr-anethole; 6, benzaldehyde; 7, fenchone; 8, byproduct of cinnamyl acetate; 9, linalool; 10, cinnamyl acetate; 11, tr-cinnamaldehyde; 12, benzyl alcohol; 13 α-terpineol; 14, β-phenyl ethyl alcohol; 15, byproduct of cinnamyl acetate; 16, eugenol.

cyanosilica column at 40°C was used to determine four substances (safranal, picrocrocin, crocetin, and crocins) at three different wavelengths [37].

3. Free and Conjugated Carbohydrates

Sugar analysis with ELSD is shown in Fig. 19. With a trimethylsilane (TMS) column and 20% modifier containing 8% water, 8 sugars were quantified in 10 min [38]. The limit of detection was 20 ng. Typical application matrices were brandy, tobacco, etc. A review of SFC of carbohydrates, highlighting the advantages of pSFC as compared with cSFC, was recently published [39]. TMS-bonded silica with ELSD was also used to analyze glycolipids (mono- and digalactosyldiacylglycerols, glucosylsitosterol, and galactocerebroside), phospholipids, and carbohydrates in a plant extract [40] (see also Chapter 4, Fig. 9).

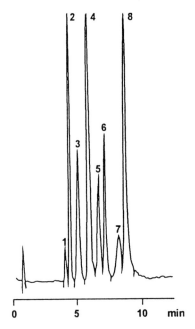

Figure 19 Analysis of sugars [38]. Detection: ELSD (Sedex 55), nebulizer at 70°C, evaporator at 60°C. Injection: 20 µL. Sample: 0.3 µg/µL of each sugar in methanol. Column: Zorbax TMS-silica, 5 µm, 4.6 × 250 mm. Mobile phase: carbon dioxide with 20% methanol-water-triethylamine 91.5:8:0.5. Flow-rate: 5 mL/min. Pressure (column outlet): 200 bar. Temperature: 60°C. Peaks: 1, ribose (aldopentose); 2, mesoerythritol (C4 polyalcohol); 3, xylose (aldopentose); 4, xylitol (C5 polyalcohol); 5, sorbose (cetohexose); 6, mannose (aldohexose); 7, glucose (aldohexose); 8, mannitol (C6 polyalcohol).

4. Waxes

Quality control of natural waxes is presented in Fig. 20 [20,41]. These chromatograms demonstrate typical fingerprints for each type of natural wax used in the cosmetics industry in order to detect abnormal compositions of commercial waxes. Saturated aliphatic hydrocarbons and fatty esters are the two most abundant functional group classes in waxes. Bare and diol-bonded silicas facilitated separations according to functional group classes but not according to the alkyl chain lengths within each class. C18-silica solved the problem. Another difficulty concerned the sample solvent. The best solvent, chloroform, was more eluting than the mobile phase, therefore peak splitting occurred. This drawback was eliminated using solventless injection (Sec. II.D).

In Fig. 20, the circled part shows a section of heavy hydrocarbons that should not be there. Such fingerprints are both much simpler and faster obtained than

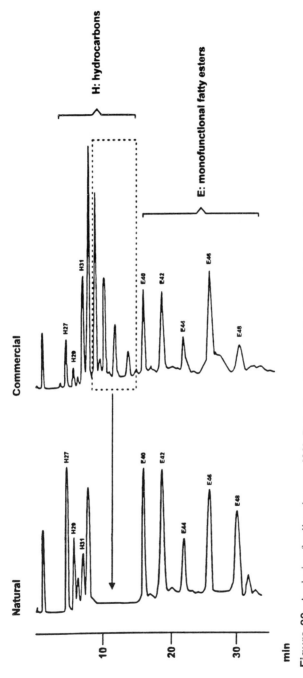

Figure 20 Analysis of yellow beewax [20]. Detection: ELSD (Sedex 55), nebulizer at 60°C, evaporator at 51°C. Injection: 10 μL. Sample: 5 μg/μL in chloroform. Column: Zorbax C18-silica, 5μm, 4.6 × 150 mm. Mobile phase: carbon dioxide with methanol, 0.1–3% in 25 min. Flow rate: 3 mL/min. Pressure (column outlet): 168 bar. Temperature: 47°C.

those produced by GC, which also may cause thermal degradation of certain components. With HPLC, two methods would be required [20].

C. Pharmaceutical Drug Substances and Products

Drug substances are active substances that may be subsequently formulated with excipients to produce drug products (tablets, lotions, creams, suppositories, etc.). In the pharmaceutical industry, the use of pSFC already started for release and stability testing, analyses of enantiomers, preparative separations of enantiomers, purification of drug candidates, and process control.

1. Release and Stability Testing

Release and stability control tests [42–44] must be carried out to analyze active substance, related compounds, and preservatives during the development and marketing stages of a drug substance or a drug product. The release tests ensure that the drug meets the specification limits after production, whereas the stability tests lead to the shelf life and storage conditions labeled on the drug container. HPLC is well established for these tests, but because pSFC has produced reliable, accurate and precise determinations of drugs [7], it is known that pSFC can also fulfill the requirements to be used as a quantitative method for drug analysis [45] (see also Chapter 10).

Release and stability testing of pharmaceutical drugs has the following statutory objective: in the presence of the active substance, related compounds (impurities, degradation products) must be quantified, identified, and qualified (for toxicity) down to the concentration of 0.1% of the mass of active substance in the finished product. This is achieved with pSFC and UV-vis detection at 210 nm and/or ELSD. Resolution is generally obtained within a few minutes using a polar column (cyano-, diol-, or aminopropyl-bonded silica, 5- or 10-μm particles), with sub- or supercritical conditions (50–150°C, 2–20% polar solvent), either constant or gradient conditions according to the sample complexity.

Crotamiton in creams and lotions [7] was one of the first pSFC methods routinely used for release and stability control at Ciba Geigy Ltd. Optimized under constant and truly supercritical conditions (cyanosilica, 3 μm, 2 × 270 mm, 1.95 mL/min, 2.6% modifier, 125 bar, and 100°C), the chromatogram quantitatively resolved seven compounds (geometric isomers of the active substance, two preservatives, two possible byproducts, and a decomposition product) in less than 5 min per sample. Four chromatographic methods were substituted in this manner: two GC, one HPLC, and a thin-layer chromatography (TLC) [46]. In addition, the preseparation of the active substance from the excipients, which is necessary before HPLC to avoid column plugging, is not necessary before pSFC. The dosage forms, cream and lotion, can be diluted in tetrahydrofuran and injected directly.

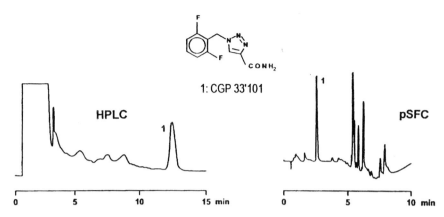

Figure 21 Analysis of a drug substance (CGP 33'101) in mice feed pellets [32]. Detection: 220 nm for HPLC and 210 nm for pSFC. Sample: 0.005 μg/μL of CGP 33'101 and 50 μg/μL of pellets in water-acetonitrile 1:1. HPLC: Injection: 10 μL. Column: Nucleosil C18-silica, 5 μm, 4.6 × 125 mm. Mobile phase: water-acetonitrile-THF 850:125:25 plus 1 g sodium pentanesulfonate, isocratic during 14 min, then composition gradient in 4 min to water-acetonitrile-THF 100:850:50, finally 6 min isocratic. Flow rate: 1 mL/min. Temperature: ambient. pSFC: Injection: 5 μL. Column: Spherisorb NH₂-silica, 2 μm, 2 × 250 mm. Mobile phase: carbon dioxide modified with methanol-acetonitrile 1:1, 6% isocratic during 3 min, then to 19% in 1 min, finally 6 min isocratic. Flow rate: 2.1 mL/min. Pressure (column outlet): 125 bar. Temperature: 55°C.

A drug substance in mouse feed pellets is shown in Fig. 21 [32]: as compared with reversed phase HPLC, this active substance (peak no. 1) is first eluted. This situation generates a similar LOD for this peak in pSFC as in HPLC (1 ppm), although in HPLC the injection volume is higher and the detector cell pathlength is longer. Whereas in HPLC, the whole cycle including column washing and reconditioning times took 45 min, in pSFC the chromatographic system was ready for the next injection after only 12 min.

Furthermore, the reverse elution order in pSFC as compared with reversed phase HPLC can also be important for resolution: Fig. 22 shows a case where peak no. 3 could only be quantified in pSFC [32].

The ELSD can also be adopted in routine methods to quantify active ingredients as well as excipients. Triglycerides in suppository drugs are shown in Fig. 23 [32]. In this case, a special low dead volume (10 μL) piezo-driven back pressure regulator was used [47] (see also Chapter 4, Figs. 4 and 5). The chromatogram is once more optimized with the standard cyano column, low amount of methanol, high temperature (150°C), and pressure gradient. Relative standard deviation of peak areas with internal standard was 3% (peaks 3–6). This method is preferred to the generally accepted one, which is more sensitive but needs a

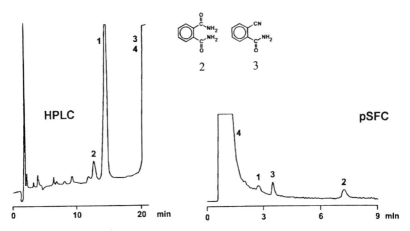

Figure 22 Analysis of a test solution spiked with two possible byproducts [32]. Detection: 215 nm for HPLC and 210 nm for pSFC. Sample: 0.002 μg/μL of 3, 0.004 μg/μL of 2, and 0.8 μg/μL of 4, in N-methylpyrrolidone-methanol 1:4. HPLC: Injection: 20 μL. Column: Nucleosil C18-silica, 5 μm, 4.6 × 125 mm. Mobile phase: water-acetonitrile 30:70 plus 2 g ammonium acetate, isocratic during 21 min, then composition gradient in 0.5 min to acetonitrile, finally 24 min isocratic. Flow rate: 1 mL/min. Temperature: ambient. pSFC: Injection: 5 μL. Column: Grom-Sil diol-silica, 3 μm, 2 × 250 mm. Mobile phase: carbon dioxide modified with 5% methanol containing 0.05% ammonium acetate. Flow rate: 2.1 mL/min. Pressure (column outlet): 125 bar. Temperature: 60°C. Peaks: 1, impurities; 2, phthalamide; 3; phthalic acid mononitrile monoamide; 4, N-methylpyrrolidone.

derivatization step (saponification followed by esterification of fatty acids) followed by GC-FID determination, which takes altogether about 90 min.

In conclusion, for release and stability control of drugs, pSFC analysis can be four to five times faster than reversed phase HPLC. A different elution order may contribute to better resolutions, unresolved cases may be resolved, and a single method may replace two or more chromatographic methods. In some cases, the sample preparation can be reduced without disturbing the separation column material. Currently, 9 validated methods are routinely used for drug stability control at Ciba Geigy Ltd, and 11 others solve analytical problems in R&D [46].

2. Biomedical Analyses

Diverse medium-polarity molecules, analytes of biomedical importance, were separated by pSFC methods at temperatures of 40–85°C. These methods were applied to biological matrices after sample preparation, to plant extracts, and to manufactured drugs.

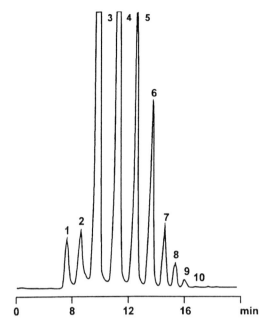

Figure 23 Analysis of triglycerides in suppository drugs [32]. Detection: ELSD (Sedex 45), nebulizer at ambient temperature, evaporator at 80°C. Injection: 5 μL, dissolved in dichloromethane. Column: Spherisorb CN-silica, 3 μm, 2 × 250 mm. Mobile phase: carbon dioxide with 2.4% methanol. Flow rate: 2.6 mL/min. Pressure (column outlet): 190 bar during 9.5 min, up to 225 bar until 14.5 min, finally isobaric. Temperature: 150°C. A special low dead-volume piezo-driven back pressure regulator was used [47]. Peaks identified by number of C atoms in fatty acid chains: 1, 10-10-2; 2, 10-12-12; 3, 12-12-12; 4, 12-12-14; 5, 12-14-14; 6, 14-14-14; 7, 14-14-16; 8, 14-16-16; 9, 16-16-16; 10, 16-16-18.

Diol-bonded silica was often the best stationary phase. It was used with UV-vis detection to carry out the resolution of morphinic alkaloids (papaverine, thebaine, codeine, and morphine) found in poppy straw extracts [48]. The resolution of 11 steroid hormones is shown in Fig. 6; they are separated in 5 min, approximately four times faster than with HPLC [9]. With ELSD, diol-bonded silica was also used for underivatized amino acids [49] and for seven bile acids [50]. The bile acids were detected down to 100 pmol (40 ng), a detection limit approximately 1000 times lower than with UV at 210 nm. Free bile acids in pharmaceutical preparations were also separated on cyano- and phenyl-bonded silica [51].

UV-vis detection and APCI-MS were used for the analysis of the four major cannabinoids (cannabidiol, Δ8- and Δ9-tetrahydrocannobinols and cannabinol) on cyano-bonded silica. The 8-min pSFC method was applied to identify these substances in resin samples and quantify them in biological matrices [24].

In all these cases, pSFC was faster and less expensive than HPLC methods considering the analysis time saved and the lower price of carbon dioxide vs. that of HPLC solvents.

3. Analyses of Enantiomers

Well known in HPLC, the analysis of enantiomers with chiral stationary phases (CSPs) is now one of the most important application fields for pSFC under subcritical conditions (subFC) [52–63]. No longer based on efficiency, resolution is governed by selectivity. Carbon dioxide can generate solute conformations (e.g., more unfolded) that do not exist in common nonpolar HPLC solvents such as hexane [57]. Resolution is typically achieved in 3–12 min, using an easy transfer from HPLC methods: same CSP column (polysaccharide-derivatized silica, Pirkle-type graftings, etc.) at the same temperature (near ambient), similar composition of mobile phase (carbon dioxide replaces the nonpolar HPLC solvent and the polar modifier is the same or, rather, similar in both techniques), and isocratic and/or isobaric conditions. About 70% of common drug racemates can be resolved on amylose carbamate (Chiralpak AD) or cellulose carbamate (Chiralcel OD) [56,59–63]. Resolution differences produced by these two columns are presented in Table 1 [63].

As compared with normal phase HPLC on the same column, chiral analysis by subFC can be three to four times faster, while the elution order is generally the same, and resolution can be better due to specific molecular interactions in carbon dioxide. As HPLC, chiral column pSFC has the potential to provide all the information about optical purity required for release and stability control on a single chromatogram.

Chapter 8 is a full review of chiral separations. Chapter 9 deals with applications of a special Pirkle-type CSP. Chapter 6 highlights applications on coupled chiral columns. Chapter 10 presents applications of a pharmaceutical company.

4. Preparative Separations

In its analytical and preparative configuration (described in Sec. II.D and applied to off-line sample preparation for lipid analysis in Sec. III.B.1), the Gilson SF3 system can produce, within a few hours, 10–500 mg of purified substance collected in liquid phase from repeated injections. These amounts are sufficient for preliminary biological testing by medicinal chemists. They are also sufficient for organic synthesis and for process research.

A prerequisite to isolate the maximum amount of substance at a given level of purity is to choose separation conditions maximizing resolution at the analytical scale. The preparative separation of propranolol racemate was investigated from the analytical conditions of Fig. 24 [53], showing a resolution of about 5 (higher than in Table 1).

Table 1 Compared Enantiomeric Resolution of Common Drug Racemates on Polysaccharide-Derivatized Silica [63]

Drug	OD	AD
Propranolol	3.6	0.0
Metoprolol	2.7	0.0
Oxprenolol	2.6	0.0
Pindolol	3.5	0.0
Nadolol	1.4	0.9
Guaifenesin	1.4	0.5
Cyclopenthiazide	1.4	0.7
Naproxen	0.0	1.5
Polythiazide	0.0	2.5
Cyclothiazide	1.4	2.4
Warfarin	4.0	5.0
trans-Stilbene oxide	2.3	9.5
Mephobarbital	0.9	19.0
Hexobarbital	0.7	14.0
Lormetazepam	1.2	9.0
Oxazepam	4.0	4.3
Mandelic acid	1.2	1.5

Column: Chiralcel OD (cellulose) and Chiralpak AD (amylose), 10 μm, 4.6 × 250 mm (Daicel Chemical Industries). Separation time: 3–12 min. Mobile phase: carbon dioxide plus 30% methanol containing 0.1% diethylamine (DEA) and 0.1% trifluoroacetic acid (TFA). Flow rate: 2 mL/min. Pressure (column outlet): 200 bar. Temperature: 30°C. Detection: 220 nm. Courtesy of A. Medvedovici, F. David, and P. Sandra [63].

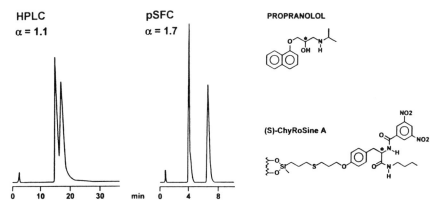

Figure 24 Analysis of propranolol racemate [52]. Detection: 224 nm. Injection: 20 μL. Sample: 0.75 μg/μL in methanol. Column: (S) thio-DNB Tyr-A on silica (ChyRoSine A), 5 μm, 4.6 × 150 mm (Sedere). Temperature: 25°C. HPLC: Hexane with 5% ethanol containing 1% *n*-propylamine, 1 mL/min. pSFC: Carbon dioxide with 10% methanol containing 1% *n*-propylamine, 4 mL/min, 200 bar.

Figure 25A shows a preparative separation using 0.2 mg per injection and a modifier gradient. Acceptable recoveries and good purity of the second fractionated enantiomer require the addition of a collection solvent (Fig. 4). The amount needed for the rinsing process (and to maintain the solutes dissolved down to the collection vials) is inversely related to the amount of mobile phase modifier used for the separation. With a flow rate of 0.25 mL/min of isopropanol, >80% recovery with 99% purity for the first enantiomer (fraction 1), and more than 90% recovery with 95% purity for the second one (fraction 3) was collected (Fig. 25B).

To increase the amount of collected fractions, the column was loaded with 1 mg (trace 1), then 4 mg (trace 2), at a constant injection volume of 20 μL (Fig. 26A). To increase the amount of injected sample, water was added to isopropanol as a sample solvent. For 4 mg, the injected solution was close to saturation (20% concentration in isopropanol-water 2:1). The separation with an increased injection volume (100 μl) for 5 mg per injection is shown in Fig. 26B.

The molar ratio of solute to chiral selector was about 6%. As compared with Fig. 26A, the lower resolution means that productivity is favored by the injection of almost saturated solutions as is also generally true in HPLC. In this case, the specific throughput (an index of preparative performance referring the mass of injected sample to the mass of stationary phase, the volume of liquid mobile phase, and the separation time [64,65]) crosses the threshold of 1 g of sample per gram of stationary phase, per liter of liquid mobile phase, and per hour of separation time (1 g/g.L.h). To the best of our knowledge, this specific throughput has not yet been reported for preparative elution chromatography, at least for moderate overload (less than 15 mg of sample per gram of stationary phase). Attempts to further increase the specific throughput of pSFC must focus on column efficiency (smaller particles) but more deliberately on selectivity, especially for enantiomers. To reach this goal, the appropriate choice of the modifier should be fruitful. This approach can be guided by the recently published determination of the eluotropic strengths of 10 solvents toward various types of anlytes under both subcritical and supercritical conditions [66].

In terms of production rate, factors favorable to pSFC as compared with normal phase HPLC on the same column can be faster separation, higher sample solubility in SF carbon dioxide than in hexane, and selectivity effects generated by carbon dioxide that do not exist in hexane. Interestingly, the last two factors increase with polar compounds. As a consequence, the same production rate can be achieved with pSFC using smaller bore columns than with normal phase HPLC. And repeated injections can be more effective for saving on expensive stationary phases.

Other preparative applications carried out with the Gilson SF3 system concern essential oils [19], active substances in a fermentation broth, purification of drug candidates and manufacturing process optimization. Chapter 14 is

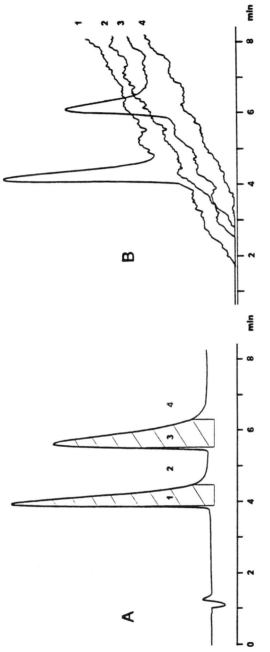

Figure 25 Preparative separation (A) and fraction analysis (B) of propranolol racemate based on detector signal amplitude [19]. Detection: 240 nm and 100 mAUFS for A, 224 nm and 10 mAUFS for B. Injections: 20 µL. Samples: 10 µg/µL (1%) of racemate in isopropanol for A (0.2 mg injected), four fractions collected from 6 preparative injections for B (1.2 mg total injected). Column and flow rate: as for Fig. 24. Mobile phase: carbon dioxide with 20–25% isopropanol (containing 1.5% diethylamine, 0.15 M) in 8 min. Pressure (column outlet): 280 bar. Temperature: 35°C. Fraction 1: Recovery: >90%; purity: 99%. Fraction 2: Recovery: >80%; purity: 95%.

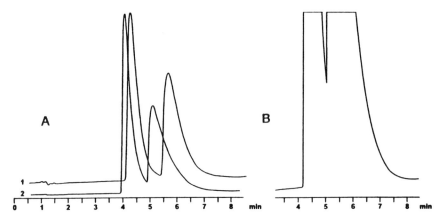

Figure 26 Preparative loading of propranolol racemate [19]. Same conditions as Fig. 25, except detection: 254 nm and 100 mAUFS. Sample solvent: isopropanol-water 2:1. Injections and samples: A.1: 20 µL, 50 µg/µL (5%, 1 mg injected); A.2: 20 µL, 200 µg/µL (20%, 4 mg injected); B: 100 µL, 50 µg/µL (5%, 5 mg injected).

dedicated to preparative SFC using different systems with larger bore columns (>5 mm diameter). See also Chapters 12 and 15.

Especially demonstrated by many purifications of enantiomers, the preparative advantages of pSFC, again compared with normal phase HPLC on the same column can be a higher production rate, a higher reproducibility, and a lower cost of mobile phase per sample (less than two thirds).

IV. CONCLUSION

pSFC, implemented on the same columns as HPLC, can offer rapid, robust, and cost-effective methods that may outperform or complement those of HPLC and are frequently superior to those of nonaqueous HPLC in terms of analysis time. These practical advantages come from three major sources: unique selectivities in carbon dioxide mixed with diverse polar solvents; the flexibility of independent programming of mobile phase pressure, composition, and flow rate; and a more systematic exploration of temperature to enhance efficiency or selectivity effects. Thanks to the commercial availability of pSFC instruments, there are an increasing number of applications for drug release control, drugs in biological matrices, and for agrochemicals, petrochemical and related products, food and extracts of natural products. New developments are anticipated, particularly in the preparative field, in off-line sample preparation, and in one-line analytical coupling of pSFC with, for example, SFE, diode array absorbance detection, MS, and the use of pSFC in industrial processes.

An important milestone in the acceptance of pSFC for the pharmaceutical industry was set when pSFC was accepted as a general method by the Swiss

Pharmacopoeia in 1994. It is anticipated that pSFC will also be soon accepted in the European Pharmacopoeia, a French contribution submitted in June 1994 should be published shortly for public review prior to final approval.

REFERENCES

1. T. L. Chester, "Missing Link or New Dimension? A Fresh Look at Supercritical Fluid Chromatography," Proceedings of Pittcon '95, New Orleans, LA, p. 1034 (1995).
2. F. Vérillon, A. Hamstra, and C. Netter, *LC-GC Int.*, 7:710 (1994).
3. K. Anton, N. Periclès, S. M. Fields, and H. M. Widmer, *Chromatographia*, 26:224 (1988).
4. A. Giorgetti, N. Periclès, H. M. Widmer, K. Anton and P. Dätwyler, *J. Chromatogr. Sci.*, 27:318 (1989).
5. K. Anton, N. Periclès, and H. M. Widmer, *J. High Resolut. Chromatogr.*, 12:394 (1989).
6. M. A. Morrissey, A. Giorgetti, M. Polasek, N. Periclès, and H. M. Widmer, *J. Chromatogr. Sci.*, 29:237 (1991).
7. K. Anton, M. Bach and A. Geiser, *J. Chromatogr.*, 553:71 (1991).
8. F. Vérillon, D. Heems, B. Pichon and J. C. Robert, "Supercritical Fluid Chromatograph with Independent Programming of Mobile Phase Pressure, Composition and Flow Rate," Proceedings of Pittcon '92, New Orleans, LA, p. 1114 (1992).
9. F. Vérillon, D. Heems, B. Pichon, K. Coleman, and J. C. Robert, *Am. Lab.*, 24:45 (1992).
10. F. Vérillon, D. Heems, T. Marin Martinod, and J. C. Robert, *Spectra 2000, 166*:57 (1992).
11. F. Vérillon, T. Marin Martinod and J. C. Robert, "Automated Pressure Control, A New Dimension of Packed Column SFC," Proceedings of the 4th International Symposium on Supercritical Fluid Chromatography and Extraction, Cincinnati, OH, pp. 9–10 (1992).
12. F. Vérillon, D. Heems, K. Coleman and B. Pichon, "Applications of a Fast and Automated Chromatographic Method with a New SFC System," Proceedings of the 4th International Symposium on Supercritical Fluid Chromatography and Extraction, Cincinnati, OH, pp. 11–12 (1992).
13. F. Vérillon, D. Heems, B. Pichon, and J. C. Robert. "Fast and Automated Chromatography using Carbon Dioxide with Pressure and Composition Gradients at the Optimum Mass Flow Rate," Proceedings of the 19th International Symposium on Chromatography, Aix-en-Provence, France, p. 22 (1992).
14. P. Sandra, private communication.
15. T. A. Berger, *J. High Resolut. Chromatogr.*, 14:312 (1991).
16. S. H. Page, J. F. Morrison, R. G. Chritensen, and S. J. Choquette, *Anal. Chem.*, 66:3553 (1994).
17. G. B. Jacobson, "Supercritical Ammonia in Synthesis, Chromatography and Extraction of ¹¹C-Labelled Compounds," thesis, Uppsala University, Sweden, 1996.
18. G. O. Cantrell, J. Blackwell, J. D. Weckwerth, P. W. Carr, and L. E. Schallinger, "Characterization of 1,1,1,2-Tetrafluoroethane as a Mobile Phase for Supercritical Fluid Chromatography," Proceedings of the 7th International Symposium on

Supercritical Fluid Chromatography and Extraction, Indianapolis, IN, p. A07 (1996).

19. K. Coleman and F. Vérillon, "Laboratory-Scale Preparative Chromatography Enhanced by Fluids Containing Carbon Dioxide under Automated Pressure Control," Proceedings of the 3rd International Symposium on Supercritical Fluids, Strasbourg, France, pp. 415–420 (1994).

20. S. Brossard, M. Lafosse, M. Dreux, and J. Bécart, *Chromatographia, 36*:268 (1993).

21. K. D. Bartle, C. D. Bevan, A. Clifford, S. A. Jafar, N. Malak, and M. S. Verrall, *J. Chromatogr., 697*:579 (1995).

22. C. K. Huynh, T. V. Duc, and M. Guillemin, "Analysis of Nitrosamines by SFC-TEA," Proceedings of the 3rd International Symposium on Supercritical Fluids, Strasbourg, France, p. 483 (1994).

23. F. Sadoun, H. Virelizier, and P. J. Arpino, *J. Chromatogr., 647*:351 (1993).

24. M. Carrott, D. Jones, G. Davidson, B. Bäckström, M. Cole, and K. Coleman, "Analysis of Cannabis using SFC-APCI/MS," Proceedings of the 6th International Symposium on Supercritical Fluid Chromatography and Extraction, Uppsala, Sweden, p. P93 (1995).

25. K. Anton, Supercritical Fluid Chromatography: Instrumentation, *Encyclopedia of Analytical Sciences* (A. Townshend, ed.), Academic Press, San Diego, CA, p. 4856 (1995).

26. M. Kawakatsu, Y. Yamamoto, H. Kotaniguchi, and F. Vérillon, "Flame Ionization Detector with Pressure-Gradient Packed-Column SFC," Proceedings of the 6th International Symposium on Supercritical Fluid Chromatography and Extraction, Uppsala, Sweden, p. P07 (1995).

27. M. Dreux, M. Lafosse, and L. Morin-Allory, *LC-GC Int., 9*:148 (1996).

28. F. Guerrero and J. L. Rocca, "Couplage de la micro-chromatographie en phase liquide avec un détecteur évaporatif à diffusion de lumière," Proceedings of the Congrès SEP 95, Chromatographies et techniques apparentées, Lyon, France, p. C25 (1995).

29. B. W. Wenclawiak and T. Hees, *J. Chromatogr., 660*:61 (1994).

30. M. Liu, A. Thienpont, M. H. Delville, G. Félix, and C. Netter, *J. High Resolut. Chromatogr., 17*:104 (1994).

31. A. Giorgetti and F. Vérillon, Atlas of chromatograms, SFC 148, *J. Chromatogr. Sci., 31*:242 (1993).

32. K. Anton, M. Bach, C. Berger, F. Walch, G. Jaccard, and Y. Carlier, *J. Chromatogr. Sci., 32*:430 (1994).

33. T. P. Lynch, *Applications of Supercritical Fluids in Industrial Analysis* (J. R. Dean, ed.), Blackie Academic, CRC Press, Boca Raton, FL, p. 188 (1994).

34. T. P. Lynch and M. P. Heyward, *J. Chromatogr. Sci., 32*:534 (1994).

35. A. Medvedovici, P. Sandra, F. David, and A. Hamstra, "SFC Analysis of Lipids in the Reversed Phase and Silver-Loaded Normal Phase Modes," Proceedings of the 7th International Symposium on Supercritical Fluid Chromatography and Extraction, Indianapolis, IN, p. B14 (1996).

36. P. Sandra, A. Medvedovici and F. David, *Chromatographia, 44*:37 (1997).

37. C. Bicchi, C. Balbo, F. Belliardo, O. Panero, and Rubiolo, "SFC-UV Analysis of Safranal, Picrocrocin, Crocetin and Crocins in Saffron," Proceedings of the 18th

International Symposium on Capillary Chromatography, Riva del Garda, Italy, p. 1746 (1996).

38. B. Herbreteau, unpublished result.
39. M. Lafosse, B. Herbreteau, and L. Morin-Allory, *J. Chromatogr.*, *720*:61 (1996).
40. B. Herbreteau, M. Lafosse, M. Dreux, V. Krzych, and P. André, *Int. J. Biochromatogr.*, *1*:301 (1996).
41. S. Brossard, M. Lafosse, and M. Dreux, *J. Chromatogr.*, *623*:323 (1992).
42. *International Conference on Harmonization of Technical Requirements for the Registration of Pharmaceuticals for Human Use (ICH)*, Stability Testing of New Drug Substances and Products, 27 October 1993.
43. J. Vessman, Pharmaceutical Analysis: Overview, *Encyclopedia of Analytical Sciences* (A. Townshend, ed.), Academic Press, San Diego, CA, p. 3798 (1995).
44. P. Lacroix, Pharmaceutical Analysis: Drug Purity Determination, *Encyclopedia of Analytical Sciences* (A. Townshend, ed.), Academic Press, San Diego, CA, p. 3808 (1995).
45. R. C. Williams, M. S. Alasandro, V. L. Fasone, R. J. Boucher, and J. F. Edwards, *J. Pharm. Biomed. Anal.*, *14*:1539 (1996).
46. K. Anton, private communication.
47. N. Periclès, Eur. Patent 0427671A1, 30 October 1990; U.S. Patent 5,224,510, 6 July 1993.
48. J. L. Janicot, M. Caude, and R. Rossets, *J. Chromatogr.*, *437*:351 (1988).
49. V. Camel, D. Thiébaut, M. Caude, and M. Dreux, *J. Chromatogr.*, *605*:95 (1992).
50. V. Villette, B. Herbreteau, M. Lafosse, and M. Dreux, *J. Liquid Chromatogr.*, in press.
51. S. Scalia and D. E. Games, *J. Pharm. Sci.*, *82*:44 (1993).
52. A. Tambuté, L. Siret, A. Begos, and M. Caude, *Chirality*, *4*:36 (1992).
53. L. Siret, N. Bargmann, A. Tambuté, and M. Caude, *Chirality*, *4*:252 (1992).
54. N. Bargmann-Leyder, A. Tambuté, A. Bégos, and M. Caude, *Chromatographia*, *37*:433 (1993).
55. N. Bargmann-Leyder, J. C. Truffert, A. Tambuté, and M. Caude, *J. Chromatogr.*, *666*:027 (1994).
56. K. Anton, J. Eppinger, L. Fredericksen, E. Francotte, T. A. Berger, and W. H. Wilson, *J. Chromatogr.*, *666*:395 (1994).
57. N. Bargmann-Leyder, C. Sella, D. Bauer, A. Tambuté, and M. Caude, *Anal. Chem.*, *67*:952 (1995).
58. L. Siret, P. Macaudière, N. Bargmann-Leyder, A. Tambuté, M. Caude, and E. Gougeon, *Chirality*, *6*:440 (1994).
59. J. Whatley, *J. Chromatogr.*, *697*:251 (1995).
60. S. M. Wilkins, D. R. Taylor, and R. J. Smith, *J. Chromatogr.*, *697*:587 (1995).
61. R. J. Smith, D. R. Taylor, and S. M. Wilkins, *J. Chromatogr.*, *697*:591 (1995).
62. P. Sandra, A. Medvedovici, A. Kot, and F. David, *Eur. Pharmaceut. Rev.*, *1*:41 (1996).
63. P. Sandra, A. Medvedovici, and F. David, *J. Chromatogr.*, in press.
64. G. Franke and F. Vérillon, *J. Chromatogr.*, *450*:81 (1988).
65. F. Vérillon and M. F. Claverie, *Lab. Robotics Autom.*, *3*:241 (1992).
66. G. O. Cantrell, R. W. Sringham, J. A. Blackwell, J. D. Weckwerth, and P. W. Carr, *Anal. Chem.*, *68*:3645 (1996).

4

Packed Column Supercritical Fluid Chromatography—Evaporative Light-Scattering Detection

J. Thompson B. Strode III and Larry T. Taylor

Virginia Tech, Blacksburg, Virginia

Klaus Anton and Monique Bach

Novartis Pharma AG, Basel, Switzerland

Nico Periclès

Ciba Specialty Chemicals, Inc., Kaisten, Switzerland

I. INTRODUCTION

The future of packed column supercritical fluid chromatography (pSFC) appears exceedingly bright as a complementary technique to high-performance liquid chromatography (HPLC). In fact, commercial instrumentation is readily available that enables the user to go back and forth between pSFC and HPLC without changing the column [1,2].

In pSFC, modifiers and additives (i.e., organic solvents, acids, and/or bases) are frequently required to efficiently elute polar analytes off the column [3] (see Chapter 2). The modifiers/additives perform this function primarily by either increasing the solubility of the analyte in the supercritical fluid (SF) or by reducing the activity of the column. Unfortunately, the introduction of modifiers/additives can interfere with the use of the most popular gas chromatographic (GC) flame ionization detector (FID) [3]. In addition, molecules with no chromophores cannot be detected by the popular HPLC ultraviolet-visible (UV-vis) detector, which is mainly used for pSFC. Therefore, beside UV-vis a detector is needed for pSFC in which the modifier does not interfere and in which detection of fewer or no absorbing molecules in combination with pressure/modifier gradients is possible. One such detector is the quasi-universal evaporative light-scattering detector (ELSD), which responds to the light scattered by nonvolatile analytes after the mobile phase has been evaporated [4–7]. Numerous developments have been made in pSFC/ELSD interfaces during the last few years, and applications of pSFC/ELSD have begun to routinely appear in the literature [8]. A review of the state of the art including some industrial applications will be presented. Comparison of pSFC to HPLC methods will also be made where data are available.

II. BASIC EVAPORATIVE LIGHT-SCATTERING DETECTOR OPERATION

The ELSD was originally developed for connection to conventional HPLC. Detection is based on the mass of a relatively nonvolatile compound rather than on the ability of that compound to absorb light of a specific wavelength. The ELSD has three important zones, as shown in Fig. 1 [12]. The first zone is the nebulizer, which in HPLC is used to form an aerosol and ultimately to produce droplets. The heated drift tube constitutes the second zone, which is used to evaporate away the mobile phase and any volatile material (i.e., buffers) from droplets formed during nebulization. As the mobile phase is evaporated away from the droplet, a particle is formed. The dry particles then drift down the heated drift tube into the third zone, a detection cell. The detection cell contains a light source (i.e., tungsten lamp or laser) and detector (i.e., photodiode). As the dry particles drift through the detection cell, they pass through the laser beam and scatter light. The scattered light intensity is then measured by the photodiode detector at a 120° angle to a beam of light produced by a tungsten source or at a 90° angle to a helium/neon laser. These zones are depicted in Fig. 1 along with a listing of the various experimental parameters that influence the quality of ELSD detection.

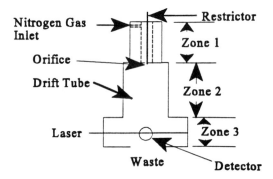

Figure 1 Schematic of ELSD with appropriate important experimental parameters. (From Ref. 12. Reproduced by permission of Preston Publications, A Division of Preston Industries, Inc., Niles, IL.)

Particles can scatter light by four main mechanisms: (1) reflection, (2) refraction, (3) Raleigh scattering, and (4) Mie scattering [4]. Reflection and refraction occur when the size of the particle approaches the wavelength of light used in detection. Reflection and refraction can be described simply as the deviation of light quanta as they encounter the boundary between phases. Mie scattering dominates when the particle size is less than the wavelength (λ) of the probe light but greater than $\lambda/2\theta$. Mie scattering can be explained by the notion that different points on the same particle are exposed to incident light, which causes induced oscillating dipoles to form. The induced oscillating dipoles then produce waves that can interfere with each other similar to Young's double-slit light experiment [9]. As a result, the scattered light can have a greater or lesser intensity depending on the angle of observation. Raleigh scattering, on the other hand, begins to dominate when the particles are much smaller than the wavelength of light ($r/\lambda < 5 \times 10^{-2}$, where r = radius of the particles). Because the particles are very small in relation to the wavelength of light, the particles behave as point sources. The light quanta induce oscillating dipoles similar to Mie scattering. The particles then radiate low-intensity light in all directions. There is no single dominating mechanism. Reflection and refraction occur most frequently, but as the particle size decreases (e.g., lower concentration and mass of sample injected) Raleigh and Mie scattering begin to dominate. Because Raleigh and Mie scattering are exponentially proportional to the intensity of light scattered, the amount of light scattered is small. Therefore, under operating conditions of most ELSDs, reflection and refraction are the dominating mechanism.

A. Nebulizer Design for pSFC

1. Fixed Restrictor

It is generally believed that the design of the ELSD nebulizer spray device significantly influences the sensitivity of the detector because changes in spray particle size and the direction and velocity of particles in the drift tube will probably be affected. Several different nebulizer designs have been utilized in pSFC. Most designs used a standard or slightly modified HPLC nebulizer with a crimped (i.e., pinched tube) stainless steel restrictor inserted into the nebulizer assembly to deliver the SF effluent to the detector with analytical columns (4.6 mm. id) [10,11]. Typical fixed restrictor types are shown in Fig. 2 [38].

The conventional Venturi nebulizer that atomizes the mobile phase in HPLC applications was thought to be unnecessary in pSFC because the restrictor that controls pressure and flow rate in capillary column SFC (cSFC) was believed to perform like a nebulizer. A tapered stainless steel tube was used first as a restrictor, but it was difficult to reproducibly control the pressure and flow rate [10]. It was replaced with a linear restrictor that was a short length of fused silica (75 μm id × 60–200 mm). The response of the detector with this restrictor

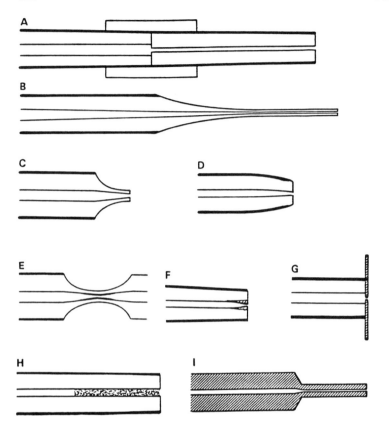

Figure 2 Schematic cross-section diagram of restrictor designs used in cSFC with FID and in pSFC and ELSD. All restrictors except (G), which represents a pinhole laser-drilled orifice, and (I), a pinched restrictor, are fabricated from fused silica tubing coated with polyimide. A, linear; B, C, and F, tapered; D, integral; E, converging-diverging; G, pinhole; H, porous frit; I, pinched. (From Ref. 38.)

was found to be highly dependent on the flow rate of the decompressed carbon dioxide. A maximum response at 2.2 L/min decompressed flow [10] was observed, but this high flow rate was not practical for most analytical scale (4.6 mm id) pSFC analyses. In other words, the chromatographic resolution of nonpolar to moderately polar analytes would suffer because the resulting liquid carbon dioxide flow rate of 4.5 mL/min (e.g., approximately 2.2 L/min decompressed carbon dioxide flow) would be much higher than the widely used pSFC value of 2.0 mL/min liquid flow [3].

Furthermore, in these early studies with a linear fused silica restrictor, the Joule–Thompson cooling effect at the high decompressed carbon dioxide flow rates caused dry ice to form, which increased the background noise and decreased the performance of the detector. To alleviate this problem, the tip of the restrictor in a second-generation interface was placed inside a heated brass ring that heated the restrictor and prevented ice formation [11]. However, ice formation could not be completely eliminated with this arrangement when the content of methanol as modifier was increased from 2.8% (w/w) to 10% (w/w). Excessive ice formation caused by the additional methanol could be reduced by increasing the drift tube temperature, but the detector's performance was also reduced by this strategy. Nizery et al. found that the ice formation and resulting noise could be minimized without affecting the performance of the detector by heating a small section of tubing located after the tip of the restrictor but before the drift tube [11]. This action reduced the noise when both 100% carbon dioxide and low concentrations of methanol-modified carbon dioxide [3% (w/w)] mobile phases were employed but at higher methanol concentrations [4.8% (w/w)] a make-up gas was necessary to aid in the evaporation of the mobile phase at high decompressed carbon dioxide flow rates (>3 L/min). A detection limit of 12 ng was reported for docosanol and octadecanol on a Lichrospher Sil column, but the mobile phase composition was not reported [11]. In these early studies, it was believed that the detector response was highly dependent on the mobile phase flow rate. A plot of peak area and flow rate yielded a maximum where it was speculated that the ascending portion of the curve was caused by losses of solutes on the walls of the drift tube at low flow rates, whereas the descending portion was explained by the formation of smaller drops at high flow rates. For these early reasons, it was deduced that pressure programming in pSFC could not be performed with ELSD. This notion has now been proven to be incorrect.

An alternate approach for pSFC/ELSD with analytical scale columns and an Alltech Mark III ELSD was shown by Strode et al. when they used an integral fused silica restrictor as a modified nebulizer to deliver the SF effluent to the detector (Fig. 3) [12]. They observed that the make-up nitrogen gas flow rate also had a large effect on the response of the Alltech Mark III detector. As the nitrogen gas flow rate was increased, the signal response of the detector decreased. The decreased signal was thought to be caused by a decreased residence time of the particles in the laser beam [13]. It was, however, believed that the total flow rate through the detector (not just the nitrogen flow) controlled the response of the analyte. The total flow rate of the detector is defined as the combination of the make-up nitrogen gas flow rate and the decompressed carbon dioxide flow rate. The total gas flow rate was plotted against the response of the detector using 5% methanol-modified carbon dioxide

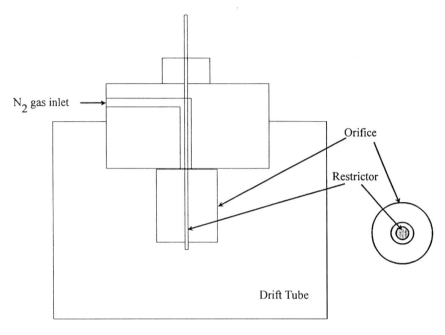

Figure 3 pSFC/ELSD interface. (From Ref. 12. Reproduced by permission of Preston Publications, A Division of Preston Industries, Inc., Niles, IL.)

wherein a total gas flow rate equal to or higher than 1000 mL/min was maintained. It was surprisingly found that the response of the detector did not change by orders of magnitude as a function of carbon dioxide flow rate as previously reported by Carraud et al. [10]. Consequently, this integral nebulizer design resulted in less dependency on the carbon dioxide flow rate. As a result, slower decompressed carbon dioxide flow rates and smaller id columns could be equally well used without sacrificing the performance of the detector. Furthermore, pressure programming with the integral restrictor wherein the flow rate would increase could be utilized. Detection limits of 10 ng or less for progesterone, testosterone, and 17α-hydroxyprogesterone at 200 bar, 700 mL/min decompressed carbon dioxide, and 2% (v/v) methanol-modified carbon dioxide were reported [12].

A miniaturized ELSD was utilized for packed capillary supercritical fluid chromatography (pcSFC) by Hoffmann et al. [14]. To accomplish this, the nebulizer and the drift tube were removed from a conventional HPLC-ELSD. Typical flow rates of 10 mL/min decompressed carbon dioxide (20 μL/min liquid) for the pcSFC were utilized. The drift tube heating block was used as the column oven. The restrictor was placed at the entrance of the detection cell. A detection limit of <5 ng for both Irgafos 168 and trimyristin using 5.6% (mol) n-propanol-modified carbon dioxide and a Novapak-4 ODS column (100 mm × 0.32 mm id) was observed [14]. The separations of glyceryl monosterate and ethylene bisstearamide were discussed. Similarly, Demirbüker et al. described a pcSFC system in which another miniaturized ELSD was utilized [15]. The restrictor (130 mm × 10 μm fused silica) was connected to the miniaturized drift tube (1/16 in. od ss tube). The drift tube was then connected to a small detection cell where the analyte particles could scatter light and be detected. They found that the lower volume ELSD did not suffer the mobile phase flow rate dependence that was observed earlier by Carraud et al. [10]. For example, the response of the detector remained constant over a range of 8–16 mL/min decompressed carbon dioxide (13–26 μL/min liquid). They also discussed the argentation chromatography of triacylglycerols using modified carbon dioxide mobile phases. A detection limit of 6 ng was determined for triolein with a packed capillary column. A recently published pSCF/ELSD design allows independent optimization of the flow rate for the pSFC separation and for the ELSD detection. Pinkston et al. [16] transfer with a "make-up solvent" (e.g., before ELSD inlet) the pSFC/ELSD into a HPLC/ELSD coupling to prevent carbon dioxide ice formation and sample precipitation during the depressurization process [8].

2. Variable Back Pressure Regulator

A new type of variable back pressure regulator also suitable for pSFC/ELSD was patented in 1989 by N. Periclès and has been for preparation of drug products (Fig. 4) [17–19]. The regulator is based on a modified needle valve whereby the needle is replaced by a piston, which can be moved with the aid of a piezo transducer according to the pressure program. The regulator was designed to be used within the following range: 100–350 bar pressure, 0.25–20.0% modifier, and 0.1–4 mL/min liquid flow rate. Because of the low internal dead volume (10 μL) packed columns from 4.6 to 1.0 mm id can be used without losing too much separation efficiency. The variable pressure regulator valve A is coupled via a stainless steel capillary B directly to the ELSD C, where a linear restrictor D allows a two-step depressurization. A comparable system was described in 1989 by Giorgetti et al. [20] and in 1992 by F. Verillon et al. [2] for a similar two step back pressure regulator design. This arrangement of a variable pressure regulator valve and a fixed restrictor in series ensures that the outlet of the variable back pressure regulator is always in the liquid mobile phase, which helps to prevent precipitation of the sample in the heated valve. The piezo transducer is driven by an electronic programmer E and the actual pressure is measured with a pressure indicator controller (PIC) F. High-speed,

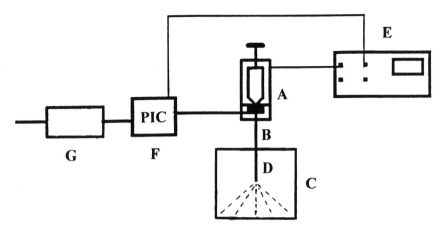

Figure 4 Scheme of the piezo-driven variable back pressure regulator. (A) Variable pressure regulator valve, (B) stainless steel capillary, (C) ELSD, (D) linear restrictor, (E) electronic programmer, (F) pressure indicator controller (PIC), (G) column. (From Ref. 19.)

high-resolution analysis with the described variable back pressure regulator and ELSD is shown in Fig. 5 whereby triglycerides for suppository mass are separated under (A) isobaric and isocratic, (B) isocratic and pressure programmed, and (C) isobaric and composition programmed pSFC conditions. This application is discussed in further detail in Chapter 3, Fig. 23. In the same chapter, Figs. 10 and 11 show a complete pSFC/ELSD system and scheme. Other commercially available one-step decompression valves (1,21) could also be coupled to ELSD but no data are found in the literature (8).

Takeuchi et al. coupled pSFC to a conventional HPLC/ELSD to permit the use of (1) smaller columns (i.e., 1–2 mm id), (2) a variable restrictor, and (3) slower flow rates [22]. They accomplished this by connecting the outlet of a variable restrictor (i.e., micrometering valve) to a mixing tee that was connected to the inlet of a conventional HPLC/ELSD nebulizer. A particle-forming solvent via the mixing tee was also found to be necessary for making a proper aerosol. This approach enabled the use of slower mobile phase flow rates [300 μL/min (liquid)] and smaller diameter columns (1.7 × 250 mm) but it diluted the molar concentration. No detection limits were reported.

3. Mobile Phase Flow Rate Ranges

The liquid carbon dioxide flow rate range for the various pSFC/ELSD studies is summarized in Fig. 6. As one can see, the liquid flow rate spanned by the instruments of Strode et al. and Pericles was the greatest.

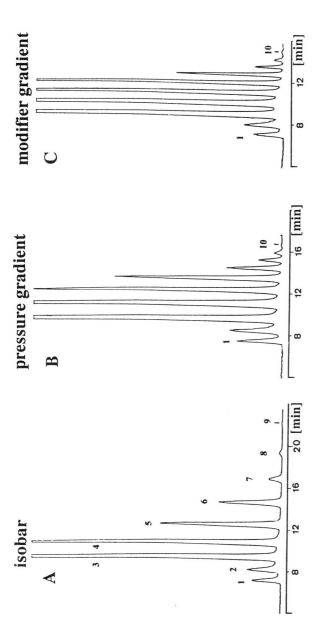

Figure 5 Separation of triglycerides (suppository mass) by pSFC/ELSD. pSFC conditions. Column CN 3 μm, 250 × 2 mm id; pressure. (A) 190 (iso)bar; (B) 9.5 min 190 (iso)bar, in 5 min linear to 225 bar, 2.5 min 225 (iso)bar; (C) 190 (iso)bar. Flow: carbon dioxide: 2.5 mL/min. MeOH: (A, B): 0.0625 mL/min isocratic, (C): 5 min isocratic 0.0625 mL/min, in 6 min linear to 0.125 mL/min, 6 min isocratic 0.125 mL/min, oven: 150°C; detector: ELSD 80°C; variable back pressure regulator. Peak identification: peaks identified by number of C atoms in fatty acid chains. 1: 10-10-12; 2: 10-12-12; 3: 12-12-12; 4: 12-12-14; 5: 12-14-14; 6: 14-14-16; 7: 14-14-16; 8: 14-16-16; 9: 16-16-16; 10: 16-16-18. (B. from Ref. 37. Reproduced by permission of Preston Publications, A Division of Preston Industries, Inc., Niles, IL.)

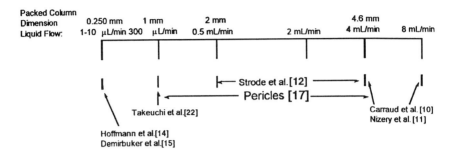

Figure 6 Summary of pSFC/ELSD studies organized by liquid carbon dioxide flow rate and column dimensions.

B. Drift Tube Temperature

The effect of drift tube temperature was the same for all of the previously discussed nebulizers. As the drift tube and/or detection cell temperature was increased, the response of the analyte was decreased. The decreased signal was attributed to solute vaporization at the higher detector temperatures [4,13]. For the analysis of progesterone, testosterone, and 17α-hydroxyprogesterone, the response of progesterone was found by Strode et al. to decrease the most with increasing drift tube temperature (Fig. 7) [12]. Above 70°C, the response of the analytes was reduced even more when methanol was used as a modifier. The effect of the drift tube temperature was reduced with progesterone when the modifier was changed from methanol to either ethanol or isopropanol (Fig. 7). A similar effect was observed when testosterone and 17α-hydroxyprogesterone were used. Since ethanol and isopropanol have higher boiling points than methanol, it was speculated that more thermal energy would be required to evaporate these solvents.

C. Solute Volatility

Nizery et al. found that the volatility of the analyte had a significant effect on the response of the detector [11]. The more volatile analytes produced lower signal responses than the higher boiling analytes. The effect of solute volatility was investigated by Strode et al. studying a homologous series of hydrocarbons (C16, C18, C20, C24, C26, C28, C30, and C32 each present at 500 ng at the ELSD) using 100% carbon dioxide and 5% methanol-modified carbon dioxide as shown in Fig. 8 [12]. The lowest member of the series detected by ELSD was C24 (Fig. 8A) whereas the FID detected all eight of the hydrocarbons including C16 with 100% carbon dioxide mobile phase (Fig. 8B). When 5% methanol

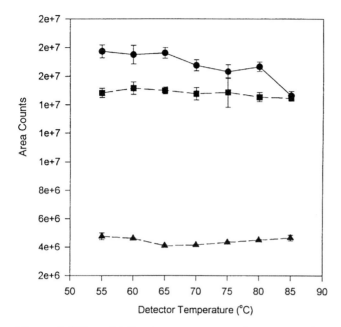

Figure 7 Effect of drift tube temperature on the response of progesterone. Conditions: postcolumn split flow = 800 mL/min decompressed carbon dioxide, 5% methanol-modified carbon dioxide, Deltabond cyanopropyl-derivatized silica (250 × 4.6 mm, dp = 5 μm), 200 bar column inlet pressure, 50°C (oven), 2 mL/min carbon dioxide (liquid flow measured at the pump), 0.4 standard liters per minute (SLPM) nitrogen make-up gas, and 0.0165 in orifice. ●, 5% (v/v) methanol-modified carbon dioxide, 5% (v/v); ■, ethanol-modified carbon dioxide; and ▲, 5% (v/v) isopropanol-modified carbon dioxide. Integral restrictor.

modified carbon dioxide was used, C24 was still the lowest hydrocarbon that could be detected by the ELSD (Fig. 8C). Although the methanol did not aid in the detection of the more volatile analytes, the ELSD improved the detection limit of higher molecular weight hydrocarbons in the presence of methanol, whereas the FID could not be used due to the presence of organic modifier. The improved detection limit, however, may be a result of the faster elution one gains with modified carbon dioxide because the modifier enhances solvating power and reduces possible reactive sites on the column (see also discussion in Sec. II.D).

The effect of solute volatility has also been studied by Berry et al. with pSFC/ELSD [23]. A homologous series of 14 monocarboxylic acids ranging from acetic acid (mpt. 16°C, bpt. 117°C) to pentadecanoic acid (mpt. 52°C, bpt.

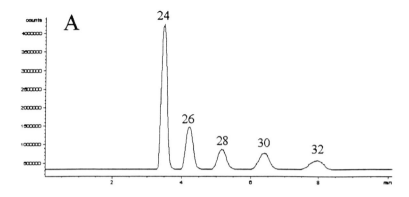

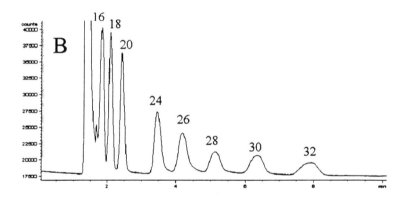

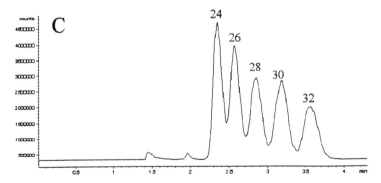

Figure 8 Effect of *n*-alkane volatility on the performance of the ELSD. Conditions: 1.5 mL/min (liquid carbon dioxide measured at the pump), 112 bar column inlet pressure, Deltabond cyanopropyl-derivatized silica, 600 mL/min decompressed carbon dioxide diverted to the ELSD by an integral restrictor, 75°C oven temperature, 0.4

340°C) was studied. The lowest member of the series detected was butanoic acid (mpt. −6°C, bpt. 162°C). The response of butanoic acid, however, was two orders of magnitude lower than pentadecanoic acid. Methanol-modified (10%) carbon dioxide served as the mobile phase. The workers stated that their results were much improved over HPLC/ELSD in which case the lowest homolog detected was octanoic acid and the overall sensitivity achieved was only 20% of that afforded by pSFC/ELSD.

D. Detector Performance

Because the response of the ELSD is a function of the four different light-scattering mechanisms, the response of the analyte has been determined to be proportional to the analyte mass by a nonlinear relationship [4]. To obtain linear plots, Eq. (1) is plotted on a log-log scale, thereby resulting in Eq. (2). With Eq. (2), linear log-log plots have been obtained over limited sample ranges.

$$y = (a)(m^b) \tag{1}$$

$$\log(y) = \log a + b \log m \tag{2}$$

where y = ELSD response
m = mass of the analyte
a = response factor
b = coefficient of regression (slope) of the response curve

. The minimum detection limit of pSFC/ELSD has progressively increased over the years. Carraud et al. originally reported detection limits at approximately 50 ng. This value was later improved to 12 ng for a signal-to-noise ration of 3:1. In general, low-nanogram detection limits have been observed with many pSFC/ELSD systems (Table 1). For example, the limit of detection (LOD) was determined for progesterone, testosterone, and 17α-hydroxyprogesterone using 2%, 10%, and 20% methanol-modified carbon dioxide under ELSD conditions of 60°C, and 200 bar column inlet pressure, on a Deltabond cyanopropyl column [12]. Calibration curves (Table 2) of peak area vs. concentration were based on an average ELSD response (n = 4) in the concentration of 100–5000 ppm. Precision appeared to be quite high regardless of the percent modifier or steroid type. The LOD of the three steroids was determined to be 10 ng or lower (s/n = 3) at all modifier concentrations.

SLPM nitrogen make-up gas flow rate, 65°C drift tube temperature, and 0.0165-in. orifice. A: ELSD, 100% carbon dioxide; B: FID, 100% carbon dioxide; C: ELSD, 5% methanol-modified carbon dioxide. Numbers over peaks correspond to the carbon number. (From Ref. 12. Reproduced by permission of Preston Publications, A Division of Preston Industries, Inc., Niles, IL.)

Table 1 Limits of Detection (LOD) for Different Nebulizer Designs

Column i.d. (mm)	Analyte	Mobile phase	ELSD decompressed flow rate (mL/min)	LOD (ng)	Ref.
4.6	Docosanol	Not reported	2700	12	10[a]
4.6	Progesterone	2% (v/v)methanol-carbon dioxide	700	9	12[a]
2	Tripalmitic triglycerides	2.4% (v/v) methanol-carbon dioxide	1000	5	37[b]
0.320	Trimyristin	5.6% (mol) n-propanol-carbon dioxide	10	5	14[a]

[a] Fixed restrictor.
[b] Variable back pressure regulator.

Table 2 Calibration Curve Results for pSFC/ELSD

	Progesterone		
*	2% CH_3OH/98% CO_2	10% CH_3OH/90% CO_2	20% CH_3OH/80% CO_2
m	1.80	1.74	1.65
i	2.00	2.48	1.65
r^2	0.998	0.994	0.9998
s_m	0.044	0.052	0.013
s_i	0.28	0.32	0.081

	Testosterone		
*	2% CH_3OH/98% CO_2	10% CH_3OH/90% CO_2	20% CH_3OH/80% CO_2
m	1.83	1.68	1.50
i	1.88	2.38	3.06
r^2	0.998	0.993	0.9994
s_m	0.035	0.023	0.021
s_i	0.25	0.16	0.12

	17α-Hydroxyprogesterone		
*	2% CH_3OH/98% CO_2	10% CH_3OH/90% CO_2	20% CH_3OH/80% CO_2
m	1.69	1.79	1.52
i	2.18	2.18	2.91
r^2	0.998	0.9996	0.995
s_m	0.033	0.017	0.057
s_i	0.22	0.098	0.36

*m is the slope, i is the intercept, r^2 is the correlation coefficient, s_m is the standard error in the slope, and s_i is the standard error in the intercept.
Source: Ref. 12.

Table 3 Limits of Detection for Steroids for pSFC/ELSD (ng)

Compound	2% CH₃OH 98% CO₂	10% CH₃OH 90% CO₂	20% CH₃OH 80% CO₂
Progesterone	9	10	4
Testosterone	7	6	5
17α-Hydroxyprogesterone	7	6	5

Source: Ref. 12.

Furthermore, the detection limits can surprisingly be improved as the modifier concentration is increased (Table 3) [11]. It was theorized that the additional methanol in the mobile phase produced larger droplets that resulted in larger particles and a larger response by the ELSD. This increase in sensitivity may also be just an effect of faster eluting compounds, so that the signal to noise increases which should result in a better detection limit. However, the possible changes in sensitivity (increase, none, decrease) with methanol-modified carbon dioxide is known [8]. It seems that the experimental design as described by Pinksten [16], whereby pSFC nebulization is transferred to HPLC nebulization, decreases the possible variations in response [39].

III. APPLICATIONS

A. Carbohydrates

Carraud et al. first reported the analysis of fructose and sucrose by pSFC/ELSD using an octadecyl silica (ODS) column and 7.5% methanol-modified carbon dioxide [10]. Later, Herbreteau et al. [24] discussed the separation of a wide range of sugars by pSFC/ELSD using different column types with different methanol-modified carbon dioxide compositions as mobile phase (Fig. 9). They concluded that for pSFC/ELSD there was a greater range of selectivity compared to HPLC/ELSD. However, the low temperature used for these pSFC applications can generate, depending on modifier concentration, subcritical conditions with a liquid phase similar to HPLC. Another example can be seen in Chapter 3, Fig. 19.

Berry et al. [23] investigated the use of β-cyclodextrin and cyano-bonded stationary phases for the separation of eight sugars in <12 min using carbon dioxide with a methanol gradient up to approximately 20%. The sugars eluted as three distinct groups (e.g., mono-, di-, and trisaccharides). Upnmoor and Brunner [25] found an improved peak shape when comparing Nucleosil 100 Si with Lichrosorb RP-18 column material at 220 bar and 40°C with different methanol composition percentage relative to carbon dioxide.

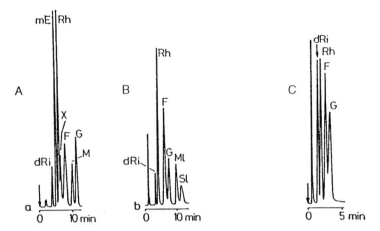

Figure 9 Separation of sugars with subFC/ELSD. subFC conditions: column: (A) Lichrospher diol, (B) RSIL NO2, (C) Zorbax CN; pressure A: 265 bar, B, C: 238 bar; total liquid flow A: 1.77 mL/min, B: 3.8 mL/min, C: 4.49 mL/min; eluent: carbon dioxide-methanol. A: (84:5:15.5, w/w); B: (87.0:13.0, w/w); C: (91.1:8.9, w/w). dRi = 2-deoxy-D-ribose, mE = mesocrythritol, Rh = rhamnose, X = xylose, F = fructose, M = mannose, G = glucose, M1 = mannitol, S1 = sorbitol. (From Ref. 24. Reproduced by permission of Elsevier Science-NL, Amsterdam, The Netherlands.)

B. Oligomers

Low molecular weight PEG 200 can be rapidly separately on a Lichrospher aminopropyl-derivatized silica column with 15% methanol-modified carbon dioxide [12]. With the same mobile phase but different composition percentage a similar resolution could be obtained on a diol column type [26]. Figure 10A', B' demonstrates decreased chromatographic run time together with increased resolution for the same type of PEGs compared to Fig. 10A, B. While in A and B the pressure is stabilized with flow rate adjustment together with the use of a linear flow restrictor at high isobaric conditions, the programmable low dead-volume back pressure regulator is used at lower isobaric conditions together with a composition gradient in A' and B'. The method in A' and B' is working with a longer column length, smaller id, and particle size together with a lower flow rate.

To increase the solvating strength of the mobile phase, an additive can be used. Brossard et al. [27] added triethylamine to the methanol-carbon dioxide mobile phase to elute the higher molecular weight PEGs from a silica and diol–derivatized silica phase with 2.5 L/min decompressed carbon dioxide and 268 bar. To increase the resolution between the oligomers, they found that water was required to reduce the activity of the column. Strode et al. [12] separated PEG

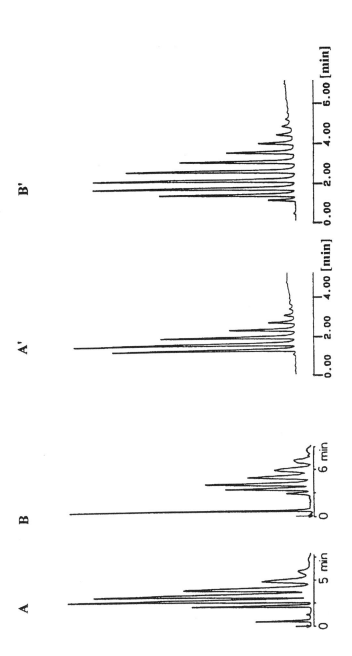

Figure 10 Separation of PEG 300 (A, A') and 400 (B, B') with subFC/ELSD and a linear restrictor (A, B) or with pSFC/ELSD (A', B') and the low dead-volume variable back pressure regulator. Conditions: column: (A, B) Lichrospher diol, 5 μm, 125 × 4 mm id, (A', B') Permacoat diol, 3 μm, 200 mm × 2 mm id; pressure: (A) 272 bar, (B) 170 bar, (A') 180 bar, (B') 190 bar; neat carbon dioxide flow: (A) 3.4 mL/min, (B) 1.8 mL/min, (A', B') 2.0 mL/min, MeOH modifier flow, (A') 0.25 min isocratic 0.2 mL/min, in 2.5 min linear to 0.35 mL/min, 1.25 min isocratic 0.35 mL/min, (B') 0.25 min isocratic 0.2 mL/min, in 4.5 min linear to 0.35 mL/min, 1.25 min isocratic 0.35 mL/min; oven temperature: (A, B) ambient temperature, (A', B') 100°C, detector: (A, B, A', B') 45°C. (A, B: From Ref. 26. Reproduced by permission of Elsevier Science-NL, Amsterdam, The Netherlands.)

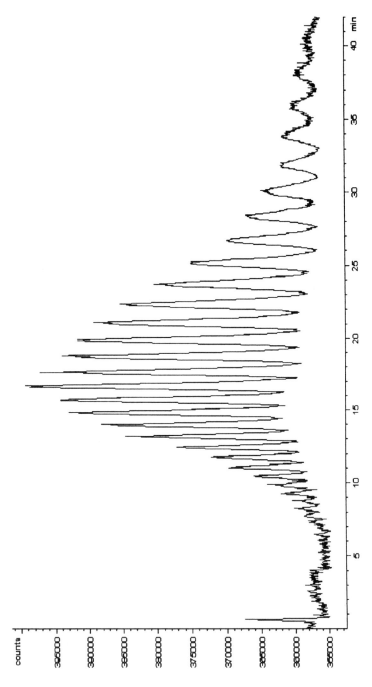

Figure 11 Separation of PEG 2000. Conditions: 20% (v/v) [1% (v/v) triethylamine, 5% (v/v) water, 14% methanol]-modified carbon dioxide, 4.0 mL/min (liquid carbon dioxide measured at the pump), 200 bar column inlet pressure, Lichrospher aminopropyl-derivatized silica, 1000 mL/min decompressed carbon dioxide diverted to the ELSD by an integral restrictor, 50°C oven temperature, 0.4 SLPM nitrogen make-up gas flow rate, 65°C drift tube temperature, and 0.0165-in. orifice (From Ref. 12. Reproduced by permission of Preston Publications, A Division of Preston Industries, Inc., Niles, IL.)

2000 on a Lichrospher amino column using a similar mobile phase as Brossard (Fig. 11). Thirty-three oligomers were observed, which was in good agreement with the open tubular capillary (cSFC-FID) separation of derivatized PEG 2000 by Pinkston et al. [28]. More importantly, these complex mobile phases did not interfere in the detection of ethoxylated alcohols. Decreased analysis time with similar resolutions can be observed with a methanol-modified carbon dioxide mobile phase, using the programmable low dead-volume back pressure regulator together with a diol 3 μm 200 mm × 2 mm id column in a pressure gradient mode. This is demonstrated in Fig. 12 where PEG 1500 is separated and about 26 oligomers were observed.

To our knowledge other separation techniques like GC-FID or cSFC-FID do not show such short retention times for PEGs as can be seen in Figs. 10A', B'

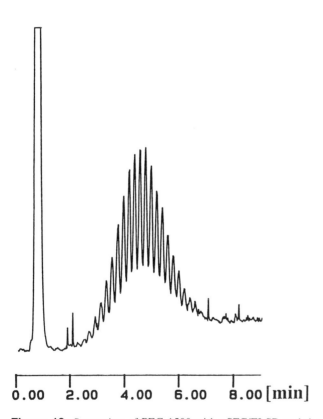

Figure 12 Separation of PEG 1500 with pSFC/ELSD and the low dead-volume variable back pressure regulator. pSFC conditions: column, carbon dioxide flow, oven and ELSD temperature as shown in Fig. 10, A', B'. Pressure 1 min 250 (iso) bar, in 8 min linear to 400 bar, 2 min 400 (iso) bar; flow methanol: 0.3 mL/min isocratic.

and 12. pSFC can also be used to fractionate and collect the individual oligomer as is shown in more detail in Chapter 12.

C. Ginkgolides

Ginkgolides are prescribed for the postponement of the symptoms of old age (i.e., forgetfulness and early dementia). The ginkgolides are extracted from the *Ginkgo biloba* leaf. Theire structure is shown in Fig. 13 [12]. The ginkgolides have a poor UV chromophore that prohibits the use of a UV-vis detector. The current assay method is usually performed using reversed phase HPLC/ELSD with an ODS column [29]. This assay method has been plagued by poor peak area reproducibility and large solvent waste. The ginkgolides could be separated and quantitated by GC with FID, but an additional silylation derivatization step was required [30]. In an effort to improve the chromatography performance of the assay and reduce solvent waste, a pSFC method was developed using a Lichrospher aminopropyl-derivatized silica column. The ginkgolide extract was separated into three peaks with 20% (v/v) methanol-modified carbon dioxide [12,31].

1. Bilobalide   2. Ginkgolide A   3. Ginkgolide B

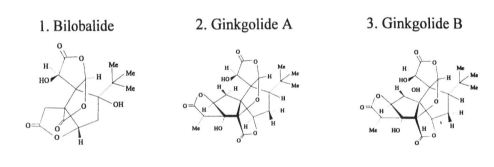

4. Ginkgolide J   5. Ginkgolide C

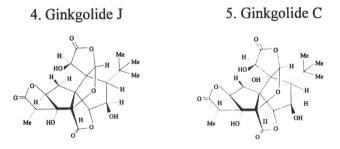

Figure 13 Structures of the five main components of the *Ginkgo biloba* leaf. (From Ref. 12.)

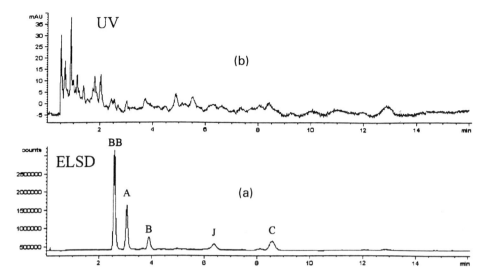

Figure 14 Separation of the ginkgolide extract from the *Ginkgo biloba* leaf. Conditions: 12% (v/v) methanol-modified carbon dioxide, 4.0 mL/min (liquid carbon dioxide measured at the pump), 280 bar column inlet pressure, Deltabond aminopropyl-derivatized silica. (a) ELSD 600 mL/min decompressed carbon dioxide diverted to the ELSD by an integral restrictor, 40°C oven temperature, 0.4 SLPM nitrogen make-up gas flow rate, 65°C drift tube temperature, and 0.0165-in. orifice. (b) UV-vis detection at 220 nm. (From Ref. 31. Reproduced by permission of Elsevier Science-NL, Amsterdam, The Netherlands.)

As the modifier concentration was reduced to 12% (v/v) methanol-modified carbon dioxide, the five expected ginkgolide peaks were observed but not resolved. The separation was theorized to occur by two opposing mechanisms. The first involved hydrogen bonding with the amino group, whereas, the second opposing mechanism involved the adsorption of the ginkgolides onto the residual silica sites. The five ginkgolide compounds were further separated on a Deltabond amino-derivatized column with 12% (v/v) methanol-modified carbon dioxide. UV detection of the ginkgolide compounds was possible at low wavelengths, but the extract contained many interfering compounds that have better chromophores than the ginkgolides (Fig. 14). Because the coeluting peaks were volatile compounds, they were not detected by the ELSD.

D. Miscellaneous

Lafosse et al. [32] reported the effective separation of three relatively nonpolar underivatized amino acids using a mixture of carbon dioxide, methanol, water, and triethylamine as a mobile phase using pSFC/ELSD. When pyridine was used instead of triethylamine, other underivatized amino acids were separated

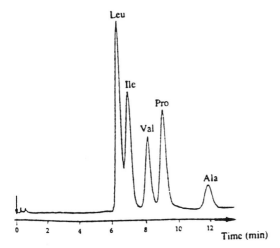

Figure 15 Separation of a standard mixture of five amino acids with subFC and a linear flow restrictor connected to ELSD. subFC conditions: column: Lichrosorb diol 5 μm; column inlet pressure: 107 bar; flow: 2.5 mL/min; oven: 30°C; composition: carbon dioxide-modifier (85:15, v/v), modifier, methanol-water-triethylamine-pyridine (87.95:7:0.05:5, v/v); detector: ELSD 45°C. (From Ref. 33. Reproduced by permission of Elsevier Science-NL, Amsterdam, The Netherlands.)

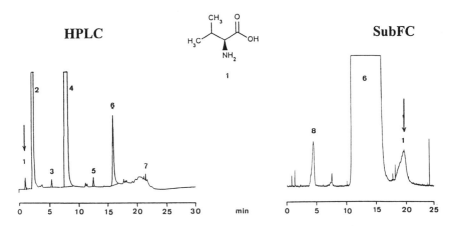

Figure 16 Separation of L-valine in an intermediate [(R)-valine-benzylester-toluene-4-sulfonate]. Comparison of HPLC/UV-vis with subFC and the programmable low dead-volume back pressure regulator connected to ELSD. *HPLC conditions*: column C18, 5 μm, 125 × 4.6 mm id; mobile phase 2.5 g/L tetramethylammonium sulfate in (A) water + ACN (9 + 1), (B) water + ACN (1 + 9), 15 min 5–35% then 5 min 35–95% then 1 min 95–5%, flow: 1.5 mL/min; oven: 35°C; detector: UV-vis 205 nm. *subFC conditions*: column Lichrospher 100 diol 5 μm, 250 × 4.6 mm id. Pressure: 300 (iso) bar, flow: carbon dioxide 1.7 mL/min, modifier 0.3 mL/min (methanol-water-triethylamine-pyridine (87.95:7:0.05:5, v/v); oven: ambient temperature; detector: ELSD 45°C. Peak 1: L-valine, peak 4: (R)-valine-benzylester-toluene-4-sulfonate, peaks 2, 3, 5–8: possible reaction products.

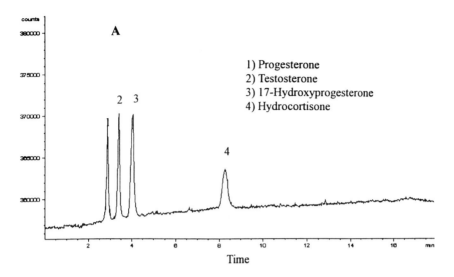

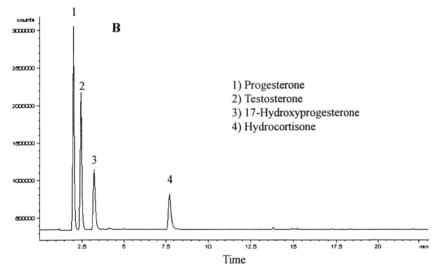

Figure 17 Effect of pressure gradient (A) and modifier gradient (B) programming on the performance of the ELSD. Deltabond cyanopropyl-derivatized silica, 858 mL/min decompressed carbon dioxide diverted to the ELSD by an integral restrictor, 60°C oven temperature, 0.4 SLPM nitrogen make-up gas flow rate, 65°C drift tube temperature, and 0.0165-in. orifice. (From Ref. 12.) A: 5% methanol-modified carbon dioxide, 2.0 mL/min (liquid carbon dioxide measured at the pump), 100 bar column inlet pressure (1.5 min hold) to 350 bar (1 min hold) at 15 bar/min. B: 1% methanol-modified carbon dioxide (0.5 min hold) to 20% methanol-modified carbon dioxide (2 min hold) at 1%/min, 280 bar column inlet pressure, 3.0 mL/min (liquid carbon dioxide measured at the pump). (Reproduced by permission of Preston Publications, A Division of Preston Industries, Inc., Niles, IL.)

under subFC conditions as shown in Fig. 15 [33]. Based on these results, a subFC/ELSD separation method was developed within several days during a drug development process to determined the content of L-valine in an intermediate [(R)-valine-benzylester-toluene-4-sulfonate] (Fig. 16). In the figure, the L-valine (peak 1) could not be determined with HPLC/UV-vis detection because of coelution with a possible reaction product (peak 8). With subFC/ELSD L-valine could be separated and quantified independent of any possible reaction products. This example clearly demonstrates how important it is to have orthogonal separation and detection methods in development laboratories.

Brossard et al. published pSFC/ELSD chromatograms for the analysis of synthetic waxes [27] and beeswax [40] (a chromatogram of yellow beeswax is published in Chapter 3, Fig. 20). Also, Carmel et al. present results for beeswax [33]. They used a diol-derivatized silica column with 2-propanol/formic acid/carbon dioxide mobile phase. Cocks and Smith discussed the separation of fatty acid methyl esters (FAMEs) by pSFC-ELSD employing 100% carbon dioxide and methanol-modified carbon dioxide with a mobile phase flow rate of 2.4 L/min decompressed carbon dioxide (35). They found that carbon dioxide could form dry ice and increase the background noise. To eliminated this noise, they suggested placing glasswool inside the heated drift tube to improve the

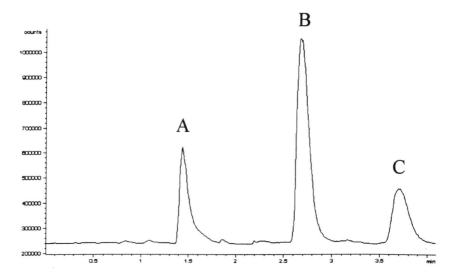

Figure 18 pSFC/ELSD of Ionophores. Flow: 3.5 mL/min (liquid CO_2), oven: 50°C; modifier: 20% (v/v) methanol-modified carbon dioxide; pressure: 200 bar; ELSD: 60°C, 0.4 SLPM nitrogen, and 0.0165-in. orifice, Column: Spherisorb amino (4.6 mm × 25 cm, dp = 3 µm). A: Monensin ($C_{36}H_{61}O_{11}Na$); B: salinomycin ($C_{42}H_{69}O_{11}Na$); and C: narasin ($C_{43}H_{71}O_{11}Na$). (From Ref. 12. Reproduced by permission of Preston Publications, A Division of Preston Industries, Inc., Niles, IL.)

heat transfer and improve the evaporation process. However, the glasswool was suggested to interfere in detection if the analytes absorb onto it.

Strode et al. [12] separated four steroids independently with pressure programming (Fig. 17A) and modifier gradient programming (Fig. 17B). Baseline resolution was obtained in each case. Pressure programming gave a more stable baseline than did modifier gradient programming.

Loran et al. [36] compare the regression data and the LODs for steroids gained with pSFC and tungsten filament light-scattering (TFLS) detection using a simple crimped stainless steel tube nebulizer design. They proposed a useful extension to the range of detectors available for pSFC. Berry et al. [23] show in an example of six steroids that ELSD can detect all compounds compared to UV-vis detection at 254 nm where only three compounds are detected. The successful separation by these same workers as well as Strode et al. [12] of ionophores has also been achieved (Fig. 18).

IV. SUMMARY

The response of ELSD interfaces was affected by restrictor type, drift tube temperature, nitrogen make-up gas flow rate, and modifier concentration. The signal was found to decrease as the detector temperature was increased. Increasing nitrogen flow rates decreased the signal for three of the interfaces. Modifier concentration is responsible for response variations. A recent HPLC-like nebulizer design seemed to stabilize the response. Sensitive detection (<10 ng) was found to be possible with each described interface type. pSFC/ELSD seems to be a promising technique for assaying foodstuffs, polymers, and drug substances and products of pharmaceuticals. It would be especially useful when organic modifiers/additives are required that prohibit the use of universal FID detection. The usefulness of orthogonal separation methods (pSFC/ELSD vs. HPLC/UV-vis) is demonstrated.

REFERENCES

1. R. Kornfeld, *Am. Lab.*, *24*:25–32 (1991).
2. F. Verillon, D. Heems, B. Pidion, K. Coleman, and J. C. Robert, *Am. Lab.*, *25*:45–53 (1992).
3. M. L. Lee and K. E. Markides, "Analytical Supercritical Fluid Chromatography and Extraction", Chromatography Conferences, Inc., Provo, UT (1990).
4. J. M. Charlesworth, *Anal. Chem.*, *50*:1414–1420 (1978).
5. G. Guiochon, A. Moysan, and C. Holley, *J. Liq. Chromatogr.*, *11*:2547–2570 (1988).
6. A. Stolyhwo, H. Colin, M. Martin, and G. Guiochon, *J. Chromatogr.*, *288*:253–275 (1984).
7. T. H. Mourey and L. E. Oppenheimer, *Anal. Chem.*, *56*:2427–2434 (1984).
8. M. Dreux and M. Lafosse, LC-GC, accepted for publication.

9. B. J. Winer, *Statistical Principles in Experimental Design*, McGraw-Hill, New York, pp. 452–464 (1971).
10. P. Carraud, D. Thiebaut, M. Caude, R. Rosset, M. Lafosse, and M. Dreux, *J. Chromatogr. Sci.*, 25:395–398 (1987).
11. D. Nizery, P. Carraud, D. Thiebaut, M. Caude, R. Rosset, M. Lafosse, and M. Dreux, *J. Chromatogr. Sci.*, 467:49–60 (1989).
12. J. T. B. Strode and L. T. Taylor, *J. Chromatogr. Sci.*, 34:261–271 (1996).
13. D. Upnmoor and G. Brunner, *Chromatographia*, 33:255–260 (1992).
14. S. Hoffmann and T. Greibrokk, *J. Microcol. Sep.*, 1:35–40 (1989).
15. M. Demirbuker, P. E. Andersson, and L. G. Blomberg, *J. Microcol. Sep.*, 5:141–147 (1993).
16. J. D. Pinkston, R. Hentschel, and T. L. Chester, "Pressure Control and Response Optimization for Evaporative Light Scattering Detection in Packed-Column SFC" 7th Inst. Symp. SFC and SFE, Indianapolis, IN, April 1996.
17. N. Periclès, Eur. Patent 0427671B1 (1988); U. S. Patent 5, 224, 510 (1993).
18. L. Frederiksen, K. Anton, and P. van Hoogevest, Eur. Patent 616801-A1 (1994).
19. L. Frederiksen, K. Anton, P. van Hoogevest, H. R. Keller, and H. Leuenberger, *J. Pharm. Sci.*, 86: 921–928 (1997).
20. A. Giorgetti, N. Periclès, H. M. Widmer, K. Anton, and P. Dätwyler, *J. Chromatogr. Sci.*, 27:318–324 (1989).
21. M. Saito, Y. Yamauchi, H. Kashiwazaki, and M. Sugawara, *Chromatographia*, 25:801–805 (1988).
22. Takeuchi et al., Advances in semi packed column SFC its hyphenation, *Hyphenated Techniques in Supercritical Fluid Chromatography and Extraction* (K. Jinno, ed.), Journal of Chromatography Library Series, Vol. 53, pp. 47–63 (1992).
23. A. Berry, E. Ramsey, M. Newby, and D. E. Games, *J. Chromatogr. Sci.*, 34:245–523 (1996).
24. B. Herbreteau, M. Lafosse, L. Morin-Allory, and M. Dreux, *J. Chromatogr.*, 505:299–305 (1990).
25. D. Upnmoor and G. Brunner, *Chromatographia*, 33:261–266 (1992).
26. M. Lafossee, P. Rollin, C. Elfakin, L. Morin-Allory, M. Martens, and M. Dreux, *J. Chromatogr.*, 505:191–197 (1990).
27. S. Brossard, M. Lafosse, and M. Dreux, *J. Chromatogr.*, 623:323–328 (1992).
28. J. D. Pinkston and R. T. Hentschel, *J. High Resolut. Chromatogr.*, 16:269–274 (1993).
29. F. F. Camponovo, J. L. Wolfender, M. P. Maillard, O. Potterat, and K. Hostettman, *Phytochem. Anal.*, 6:141–145 (1995).
30. A. Hasler and B. Meier, *Pharm. Pharmacol. Lett.*, 187:187–190 (1992).
31. J. T. B. Strode, L. T. Taylor, and T. van Beek, *J. Chromatogr.*, 738:115–122 (1996).
32. M. Lafosse, C. Elfakin, L. Morin-Allory, and M. Dreux, *J. High Res. Chromatogr.*, 15:312–318 (1992).
33. V. Camel, D. Thiebaut, M. Caude, and M. Dreux, *J. Chromatogr.*, 605:95–101 (1992).
34. S. Brossard, M. Lafosse, M. Dreux, and J. Becart, *Chromatographia*, 36:268 (1993).
35. S. Cocks and R. M. Smith, *Anal. Proc.*, 28:11–12 (1991).
36. J. S. Loran and K. D. Cromie, *J. Pharm. Biomed. Appl.*, 8:607–611 (1994).

37. K. Anton, M. Bach, C. Berger, F. Walch, G. Jaccard, and Y. Carlier, *J. Chromatogr. Sci.*, *32*:430–438 (1994).
38. B. W. Wright and R. D. Smith, *Modern Supercritical Fluid Chromatography* (C. M. White, ed.), Huethig, Heidelberg, p. 189 (1988).
39. M. Dreux, private communication.
40. S. Bossard, M. Lafosse, M. Dreux, and J. Böcart, *Chromatographia, 36*:268 (1993).

5

Packed Capillary Column Supercritical Fluid Chromatography Using Neat Carbon Dioxide

Yufeng Shen and Milton L. Lee

Brigham Young University, Provo, Utah

I. INTRODUCTION

Packed column supercritical fluid chromatography (pSFC) has been performed using columns with a wide range of internal diameters, depending on whether the separation is preparative or analytical in nature. The advantages of using packed capillary columns (50–320 μm id packed with 1.5- to 40-μm particles) are as follows: (1) higher column efficiencies are normally obtained because columns with smaller aspect ratios (column diameter/particle diameter) have higher permeabilities and therefore longer columns can be used; and (2) lower mobile phase flow rates result in better compatibility with various detection systems, less dilution and therefore higher concentration of resolved compounds for detection, and more effective implementation of temperature programming. The small amounts of stationary and mobile phases in a packed capillary column favor rapid heat transfer during temperature programming.

Fused silica capillary tubing is the most commonly used material for packed capillary SFC (pcSFC). The well-developed technology to prepare fused silica capillary tubing allows this material to be flexible and have a uniform and chemically inert inner wall surface.

Compared with conventional packed columns, packed capillary columns have lower sample loadability and sample capacity. This limits the use of pcSFC for the analysis of small-volume samples. Also, extracolumn contributions to peak broadening are more serious, and more skillful operators are required.

Currently, the packing materials used for pSFC are primarily those developed for HPLC, especially silica-based microparticles. However, when using neat carbon dioxide as the mobile phase, the inherent surface activity of these particles requires the use of polar modifiers in the mobile phase to chromatograph polar solutes. In order to successfully perform pcSFC separations of polar compounds, special strategies must be developed to further deactivate these particles. Furthermore, a variety of selectivities are required for the different applications.

In this chapter, we first review the methods used for preparing packed capillary columns for pcSFC and then describe various applications. The preparation of well-deactivated packing materials for the analysis of polar compounds is emphasized.

II. OVERVIEW OF THE PREPARATION OF PACKED CAPILLARY COLUMNS

Packed capillary columns can be prepared using dry or slurry packing methods [1]. In the dry packing method, particles are entrained and carried into the capillary column (*e.g.*, 250 μm id fused silica) with gases such as N_2 [2,3]. Up to 70 cm of packed capillary length and minimum reduced plate heights (h_{min}) of approximate 2.2 can be obtained using 5-μm particles. A small amount of solvent such as ethyl alcohol is used to reduce static charge and aggregation prior to packing the column. During packing, the column is vibrated in an ultrasonic bath to obtain a uniform packed bed. The advantage of this method is that with relatively simple operation and short time, a reasonably good packed capillary column can be produced. However, the suitability of this method to pack capillary columns of different inner diameters and with various particle sizes must be more thoroughly investigated.

Slurry packing methods are more commonly used in the preparation of packed capillary columns [4–11]. Liquid slurries are prepared by suspending microparticulate packings in selected organic solvents such as acetonitrile prior to packing. This method can be used for the preparation of capillary columns of inner diameters ranging from 25 to 250 μm with column lengths of up to approximately 1 m. A minimum reduced plate height of approximate 2.2 can be obtained using 5-μm particles. By decreasing the ratio of column diameter to particle size (ρ), h_{min} values as low as 0.88 can be obtained [8].

Recently, a carbon dioxide slurry packing method was introduced to prepare packed capillary columns [10,11]. The advantages of this method lie in the low viscosity and adjustable density of the carbon dioxide carrier. The low viscosity makes it possible to prepare long packed capillary columns of up to 10 m because a relatively low pressure drop is produced; this can sustain the necessary flow rate of the slurry during packing. Vibration of the column during packing is critically important in order to obtain uniform packed beds. The density of carbon dioxide can be controlled in the range from 0.3 to 0.8 g/mL, which is in agreement with the density of most porous silica-based packing materials; this favors the formation of uniform slurries of particles in carbon dioxide. However, carbon dioxide has little ability to reduce static charge, and the applicability of the carbon dioxide slurry method for particles having significant static charge must be further investigated. Although various packing methods have been used for the preparation of packed capillary columns, similar column performance has been reported. Minimum reduced plate heights of 2.3 ± 0.3 can be obtained for 5-μm particles using the methods mentioned above, which suggests that uniform packed beds in short columns can be obtained using any of these methods.

III. GROUP-TYPE SEPARATION OF PETROLEUM HYDROCARBONS USING CAPILLARY COLUMNS PACKED WITH POROUS SILICA PARTICLES

Using neat carbon dioxide as mobile phase and silica particles as stationary phase, only nonpolar compounds, such as hydrocarbons, can be chromatographed. The group-type separation of hydrocarbons is a significant application of this type of pcSFC.

Fuel properties that specify fuel quality, such as boiling range, density, pour point, cloud point, and diesel index, are dependent on fuel composition [12]. Several methods for group-type analysis have been developed, and the advantages of pcSFC over other methods are mainly due to fast separation and ease of quantitation when using the flame ionization detector (FID) [13] (see also Chapter 3). Figure 1 shows a pcSFC chromatogram of diesel sample obtained using porous silica particles as stationary phase. The aliphatic hydrocarbons eluted first, all grouped in the first peak. The monoaromatics eluted between toluene and naphthalene, and the polyaromatics eluted after naphthalene. There was a slight overlap between the saturated hydrocarbons and the monoaromatics. Five additional diesel samples were analyzed (Table 1) and the results were compared with those obtained from other research groups [14].

The selectivity of the silica particles for aromatics results from interactions between the polar silanol groups on the silica surface and polarizable aromatics. These interactions are strongly affected by temperature. Therefore, changing the temperature produces great changes in selectivity and resolution, as illustrated in Fig. 2. At approximately 45°C, maximum resolution was obtained. It was found that the pressure of supercritical carbon dioxide had little influence on the selectivity [13].

The effect of silica pore size on petroleum group–type separation was also investigated. Particles with smaller pores (60 Å) produced greater resolution than those with larger pores (150 and 300 Å). This result was credited to the fact that particles with smaller pores have larger specific surface area and, therefore, a larger number of silanol groups.

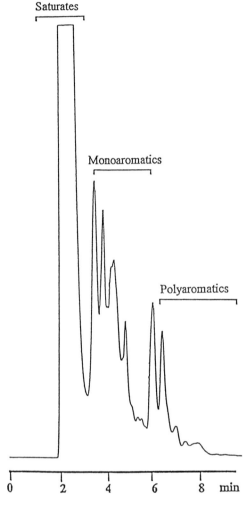

Figure 1 Typical pcSFC group–type separation of a diesel fuel. Conditions: Lee Scientific Model 50/SFC system, helium-actuated automatic Valco injector with a 0.2-μL sample loop; 98 cm × 200 μm id capillary column packed with 10 μm (60 Å) Keystone silica material; 40 cm × 10 μm id linear restrictor; CO_2; 45°C; column inlet pressure of 250 atm (1 atm = 1.013 bar); FID. (From Ref. 13, copyright 1995, American Chemical Society.)

Table 1 Group-Type Components in Diesel Sample Separated Using pcSFC[a]

Anal. no.	Hydrocarbon type (wt%)			
	Saturated	Monoaromatic	Polyaromatic	Total aromatic
1	68.3	23.6	8.1	31.7
2	69.1	23.6	7.2	30.8
3	69.2	22.9	7.8	30.7
4	68.7	23.1	8.1	31.2
5	68.6	23.0	8.4	31.4
Average	68.9	23.2	7.9	31.2
SD	0.33	0.30	0.40	0.37
RSD(%)	0.5	1.2	5.1	1.2

[a]Conditions: 52 cm × 200 μm id fused silica column packed with Keystone 10 μm, 60 Å silica; 45°C; 152 bar; CO_2; FID; timed-split injection.

Source: From Ref. 13, copyright 1995, American Chemical Society.

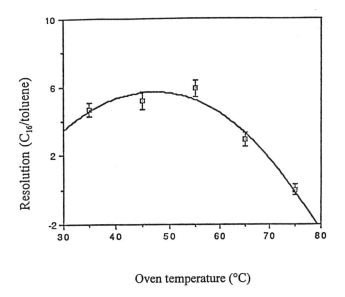

Oven temperature (°C)

Figure 2 Effect of oven temperature on resolution of *n*-hexadecane and toluene. Conditions: Lee Scientific Model 501SFC system, helium-actuated automatic Valco injector with a 0.2-μL sample loop; 54 cm × 200 μm id capillary column packed with 10 μm (60 Å) Keystone silica material, 50 cm × 10 μm id linear restrictor; CO_2; column inlet pressure of 150 atm (1 atm = 1.013 bar); FID. (From Ref. 13, copyright 1995, American Chemical Society.)

IV. INERT PACKING MATERIALS AND THE
SEPARATION OF POLAR COMPOUNDS

A. Characteristics of the Silica Surface

The specific surface area, pore size distribution, and particle size are parameters used to characterize the physical properties of silica packing materials (see also Chap. 2). Under SFC conditions (carbon dioxide mobile phase), only inert solutes, such as hydrocarbons, can be chromatographed, and these physical parameters mainly affect the chromatographic efficiency, as discussed above.

Chemical properties, *e.g.*, silanol chemistry, determine pSFC performance for polar compounds, and this is the reason that most pSFC separations of polar compounds require polar modifiers in the mobile phase (see Chapter 4). Silica particles are composed of siloxane ($\equiv$Si-O-Si$\equiv$) and silanol (Si-OH) groups. It was found that the concentration of silanols on a fully hydroxylated silica surface was $4.6 \pm 0.5/100$ Å^2, and this number was independent of the origin and physical properties of the silica particles [15,16]. However, an even larger density of $8/100$ Å^2 has also been reported [15].

Based on the fact that silica can undergo cation exchange not only in neutral but also in acidic (pH = 2–4) solutions [15], the surface must have very strong acidic centers because the dissociation constant of Si-OH is only about 9.9 in Si-O tetrahydrates. These acidic centers are responsible for the difficulties encountered in the separation of polar compounds (especially basic compounds) using silica-based packings and neat carbon dioxide. Several strategies have been developed for the deactivation of silica particles for pcSFC. These strategies include the elimination of silanol groups using a suitable silylation reagent, hydrogen–bond interaction, and acid–base neutralizing methods.

B. Elimination of Silanol Groups Using
Polymethylhydrosiloxane

From considerations of silanol group density and reagent group spacing, it has been found that polymethylhydrosiloxane (PS) has an advantage of similar distance between reactive groups in the polymer compared to silanol distribution on the silica surface, which should facilitate more complete elimination of silanol groups on the silica surface. Furthermore, compared to alkylsilane reagents, which bond onto the particle surface at one end of the reagent to form a "brush" structure, PS reactive groups (silicon hydrides) are distributed uniformly along the molecule chain, and many reactive sites in one molecule are bonded onto the particle surface. The terminology "mat" structure was used to describe this bonded layer (Fig. 3). This "mat" layer forms a flat, highly crosslinked surface on the deactivated particles and can more effectively cover any residual silanols [17].

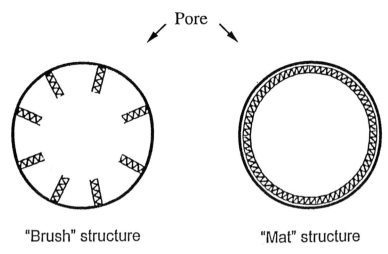

"Brush" structure   "Mat" structure

Figure 3 General bonded phase structure. (From Ref. 17, copyright 1995, Vieweg Publishing.)

Using a hydrosiloxane polymer as a silylation reagent, the optimum temperature for the dehydrocondensation reaction was reported to be between 200°C and 300°C [18]. The activity of the deactivated particles was investigated by SFC separation of polar test compounds. Figure 4 shows pcSFC chromatograms of these compounds using capillary columns packed with PS deactivated silica particles and a commercial C18-bonded phase. Better peak shapes and shorter retention times were obtained on the column containing PS-deactivated particles than on that containing a commercial C18-bonded phase. This illustrates that a more inert surface was obtained using PS as a deactivation reagent than when using sterically limited octadecylsilane reagent.

Experiments showed that alcohols, phenols, and diols eluted with relatively good peak shapes on the column packed with PS-deactivated silica particles [17]. Although the nitro-substituted phenols have relatively strong acidity because of the electron-withdrawing effect of the nitro group(s) on the benzene ring, they eluted with relatively symmetrical peaks on the packed column, even when using neat carbon dioxide as mobile phase. These polar compounds were also separated on the column packed with the C18-bonded phase; however, longer retention and severe peak tailing was obtained. Alkylamines ($pK_b \sim 4$), in which the nitrogen is attached to a saturated carbon, are stronger bases than arylamines ($pK_b \sim 9$), where the nitrogen is attached to an unsaturated carbon, such as in a benzene ring. N-Substituted amines have weaker polarity than unsubstituted amines. On the column containing PS-deactivated silica particles, aniline eluted before *N,N*-dimethylaniline, whereas on the column containing

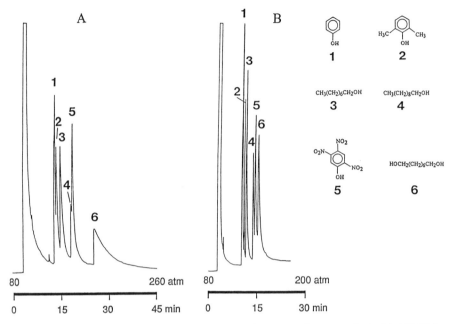

Figure 4 pcSFC chromatograms of polar test solutes on columns packed with (A) C18-bonded and (B) PS-deactivated phases. Conditions: Lee Scientific Model 600 SFC/GC system, manual valve with a 0.2-μL sample loop; 40 cm × 250 μm id capillary columns packed with (A) a commercial C18-bonded phase and (B) PS-deactivated silica particles (10 μm); CO_2; column inlet pressure programmed (A) from 80 atm to 260 atm at 4 atm/min and (B) from 80 atm to 200 atm at 4 atm/min (1 atm = 1.013 bar); 60°C; FID. (From Ref. 17, copyright 1995, Vieweg Publishing.)

the C18-bonded phase, *N,N*-dimethylaniline eluted first. This means that on the PS-deactivated silica particles, arylamines elute according to molecular weight, whereas on the C18-bonded phase, they elute according to polarity. Much better peak shapes were obtained on the column containing PS-deactivated particles than on that containing a C18-bonded phase, which again verifies that the PS-deactivated particles have a more inert surface than the commercial C18-bonded phase. However, the unacceptable peak shapes of more strongly basic alkylamines and free carboxylic acids reveal that additional work must be done to further decrease the interaction between the stationary phase and these solutes.

C. Hydrogen Bonding of Residual Silanols

It is impossible to eliminate all of the silanol groups on the silica surface because of steric hindrance. Even with extensive deactivation using hydrosiloxane polymers, a certain degree of surface activity still remains on the deactivated surface. Such deactivated packing materials are still far from satisfactory for use in separating strongly polar compounds such as bases and acids, and it is necessary to find other ways to further deactivate the silica surface.

It is common knowledge that a small amount of polar modifiers in the supercritical carbon dioxide mobile phase can dynamically deactivate the active sites on the surface of the stationary phase [19,20]. Furthermore, it has been reported that certain polar functional groups in the bonded phases interact with residual silanol groups on the bonded phase surface through hydrogen–bond interactions [21,22]. If these interactions exist in SFC, they can be used to further deactivate the residual silanol groups on the particle surface and favor the separation of polar compounds. Of course, a difference exists between dynamic deactivation with polar modifiers in the mobile phase and static deactivation with polar groups in the bonded phase. The excess, noninteracting polar modifiers in the mobile phase can more or less improve the solvating power of the mobile phase for polar compounds, while the noninteracting functional groups in the bonded phase will interact with the polar solutes and produce peak tailing or an increase in retention. Therefore, the concentration of polar groups in the bonded phase must be optimized.

3-Glycidoxypropyltrimethoxysilane (GPTMS) is a typical reagent used to prepare polar diol-bonded phases. Figure 5A shows the possible hydrogen bonding interactions between the GPTMS-bonded phase and the residual silanols on the silica surface. The residual silanols on the silica surface can interact with either the silanol (interaction **I**) or the diol group (interaction **II**) of the GPTMS-bonded phase. In order to investigate the interactions between the polar groups in the GPTMS-bonded phase and the residual silanols on the silica surface, three other bonded surface structures were designed, as illustrated in Fig. 5B–D. In Fig. 5B, the glycidoxypropyl group in GPTMS was substituted by the inert methyl group, and only interaction **I** was possible to produce a methyltriethoxysilane. In Fig. 5C, only interaction **II** can exist to produce a glycidoxypropyldimethylethoxysilane. Figure 5D shows the structure of the hexamethyldisilazane (HMDS)–deactivated surface, for which no hydrogen bonding interactions can occur [23].

From steric hindrance considerations, HMDS deactivation should produce a more inert surface and give better separation of polar compounds because of the smaller steric volume of the trimethylsilyl group. However, it was found that the GPTMS-deactivated and HMDS end-capped particles produced a more inert surface than when using only HMDS. This result suggests that the polar groups in the GPTMS-bonded phase (silanol and diol groups) further deactivate the surface, as illustrated in Fig. 5A. Experiments showed that both the silanols and the vicinal diol groups in the GPTMS-bonded phase interact with the residual silanols on the silica surface and these interactions provide the best deactivation of the residual silanols on the silica surface [23].

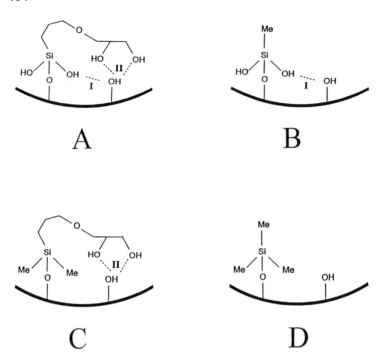

Figure 5 Structures and interactions of diol-bonded phases. (A) GPTMS-bonded surface, (B) MTES-bonded surface, (C) GPDMES-bonded surface, and (D) HMDS-bonded surface. (From Ref. 23, copyright 1996, John Wily & Sons, Inc.)

The suitability of the GPTMS-bonded and HMDS end-capped particles for the separation of polar organic compounds in SFC was investigated. At relatively low temperature (60°C), hydroxyl-containing and carboxyl-containing compounds, including alcohols, phenols, diols, aldehydes, ketones, and esters, can be separated with symmetrical peaks, except for some peak tailing of the aliphatic diol and strongly polar trinitrophenol, as illustrated in Fig. 6. These results are better than those obtained using hydrosiloxane polymer–deactivated particles [17], and with increasing temperature the peak shapes were improved. It was found that these GPTMS-bonded phases were especially suitable for the separation of hydroxyl-containing compounds.

Figure 7 shows the baseline separation of a commercial moderately polar alkylene glycol polymer, UCON LB-135. Significantly worse results were obtained when nonpolar SE-54-encapsulated particles [24] were used for this application. Free fatty acids ($pK_a \sim 4$) can be separated at relatively low temperature with only slight peak tailing, as illustrated in Fig. 8. For comparison, these free acids were also separated at 170°C by pcSFC using polymeric particles

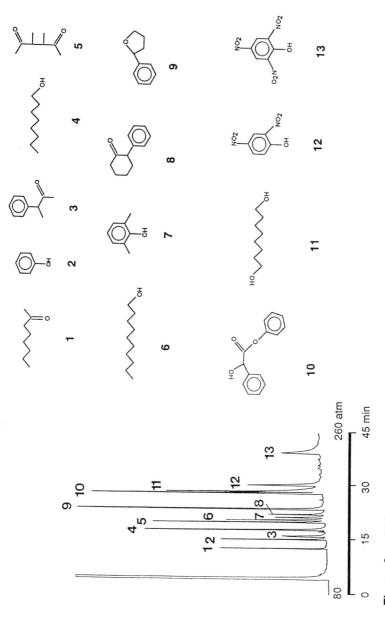

Figure 6 pcSFC chromatogram of hydroxyl- and carboxyl-containing compounds. Conditions: Lee Scientific Model 600 SFC/GC system, manual valve with a 0.2-μL sample loop; 40 cm × 250 μm id capillary column packed with GPTMS-bonded and HMDS end-capped silica particles (10 μm); CO_2; column inlet pressure programmed from 80 atm to 260 atm at 4 atm/min (1 atm = 1.013 bar); 60°C; FID. (From Ref. 23, copyright 1996, John Wiley & Sons, Inc.)

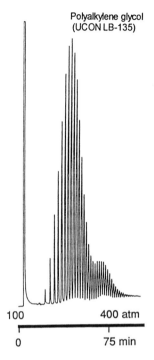

Polyalkylene glycol
(UCON LB-135)

100   400 atm

0   75 min

Figure 7 pcSFC chromatogram of UCON LB-135. Conditions: Lee Scientific Model 600 SFC/GC system, manual valve with a 0.2-μL sample loop; 40 cm × 250 μm id capillary column packed with GPTMS-bonded and HMDS end-capped silica particles (10 μm); CO_2; column inlet pressure programmed from 100 atm to 400 atm at 4 atm/min and held at 400 atm for 30 min (atm = 1.013 bar); 60°C; FID. (From Ref. 23, copyright 1996, John Wiley & Sons, Inc.)

[25]. Gas chromatography (GC) can be used for the separation of these compounds, but temperatures greater than 250°C are needed. Weakly basic arylamines ($pK_b \sim 9$) were separated on the encapsulated silica particles with relatively symmetrical peaks at 60°C according to polarity. This means that a certain degree of polarity in the stationary phase does not affect the peak shapes of moderately polar compounds in pcSFC with neat carbon dioxide. Strongly basic alkylamines ($pK_b \sim 4$) cannot be eluted from the column.

D. Acid–Base Neutralizing Interactions

In order to separate strongly basic compounds using neat supercritical carbon dioxide as mobile phase, most of the proton donor groups must be eliminated. Coating the surface with basic polymers is one way to accomplish this. The strong basity and polarity of polyethylene imines (PEIs) make them hydrophilic

n-CH$_3$(CH$_2$)$_n$COOH

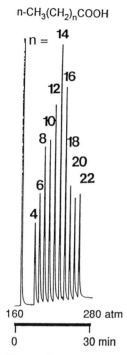

Figure 8 pcSFC chromatogram of free fatty acids. Conditions: Lee Scientific Model 600 SFC/GC system, manual valve with a 0.2-μL sample loop; 40 cm × 250 μm id capillary column packed with GPTMS-bonded and HMDS end-capped silica particles (10 μm); CO$_2$; column inlet pressure programmed from 160 atm to 280 atm at 4 atm/min (1 atm = 1.013 bar); 75°C; FID. (From Ref. 23, copyright 1996, John Wiley & Sons, Inc.)

and soluble in water, and insoluble in some organic solvents such as methylene chloride, chloroform, and acetone. In HPLC, PEI-encapsulated particles were used for the separation of biomolecules according to an ion exchange mechanism [26–29]. The water solubility of PEI required the immobilization of the coated PEI. The crosslinkers used were multifunctional molecules and, generally, proton donor compounds. The crosslinking reactions changed the basic properties of the PEI-coated surface, and the unreacted crosslinkers produced new acidic centers on the surface. Under SFC conditions using neat carbon dioxide the high molecular weight PEI coating had low solubility, and crosslinking reactions were less critical.

Figure 9 shows the possible deactivation interactions that can occur on the PEI-coated surface using diol-bonded silica particles [30]. PEI with an average molecular weight of 750,000 provided a stable coating on the diol-bonded silica

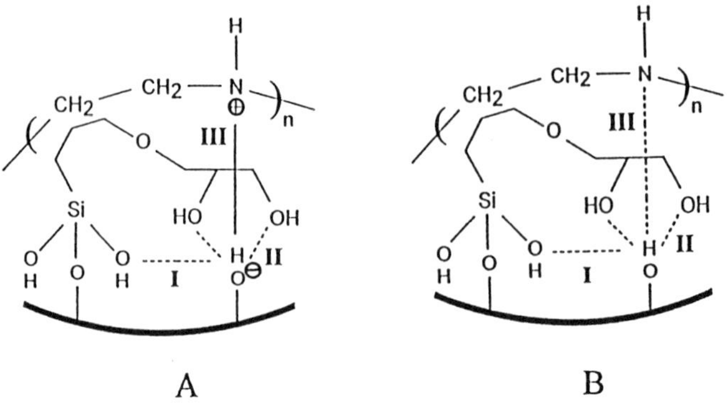

Figure 9 Silica surface interactions with diol-bonded and PEI-coated polymers. (A) Protein transfer, (B) hydrogen–bond interactions. (From Ref. 30, John Wiley & Sons, Inc.)

particles. The strong polarity and basicity of PEI result from a large number of N-H groups in the repeatable units of the PEI molecule, which gives an aqueous pH of approximate 12. The silanol groups on the silica surface have stronger acidity ($pK_a \sim 4$) than water and should facilitate proton transfer from the acidic silanol groups to the basic nitrogen atoms of the PEI, as illustrated in Fig. 9A. This neutralizing interaction represses the activity of the acidic silanol groups on the surface. Of course, some interactions between the residual silanol groups and the basic centers of the PEI can only result in hydrogen bond interactions, as illustrated in Fig. 9B. Deactivation of any remaining free amine groups in the PEI chains further lowers the activity of the packing material.

Basic anilines and alkylamines can elute from the column containing PEI-coated particles with relatively symmetrical peaks, as illustrated in Fig. 10. Recently, using spectroscopic methods, it was found that the amino groups bonded on the particle surface could react with carbon dioxide under pSFC conditions as follows [31]:

$$R-NH_2 + CO_2 \rightleftharpoons R-NHCOOH$$

Unstable carbamoyl compounds result. If it is true that primary amine solutes react with carbon dioxide to produce carbamoyl compounds under SFC conditions, the resultant carbamoyl compounds are soluble in neat carbon dioxide, and the basic particle surface is necessary to elute these unstable compounds.

The basic PEI-coated surface allows the analysis of strongly basic compounds, such as basic nitrogen-containing drugs, using pcSFC with neat carbon dioxide. Figure 11 shows a pcSFC chromatogram of four pharmaceutical

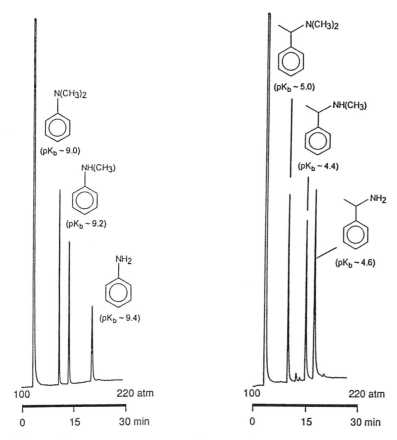

Figure 10 pcSFC chromatogram of amines. Conditions: Lee Scientific Model 600 SFC/GC system, manual valve with a 0.2-µL sample loop; 40 cm × 250 µm id capillary column packed with diol-bonded and PEI-coated silica particles (10 µm); CO_2; column inlet pressure programmed from 100 atm to 220 atm at 4 atm/min (1 atm = 1.013 bar); 80°C; FID. (From Ref. 30, John Wiley & Sons, Inc.)

compounds using a capillary column packed with Si-diol-HMDS-PEI-HMDS particles [30].

The suitability of the PEI-coated particles for the separation of other compounds was also evaluated [30]. The test compounds included hydrocarbons, polycyclic aromatic hydrocarbons (PAHs), esters, ketones, alcohols, phenols, and free carboxylic acids. Inert compounds, such as hydrocarbons and PAHs, and weakly polar compounds such as ketones and esters, produced symmetrical peaks. Proton donor alcohols ($pK_a \sim 15$) produced fronting peaks,

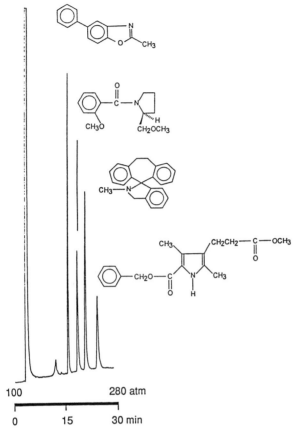

Figure 11 pcSFC chromatogram of pharmaceutical compounds. Conditions: Lee Scientific Model 600 SFC/GC system, manual valve with a 0.2-μL sample loop; 40 cm × 250 μm id capillary column packed with diol-bonded and PEI-coated silica particles (10 μm); CO_2; column inlet pressure programmed from 100 atm to 280 atm at 6 atm/min (1 atm = 1.013 bar); 80°C; FID. (From Ref. 30, John Wiley & Sons, Inc.)

and phenols ($pK_a \sim 10$) produced severe peak broadening. The reason for this is not clear. Strong proton donor carboxylic acids ($pK_a \sim 4$) could not elute from the column because of the strong interaction between the acids and the basic centers on the particle surface. After several injections of carboxylic acids, alkylamines could not elute from the column. This resulted from dynamic modification of the particle surface by the acidic samples injected.

V. APPLICATIONS USING SELECTIVE POLYMER-COATED AND ENCAPSULATED PARTICLES

A. Chromatographic Effects of Coated and Encapsulated Particles

Coating of the column packing materials with polymers has several different effects on the performance of pcSFC. These effects include (1) a decrease in the surface activities of the particles, (2) a decrease in the porosity of the particles, and (3) a change in chromatographic selectivity. However, no polymer encapsulated particles are commercially available for pcSFC.

By encapsulating the particles with a polymer coating, the residual silanols are covered and the particles become more inert. Figure 12 shows this effect on

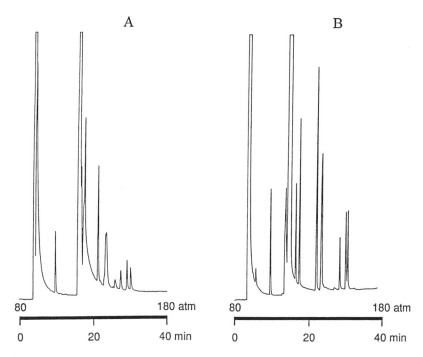

Figure 12 pcSFC chromatograms of *cherry flavor*. Conditions: Lee Scientific Model 600 SFC/GC system, manual valve with a 0.2-µL sample loop; 40 cm × 250 µm id capillary columns packed with (A) PS-deactivated and (B) PS-deactivated and SE-54-encapsulated silica particles (10 µm); CO_2; column inlet pressure programmed from 80 atm to 180 atm at 2 atm/min (1 atm = 1.013 bar); 85°C; FID. (From Ref. 17, copyright 1995, Vieweg Publishing.)

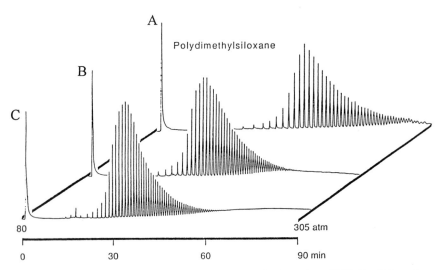

Figure 13 Effect of polymer coating on retention of solutes in pcSFC. Conditions: Lee Scientific Model 600 SFC/GC system, manual valve with a 0.2-µL sample loop; 40 cm × 320 µm id capillary columns packed with (A) untreated, (B) PS-deactivated, and (C) PS-deactivated and SE-54-encapsulated silica particles (10 µm, 80 Å); CO_2; column inlet pressure programmed from 80 atm to 305 atm at 2.5 atm/min (1 atm = 1.013 bar); 75°C; FID. (From Ref. 33.)

the separation of a commercial *cherry flavor* [17]. It is noteworthy that this encapsulation deactivation effect is limited, and the elimination of the majority of silanol groups is necessary prior to coating with the polymers [32]. The porosity of the particle surface is decreased by filling the pores with the polymer; experiments have shown that this effect is manifested by a decrease in the retention of solutes. Figure 13 shows this effect on the separation of nonpolar polydimethylsiloxanes [33].

Polysiloxane polymers have been the most widely used stationary phases in SFC. However, other polymers, such as polyethylene oxides (PEOs), are moderately polar and have a certain degree of hydrophilicity. Hydrophilic/hydrophobic properties probably play a significant role in SFC because of the relatively low operating temperature.

The properties of PEO can be varied by changing the polymer end groups. Typical terminal groups of commercially available PEO include hydroxyl, ether, esters, amino-substituted ether, etc. Three PEOs, containing hydroxyl, ether, or amino terminal groups, have been used for the preparation of PEO-coated particles for use in pcSFC [34]. Prior to coating with these polymers, the silica particles were partially deactivated using a diol phase. Figure 14 shows the possible deactivation interactions that exist on two different PEO-coated particle

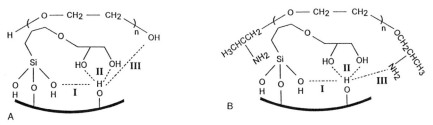

Figure 14 Surface interactions with diol-bonded and PEO-coated polymers. (A) Hydroxyl–terminated ether terminated PEO and (B) amino-substituted PEO-coated silica particles. (From Ref. 34.)

surfaces. On the hydroxyl-terminated PEO-coated surface, an excess number of proton donor groups exist, whereas on the amino-terminated PEO surface, basic centers exist. It was found that the hydroxyl-terminated PEO-coated silica particles are suitable for the separation of proton donor compounds, such as alcohols, phenols, and acids, whereas the amino-terminated PEO-coated silica surface is suitable for the separation of proton acceptor compounds, such as amines. However, strongly basic compounds such as alkylamines ($pK_b \sim 4$) cannot elute from the column packed with hydroxy-terminated PEO-coated particles, and strongly acidic compounds such as free carboxylic acids cannot elute from the column containing amino-terminated PEO coated particles [34].

Although GC can be used for the separation of some acidic and basic compounds, pcSFC can extend the molecular weight range of samples that can be chromatographed using relatively low temperature (around 100°C). Figure 15 shows separations of basic and acidic drugs using PEO-coated particles as packing materials and neat carbon dioxide as the mobile phase [34].

B. Cyanobiphenyl (CBP)-Substituted Polysiloxane-Encapsulated Silica Particles

CBPs exhibit some liquid crystalline properties and possess shape/size selectivity [35]. Figure 16 shows the structure of the 25% CBP stationary phases. There are nine isomers depending on the position of substitution on the biphenyl ring system (*p,p*-, *m,p*-, *o,p*-, *p,m*-, *m,m*-, *o,m*-, *p,o*-, *m,o*-, and *o,o*-) [36]. Table 2 lists comparative selectivities of capillary columns packed with untreated porous (80 Å), CBP-coated, and SE-54-coated silica particles for dodecylxylene isomers under pcSFC conditions [36]. All nine CBP-coated particles showed greater selectivity for the positional isomers of the benzene derivatives than SE-54-coated particles or untreated porous silica particles. The *o,p*-CBP-coated phase gave the greatest selectivity among all nine CBP coated particles [36].

The polar cyano group has been found to provide selective interaction with double bonds [37,38]. Table 3 lists the selectivities of CBP phases to the double

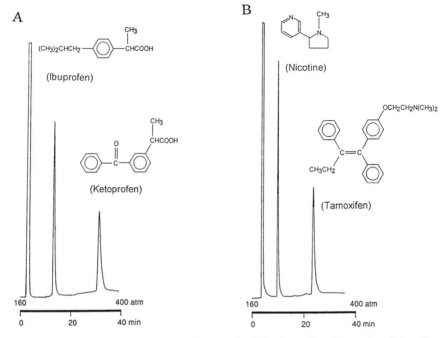

Figure 15 pcSFC chromatograms of basic and acidic drugs. Conditions: Lee Scientific Model 600 SFC/GC system, manual valve with a 0.2-μL sample loop; 40 cm × 320 μm id capillary columns packed with diol-bonded and (A) hydroxyl group–terminated and (B) amino-substituted ether–terminated PEO-coated silica particles (10 μm); CO_2; column inlet pressure programmed from 160 atm to 400 atm at 6 atm/min (1 atm = 1.013 bar); 80°C; FID. (From Ref. 34.)

bonds of selected fatty acid methyl esters (FAMEs) [36]. The column containing SE-54-coated particles showed no selectivity for double bonds, and the column containing untreated silica particles produced serious peak tailing for the FAMEs when using neat carbon dioxide as the mobile phase. The o,p-CBP-coated phase produced the greatest selectivity for double bonds compared to the other CBP-coated particles. The saturated stearic acid methyl ester could not be separated from the monoenic oleic acid methyl esters using capillary columns packed with CBP-coated particles, which suggests that the CBP-coated particles are medium polar stationary phases.

Essential oils are composed of both polar and nonpolar isomeric compounds, often differing by the position(s) of double bonds. CBP-coated particles are ideally suited for this type of application. Figure 17 shows a pcSFC chromatogram of lime oil using a capillary column packed with o,p-CBP-coated silica particles [36].

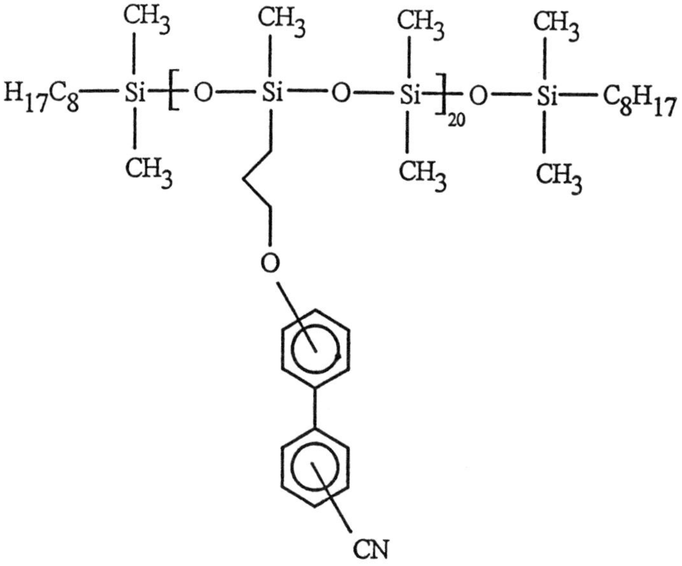

Figure 16 Structures of cyanobiphenyl polysiloxane stationary phases. (From Ref. 36, copyright 1995, John Wiley & Sons, Inc.)

Table 2 Selectivities of Cyanobiphenylpolysiloxane-Encapsulated Packing Materials to Benzene Derivatives Under pcSFC Conditions[a]

Stationary phase	$\alpha_{4/1}$[b]	$\alpha_{3/2}$[b]	$\alpha_{3/1}$[b]
p,p-CBP	1.183	1.026	1.139
o,p-CBP	1.208	1.037	1.147
m,p-CBP	1.152	1.017	1.103
p,m-CBP	1.199	1.025	1.140
o,m-CBP	1.184	1.019	1.131
m,m-CBP	1.189	1.027	1.128
p,o-CBP	1.185	1.025	1.127
o,o-CBP	1.170	1.024	1.108
m,o-CBP	1.161	1.018	1.106
SE-54	1.117	1.000	1.080
Silica	1.418	1.000	1.130

[a]Conditions: 40 cm × 320 μm id fused silica column: CO_2; 85°C; 172 bar. The dead time was measured using methane.
[b]Compounds used for the determination of selectivity were (1) 3,5-dimethyldodecyl-benzene, (2) 2,4-dimethyldodecylbenzene, (3) 2,3-dimethyldodecylbenzene, and (4) 1,5-dimethyldodecylbenzene.
Source: From Ref. [36], copyright 1995, John Wiley & Sons, Inc.

Table 3 Selectivities of Cyanobiphenylpolysiloxane-Encapsulated Packing Materials to Fatty Acid Methyl Esters (FAMEs) Under pcSFC Conditions[a]

Stationary phase	$\alpha_{2/1}$[b]	$\alpha_{3/2}$[b]
p,p-CBP	1.015	1.050
o,p-CBP	1.052	1.085
m,p-CBP	1.013	1.047
p,m-CBP	1.050	1.075
o,m-CBP	1.046	1.072
m,m-CBP	1.033	1.062
p,o-CBP	1.042	1.072
o,o-CBP	1.038	1.060
m,o-CBP	1.038	1.057
SE-54	1.000	1.000

[a]Conditions are the same as in Table 2.
[b]Compounds used for the determination of selectivity were (1) methyloleate, (2) methylionoleate, and (3) methyllinolenate.
Source: From Ref. [36], copyright 1995, John Wiley & Sons, Inc.

C. Free and Silver-Complexed Dicyanobiphenyl (DCBP)– Substituted Polysiloxane-Encapsulated Silica Particles

The selectivity of the CBP-coated particles to double bonds is ineffective for the separation of monoenic isomers because the phases have a relatively low nitrile content in the stationary phase. By increasing the cyano content in the polymer, the selectivity can be increased. Figure 18 shows the structures of two DCBP phases [39].

Experiments have shown that the asymmetric isomer, *o,o,p*-DCBP, provided higher selectivity than the symmetric isomer, *m,m,p*-DCBP, and FAMEs could be separated according to the number and position of double bonds in the compounds [39]. However, they cannot resolve monoenic cis/trans isomers. By complexing with certain metals, the selectivity to cis/trans isomers can be greatly improved. Figure 19 shows a pcSFC chromatogram of monoenic *cis/trans*-FAMEs using a capillary column packed with silica particles that have been deactivated, coated with *o,p,p*-DCBP, and silver-complexed. Although silver ions were introduced into the stationary phase, symmetrical peaks were obtained under pcSFC conditions with neat carbon dioxide [39].

Figure 20 shows the effect of temperature on selectivity for carbon number, degree of unsaturation, and position of geometric configuration around the double bond(s) for FAMEs using a column containing *o,p,p*-DCBP-coated and silver-complexed silica particles. By increasing the temperature, the separation of groups containing different carbon numbers improved, but the separation of

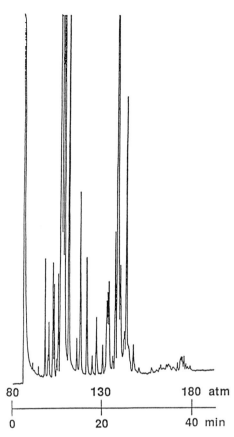

Figure 17 pcSFC chromatogram of *lime oil*. Conditions: Lee Scientific Model 600 SFC/GC system, manual valve with a 0.2-μL sample loop; 40 cm × 320 μm id capillary column packed with PS-deactivated and *o,p*-CBP-encapsulated silica particles (10 μm); CO_2; column inlet pressure programmed from 80 atm to 180 atm at 2.5 atm/min (1 atm = 1.013 bar); 85°C; FID. (From Ref. 36, copyright 1995, John Wiley & Sons, Inc.)

components of the same carbon number decreased. Overlap of peaks 9, 10, and 11 can be avoided at either higher or lower temperature than 85°C. The overlap of peaks 6 and 7 cannot be improved by changing the temperature because their interaction with the stationary phase (degree of double bonds) is the same. Figure 21 shows a pcSFC chromatogram of FAMEs in a commercial fish oil (CPL-30) using a capillary column packed with *o,p,p,*-DCBP-coated particles. Successful separation was obtained at relatively low temperature using the FID as detector [39].

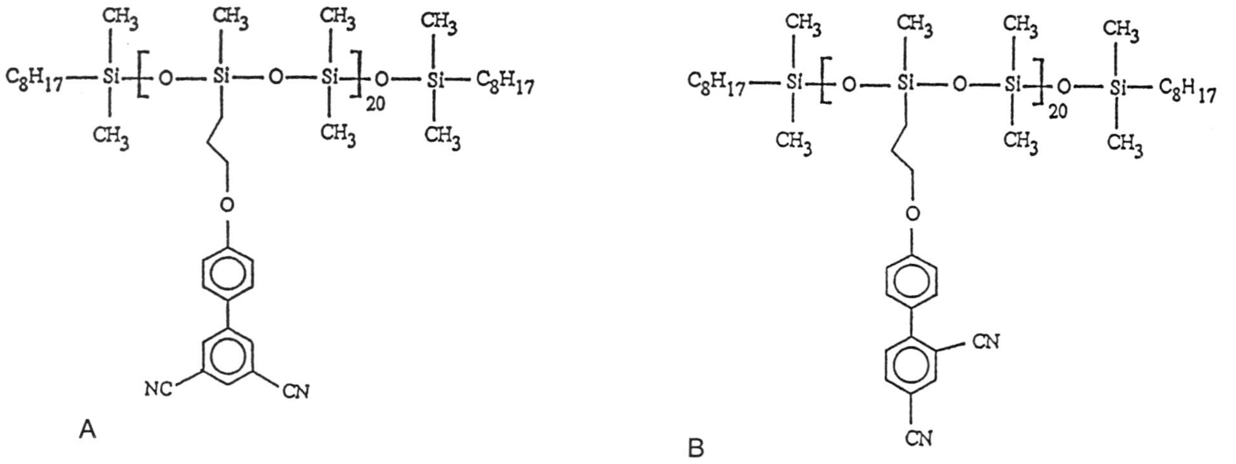

Figure 18 Structures of DCBP phases. (From Ref. 39, copyright 1995, John Wiley & Sons, Inc.)

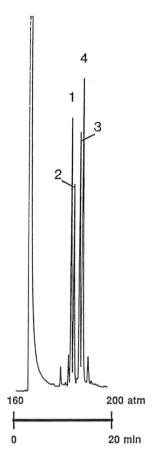

160   200 atm
├──────────────────┤
0   20 min

Figure 19 pcSFC chromatogram of FAMEs according to cis/trans structures. Conditions: Lee Scientific Model 600 SFC/GC system, manual valve with a 0.2-μL sample loop; 50 cm × 250 μm id capillary column packed with cyanopropyl polymethylhydrosiloxane–deactivated, *o,p,p*-DCBP-encapsulated, and silver-complexed silica particles (10 μm); CO_2; column inlet pressure programmed from 160 atm to 200 atm at 2 atm/min (1 atm = 1.013 bar); 45°C; FID. Peak identifications: (1) methyl palmitelaidate, (2) methyl palmitoleate, (3) methyl elaidate, and (4) methyl oleate. (From Ref. 39, copyright 1995, John Wiley & Sons, Inc.)

D. Cyclodextrin (CD)–Substituted Polysiloxane-Encapsulated Silica Particles

The use of chiral stationary phases (CSPs) in chromatography is the most convenient approach to separating enantiomers. CDs are usual chiral selectors for the separation of enantiomers in chromatography [40–44] (see also Chapter 8).

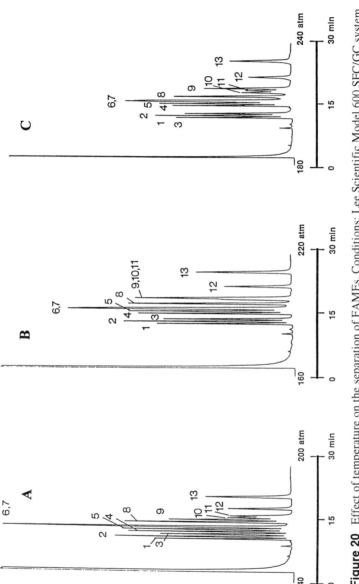

Figure 20 Effect of temperature on the separation of FAMEs. Conditions: Lee Scientific Model 600 SFC/GC system, manual valve with a 0.2-µL sample loop; 50 cm × 250 µm id capillary column packed with cyanopropyl polymethyl-hydrosiloxane–deactivated, o,p,p-DCBP-encapsulated, and silver-complexed silica particles (10 µm); CO$_2$; FID: (A) 60°C; column inlet pressure programmed from 140 atm to 200 atm at 2 atm/min (1 atm = 1.013 bar), (B) 85°C: column inlet pressure programmed from 160 atm to 220 atm at 2 atm/min, and (C) 110°C; column inlet pressure programmed from 180 atm to 240 atm at 2 atm/min. Peak identifications: (1) methyl palmitate, (2) methyl palmitelaidate, (3) methyl palmitoleate, (4) methyl palmitelaidate, (5) methyl stearate, (6) methyl oleate, (7) methyl elaidate, (8) methyl linolelaidate, (8) methyl linoleate, (9) methyl gondoate, (10) γ-methyl linolenate, (11) methyl linolenate, (12) cis-11,14,17-methyl eicosatrienoate, and (13) cis-7,10,13,16-methyl docosatetraenoate. (From Ref. 39, copyright 1995, John Wiley & Sons, Inc.)

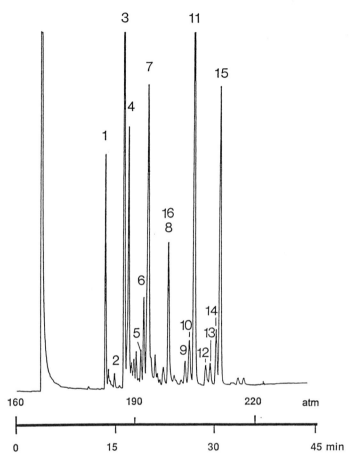

Figure 21 pcSFC chromatogram of FAMEs in *fish oil*. Conditions: Lee Scientific Model 600 SFC/GC system, manual valve with a 0.2-μL sample loop; 50 cm × 250 μm id capillary column packed with cyanopropyl polymethylhydrosiloxane–deactivated, *o,p,p*-DCBP-encapsulated silica particles (10 μm); CO_2; column inlet pressure programmed from 160 atm at 1.5 atm/min (1 atm = 1.013 bar); 85°C; FID. Peak identifications: (1) C14:0, (2) C15;0, (3) C16:0, (4) C16:1 n-7, (5) C16:4 n-4, (6) C18:0, (7) C18:1 n-9, (8) C18:4 n-3, (9) C20:4 n-6, (10) C22:1 n-11, (11) C20:5 n-3, (12) C22:4 n-6, (13) C22:4 n-3, (14) C22:5 n-3, (15) C22:6 n-3, (16) C20::1 n-9. (From Ref. 39, copyright 1995, John Wiley & Sons, Inc.)

In order to perform pcSFC separation of polar enantiomers using neat carbon dioxide as the mobile phase, inert, enantioselective stationary phases must be prepared.

Chiral compounds of different polarities, including hydrocarbons, ketones, esters, alcohols, diols, lactones, and amines (derivatives), have been chromatographed using CD-containing polymer- (see Fig. 22 [45]) encapsulated

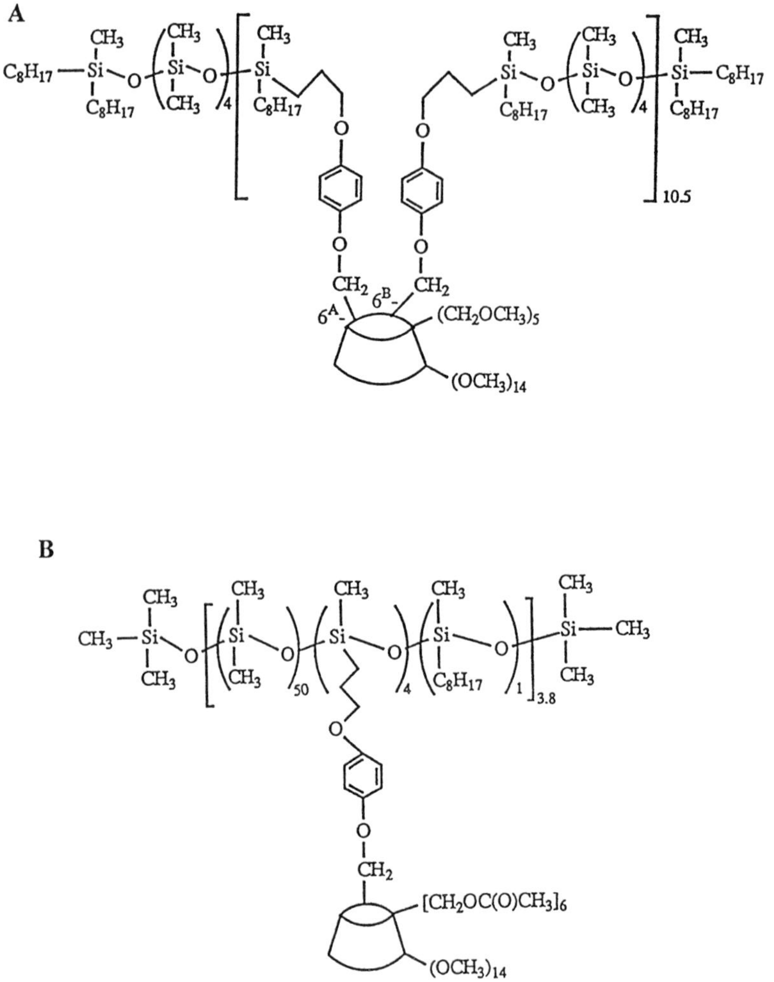

A

B

Figure 22 Structures of CD-containing polymers. (From Ref. 45, copyright 1996, John Wiley & Sons, Inc.)

particles as stationary phases and neat carbon dioxide as the mobile phase. Figure 23 shows a variety of such separations. Although the reasons are not known, it is interesting to observe that different CD polymers (side chain–substituted or copolymer) can differ greatly in their effectiveness for the separation of different enantiomers [45].

A comparison was made between packed and open tubular column SFC because open tubular columns were considered to be completely inert and void of any surface contribution to chiral selectivity. Table 4 lists the selectivity data for both open tubular and packed capillary columns containing polymer A

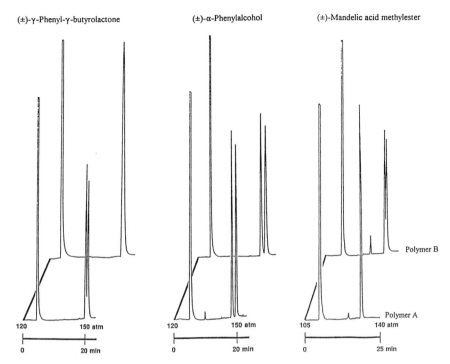

Figure 23 pcSFC chromatograms of enantiomers on capillary columns packed with CD-substituted polymer-encapsulated silica particles. Conditions: Lee Scientific Model 600 SFC/GC system, manual valve with a 0.2-µL sample loop; 40 cm × 250 µm id capillary columns packed with PS-deactivated and CD-substituted polysiloxane A and B–encapsulated silica particles (10 µm); CO_2; column inlet pressure programmed from 120 atm to 150 atm at 1.5 atm/min (1 atm = 1.013 bar); 45°C; FID. (From Ref. 45, copyright 1996, John Wiley & Sons, Inc.)

stationary phase. It can be seen that the selectivities were approximately the same from different enantiomers. These results suggest that the residual activity of the polymer-encapsulated particles is minimal and has little effect on the enantiomeric selectivity for the chiral compounds investigated [45].

The effects of temperature and pressure on the enantiomeric selectivity were investigated [45]. The adsorption characteristics of the particle surface are affected by the column temperature. Increasing the temperature can decrease the adsorption of polar analytes and improve peak shapes. However, with approximately the same analysis time, an increase in column temperature reduces the enantiomeric selectivity. Lower temperature favors the separation of enantiomers, which is also an advantage of SFC over GC for this application. The effect of supercritical fluid mobile phase pressure on enantiomeric selectivity is important because the elution of analytes in SFC is dependent on the pressure of the mobile phase. Several research groups have reported

Table 4 Effect of Pressure on Chiral Selectivity in pcSFC with Cyclodextrin A Polymer Stationary Phase

Analyte	Packed column[a]			Open tubular[b]	
	P (bar)	α	k_2'	α	k_2'
(±)-α-Pinene	81	1.02	56.19	—[c]	0.726
	91	1.02	22.14		
(±)-1-Phenylalcohol	101	1.11	49.63	1.09	0.830
	111	1.11	16.49	1.09	0.312
	122	1.11	10.06		
	132	1.11	7.65		
	142	1.11	6.45		
	152	1.11	5.57		
(±)-γ-Phenyl-γ-butyrolactone	111	1.06	24.02	—	0.448
	122	1.06	12.57		
	132	1.06	8.73		
	142	1.06	6.78		
(±)-2-Phenylcyclohexanone	101	1.05	75.91	1.03	1.183
	111	1.05	18.80		
	122	1.05	9.85		
	132	1.05	7.95		
(±)-t-2,4-Pentanediol	111	1.11	15.31	1.09	0.252
	132	1.10	7.68		
	142	1.10	6.55		
	152	1.10	5.69		
(±)-1-Cyclohexylethamine	96	1.04	46.06	—	0.651
(derivatized with TFA)	101	1.04	20.33		
	111	1.04	5.06		

[a] 40 cm × 250 µm id capillary column packed with phase A–encapsulated silica particles (10 µm); CO_2; 45°C; t_0 was determined using methane; P = column inlet pressure.
[b] 10 m × 50 µm id capillary column coated with 0.25 µm phase A; CO_2; 45°C; t_0 was determined using methane.
[c] Indicates that the separation was too small to determine accurately.
Source: From Ref. [45], copyright 1996, John Wiley & Sons, Inc.

different results for the effect of pressure on enantiomeric selectivity in cSFC [46,47]. Table 4 lists the effect of supercritical carbon dioxide pressure on enantiomeric selectivity in pcSFC. From this table, it is clearly seen that although the retention factors can change as much as 10-fold within a certain range of pressure, the chiral selectivity remains unchanged.

E. Liquid Crystal Polysiloxane–Coated Silica Particles

The recognition of molecular shape or special molecular interactions between the stationary phase and certain functional groups in the solutes is fundamental to selectivity. Liquid crystal stationary phases are the most effective for shape-selective separations in GC [48,49]. Nonpolar compounds, such as polycyclic aromatic hydrocarbons (PAHs), and polar compounds, such as steroids, have been separated using liquid crystalline open tubular columns in SFC. An advantage of using liquid crystalline phases in SFC over GC is that the lower operating temperature of SFC leaves the liquid crystalline phases more ordered and they provide higher selectivity [50,51].

Fat-soluble vitamins have been separated using various forms of chromatography. However, it is difficult to analyze all of them using one technique because of the obvious differences in their physical and chemical properties [52–60]. Fat-soluble vitamins have different polarities, molecular weights, and molecular structures as illustrated in Fig. 24 [32]. The difference between vitamins K_1 (MW = 451) and K_2 (MW = 445) is that they contain a different number of double bonds. The difference between vitamin D_2 (MW = 397) and D_3 (MW = 385) is that D_2 has one more double bond and methyl group than D_3. In addition, the D vitamins are polar compounds containing one hydroxyl group, and they have been considered to be the most difficult pair of vitamins to separate using chromatographic methods. Vitamins A (MW = 286) and E (MW =

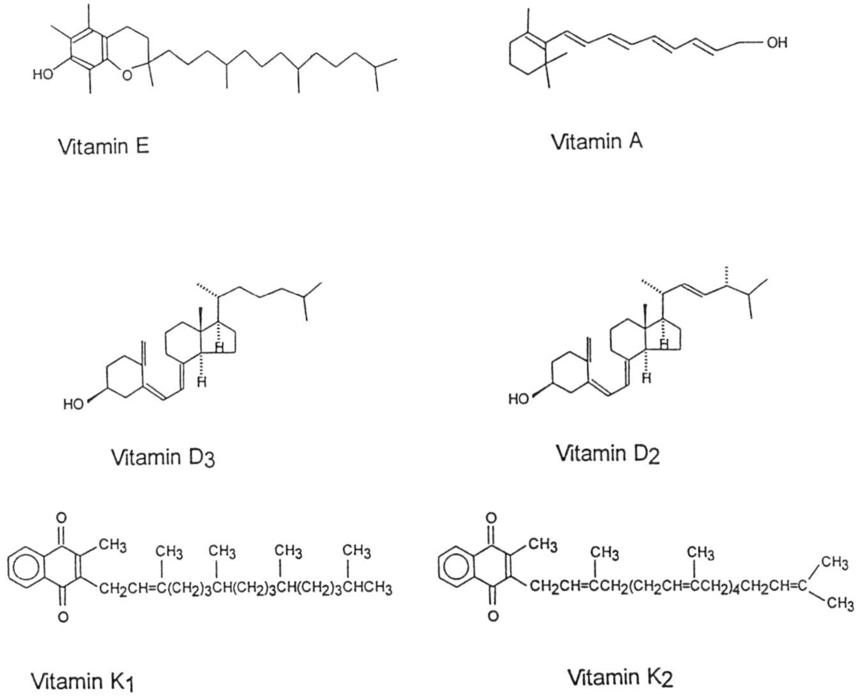

Vitamin E

Vitamin A

Vitamin D3

Vitamin D2

Vitamin K1

Vitamin K2

Figure 24 Structures of fat-soluble vitamins. (From Ref. 32, copyright 1996, Vieweg Publishing.)

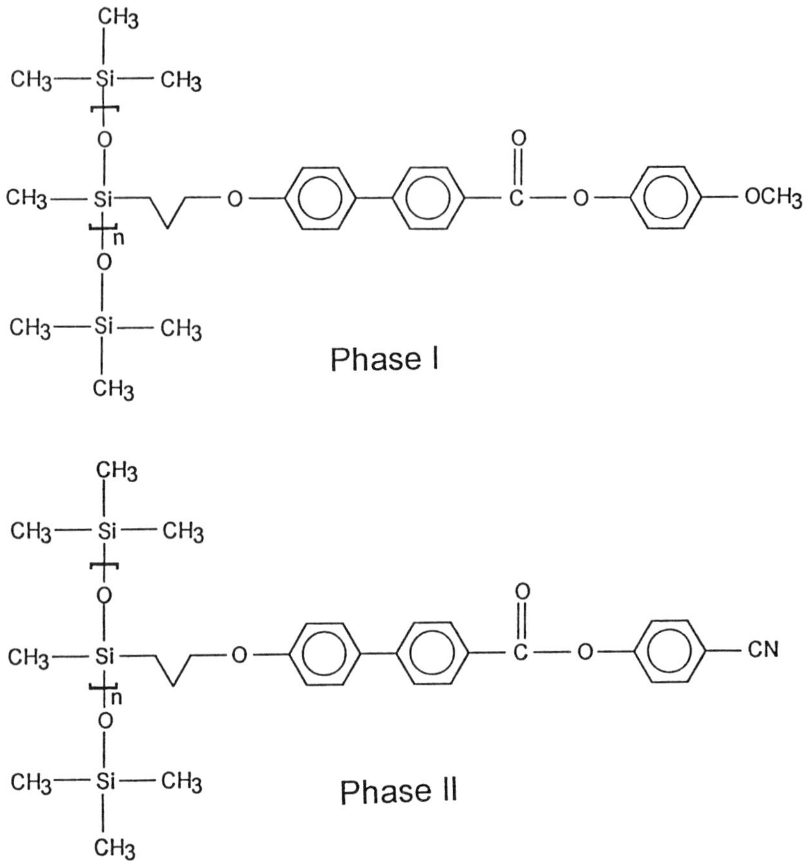

Figure 25 Structures of liquid crystalline polysiloxanes. (From Ref. 61, copyright 1986, Elsevier Scientific Publishing Company.)

431) are relatively easy to separate because they have obvious differences in molecular weight and polarity.

Using pcSFC packed with liquid crystalline polysiloxane–coated particles (see Fig. 25 [61]), these six fat-soluble vitamins could be separated at relative low temperature (70°C) within 40 min as illustrated in Fig. 26 [32]. On the column containing phase I–coated particles, the elution order was vitamins E, K_1, K_2, D_2, D_3, and A. From the structures of the vitamins, it can be seen that vitamin A has a linear structure (all trans double bonds) and eluted last, although it is polar (one hydroxyl group) and has the lowest molecular weight among these vitamins. Vitamin E has a long chain of single bonds with several methyl groups. These single bonds are very flexible; therefore, vitamin E eluted

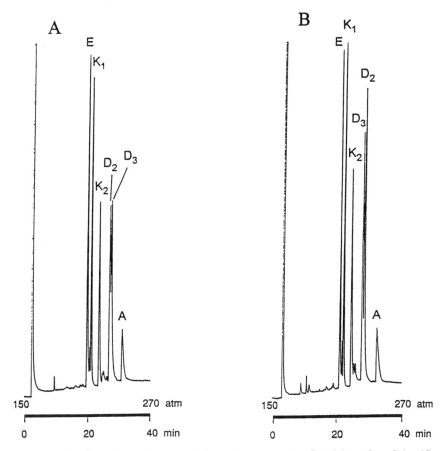

Figure 26 pcSFC chromatograms of fat-soluble vitamins. Conditions: Lee Scientific Model 600 SFC/GC system, manual valve with a 0.2-μL sample loop; 40 cm × 250 μm id capillary column packed with PS-deactivated and (A) phase I– and (B) phase II– coated silica particles (10 μm); CO_2; column inlet pressure programmed from 150 atm to 270 atm at 3 atm/min (1 atm = 1.013 bar); 70°C; FID. (From Ref. 32, copyright 1996, Vieweg Publishing.)

first. Vitamin K_1 has a long chain similar to E, but has a higher molecular weight than E, and it was eluted after E. Vitamin K_2 has several double bonds and is more rigid and linear than K_1; therefore, it eluted after K_1. Although the difference in molecular weight between vitamins K_1 and K_2 is only 4, they still can be separated very well. The elution of vitamin D_2 before D_3 results from the methyl groups in D_2. This minor difference in molecular structure causes D_2 to elute before D_3, although D_2 has a greater molecular weight.

From these results it can be seen that neither molecular weight nor polarity of these vitamins is the critical factor in determining the elution order on the column containing the liquid crystal polysiloxane–coated particles, whereas molecular shape is a major factor. Using the cyano-substituted liquid crystal phase II–coated particles as packing material, all elution orders remained the same except for a reversal in D_2 and D_3. These results indicate that for the cyano-substituted liquid crystalline polysiloxane–coated particles, the long hydrocarbon chains in the vitamins affect the selectivity.

One unique advantage of pcSFC over HPLC is the convenience of universal detection. The FID is a universal detector and can be used for the detection of fat-soluble vitamins. The FID relative response factors for vitamins K_1, K_2, A, E, D_2, and D_3 under pcSFC conditions were found to be 1.00, 1.05, 0.82, 0.90, 0.95, and 0.93, respectively, showing that they give approximately the same response (within 100% difference), except for vitamin A. The relatively low response factor of vitamin A probably results from the low purity of the standard compound (the purity marked was approximate 70%). The FID is also a relatively sensitive detector and can be used to detect trace amounts of analytes. An alternative nearly universal detector for SFC is the evaporative light-scattering detector (ELSD) (see Chapter 4).

VI. CONCLUSIONS

Although neat supercritical carbon dioxide is a polarizable mobile phase, successful deactivations of packing materials allow the use of this mobile phase for separations of polar compounds including free acids and bases. Versatile deactivation strategies make it possible to use silica particles for this propose, taking advantage of their excellent properties of mechanical strength and uniform particle size. Coating or encapsulating these deactivated particles with polymers can produce inert and selective packing materials that provide selective separations for a variety of applications using neat carbon dioxide. However, these columns are not currently commercially available. The production of these various inert and selective packed capillary columns designed for pcSFC can extend the development of SFC applications in industry.

REFERENCES

1. Y. Hirata, *J. Microcol. Sep.*, 2:214 (1990).
2. G. Crescentini, F. Bruner, F. Mangani, and G. Yafeng, *Anal. Chem.*, *60*:1659 (1988).
3. G. Crescentini and A. R. Mastrogiacomo, *J. Microcol. Sep.*, *3*:539 (1991).
4. F. J. Yang, *J. Chromatogr.*, *236*:265 (1982).
5. J. C. Gluckman, A. Hirose, V. L. McGuffin, and M. Novotny, *Chromatographia*, *17*:303 (1983).

6. K. E. Karlsson and M. Novotny, *Anal. Chem.*, *60*:1662 (1988).
7. H. J. Cortes, C. D. Pfeiffer, B. E. Richter, and T. S. Stevens, *J. High Resolut. Chromatogr.*, *10*:446 (1987).
8. R. T. Kennedy and J. W. Jorgenson, *Anal. Chem.*, *61*:1128 (1989).
9. L. J. Cole, N. M. Schultz, and R. T. Kennedy, *J. Microcol. Sep.*, *5*:433 (1993).
10. A. Malik. W. Li, and M. L. Lee, *J. Microcol. Sep.*, *5*:361 (1993).
11. D. Tong, K. D. Bartle, and A. A. Clifford, *J. Microcol. Sep.*, *7*:433 (1995).
12. D. J. Cookson, C. P. Lloyd, and B. E. Smith, *Energy & Fuels*, *2*:854 (1988).
13. W. Li, A. Malik, M. L. Lee, B. A. Jones, N. L. Porter, and B. E. Richter, *Anal. Chem.*, *67*:647 (1995).
14. *Data Report from Dionex Corporation*, Technical Center, Salt Lake City, UT, 1993.
15. K. K. Unger, *Porous Silica*, Elsevier, Amsterdam (1979).
16. E. F. Vansant, P. Van Der Voort, and K. C. Vrancken, *Characterization and Chemical Modification of the Silica Surface*, Elsevier, Amsterdam (1995).
17. Y. Shen and M. L. Lee, *Chromatographia*, *41*:665 (1995).
18. K. M. Payne, B. J. Tarbet, K. E. Markides, and M. L. Lee, *Anal. Chem.*, *62*:1379 (1990).
19. A. L. Blilie and T. Greibrokk, *Anal. Chem.*, *57*:2239 (1985).
20. J. M. Levy and J. P. Guzowski, *J. Chromatogr. Sci.*, *26*:194 (1988).
21. S. M. Staroverov, A. A. Sterdan, and G. V. Lisichkin, *J. Chromatogr.*, *367*:337 (1986).
22. N. K. Shonia, S. M. Staroverov, Yu. S. Nikitin, and G. V. Lisichkin, *Zh. Fiz. Khim*, *58*:702 (1984).
23. Y. Shen and M. L. Lee, *J. Microcol. Sep.*, *8*:413 (1996).
24. Y. Shen, A. Malik, W. Li, and M. L. Lee, *J. Chromatogr.*, *703*:303 (1995).
25. F. J. Yang, *SFC Applications*, (K. E. Markides and M. L. Lee, eds.), Snowbird, Utah (1989).
26. A. J. Alpert, and F. E. Regnier, *J. Chromatogr.*, *185*:375 (1979).
27. G. Vanecek and F. E. Regnier, *Anal. Biochem.*, *109*:345 (1980).
28. D. Pearson and F. E. Regnier, *J. Chromatogr.*, *225*:137 (1983).
29. M. Flashner, H. Ramsed, and L. Crane. *Anal Biochem.*, *135*:340 (1983).
30. Y. Shen and M. L. Lee, *J. Microcol. Sep.*, *8*:519 (1997).
31. S. Zhang, G. Nicholson, B. Schindle, and E. Bayer, *Proc. 18th International Symposium on Capillary Chromatography*, Riva del Garda, Italy, p. 1785 (1996).
32. Y. Shen, J. S. Bradshaw, and M. L. Lee, *Chromatographia*, *43*:53 (1996).
33. Y. Shen and M. L. Lee, unpublished results.
34. Y. Shen and M. L. Lee, *Chromatographia*, *43*:373 (1996).
35. A. Malik, I. Ostrovsky, S. R. Sumpter, S. L. Reese, S. Morgan, B. E. Rossiter, J. S. Bradshaw, and M. L. Lee, *J. Microcol. Sep.*, *4*:529 (1992).
36. Y. Shen, W. Li, A. Malik, S. L. Reese, B. E. Rossiter, and M. L. Lee *J. Microcol. Sep.*, *7*:411 (1995).
37. R. E. Merrifield and W. D. Phillips, *J. Am. Chem. Soc.*, *80*:2778 (1958).
38. R. H. Bauer, *Anal. Chem.*, *35*:107 (1963).
39. Y. Shen, S. L. Reese, B. E. Rossiter, and M. L. Lee, *J. Microcol. Sep.*, *7*:58 (1995).
40. V. Schurig and H. P. Nowotny, *Angew. Chem. Int. Ed. Engl.*, *29*:939 (1990).
41. V. Schurig, Z. Juvancz, G. J. Nicholson, and D. Schmalzing, *J. High Resolut. Chromatogr.*, *14*:58 (1991).

42. D. W. Armstrong, Y. Tang, T. Ward, and M. Nichols, *Anal Chem.*, *651*:1114 (1993).

43. Y. Tang, Y. Zhou, and D. W. Armstrong, *J. Chromatogr.*, *666*:147 (1994).

44. G. Yi, J. S. Bradshaw, B. E. Rossiter, A. Malik, W. Li, H. Yun, and M. L. Lee, *J. Chromatogr.*, *673*:219 (1994).

45. Y. Shen, Z. Chen, N. L. Owen, W. Li, J. S. Bradshaw, and M. L. Lee, *J. Microcol. Sep.*, *8*:249 (1996).

46. X. Lou, Y. Shen, and L. Zhou, *J. Chromatogr.*, *514*:253 (1990).

47. G. Lai, G. J. Nicholson, U. Mühleck, and E. Bayer, *J. Chromatogr.*, *540*:217 (1991).

48. G. M. Janini, G. M. Muschik, and W. L. Zielinski, Jr., *Anal. Chem.*, *48*:1879 (1976).

49. K. E. Markides, M. Nishioka, B. J. Tarbet, J. S. Bradshaw, and M. L. Lee, *Anal. Chem.*, *57*:1296 (1985).

50. H. C. Chang, K. E. Markides, J. S. Bradshaw, and M. L. Lee, *J. Microcol. Sep.*, *1*:131 (1989).

51. H. C. Chang, K. E. Markides, J. S. Bradshaw, and M. L. Lee, *J. Chromatogr. Sci.*, *26*:280 (1988).

52. G. F. M. Ball, *J. Micronutr.*, *4*:255 (1988).

53. A. Rizzolo and S. Polesello, *J. Chromatogr.*, *624*:103 (1992).

54. R. Wyss, *J. Chromatogr.*, *531*:481 (1990).

55. M. P. Labadie and C. E. Boufford, *J. Assoc. Off. Anal. Chem.*, *71*:1168 (1988).

56. A. P. De Leenheer, H. J. Nelis, W. E. Lambert, and R. M. Bauwens, *J. Chromatogr.*, *429*:3 (1988).

57. U. Singh and J. H. Bradbury, *J. Sci. Food Agric.*, *45*:87 (1988).

58. M. M. D. Zamarreno, A. S. Perez, and J. H. Mendez, *J. Chromatogr.*, *623*:69 (1992).

59. H. Hasegawa, *J. Chromatogr.*, *605*:215 (1992).

60. C. R. Smidt, A. D. Jones, and A. J. Clifford, *J. Chromatogr.*, *434*:21 (1988).

61. J. S. Bradshaw, C. Schregenberger, H. C. K. Chang, K. E. Markides, and M. L. Lee, *J. Chromatogr.*, *358*:95 (1986).

6

Selectivity Tuning in Packed Column Supercritical Fluid Chromatography

Pat Sandra

University of Ghent, Ghent, Belgium

Andrei Medvedovici

University of Bucharest, Bucharest, Romania

Agata Kot

Technical University of Gdańsk, Gdańsk, Poland

Frank David

Research Institute for Chromatography, Kortrijk, Belgium

I. INTRODUCTION

Supercritical fluid chromatography (SFC) is an intermediate technique between gas chromatography (GC) and high-performance liquid chromatography (HPLC).

In chromatographic techniques, the importance of selectivity and selectivity tuning can be deduced from the master equation of chromatography [1]. The resolving power of a column is given by:

$$R_s = \frac{\sqrt{N}}{4} \; \frac{\alpha - 1}{\alpha} \; \frac{k}{k + 1} \tag{1}$$

in which N is the plate number, α is the selectivity or separation factor, and k is the retention factor.

The plate number N is related to the height equivalent to one theoretical plate (H):

$$N = \frac{L}{H} \tag{2}$$

where L is the column length. The efficiency can experimentally be calculated from the chromatogram by

$$N = 5.54 \left(\frac{t_R}{w_h} \right)^2 \tag{3}$$

in which t_R is the retention time and w_h is the peak width at half height of a peak with k larger than 5. On the other hand, the maximum attainable efficiency can be calculated from the characteristics of the column. In separation systems in which packing material is applied:

$$N \cong \frac{L}{2d_p} \tag{4}$$

in which d_p is the particle diameter. A packed column in pSFC of 25 cm in length with 5-μm particles will give under optimal conditions roughly 25,000 plates. Note that in pSFC the efficiency is independent of the internal diameter of the column.

The separation factor α, which describes the interplay of the solutes with the stationary and the mobile phases, is defined by:

$$\alpha = \frac{k_2}{k_1} \tag{5}$$

with k_1 and k_2 the retention factors of the first and the second peak of a critical pair, respectively. The k value gives the residence time of a solute in the stationary phase:

$$k = \frac{t_R - t_M}{t_M} \tag{6}$$

in which t_R is the measured retention time and t_M is the retention time of an unretained solute or the residence time of the solutes in the mobile phase.

From the master equation [Eq. (1)], it is clear that the separation factor α is the most important parameter to enhance on resolution. This is illustrated in Fig. 1.

The peak resolution of two solutes ($k = 5$) with α of 1.05 on a column with a plate number of 20,000 is 1.4. This value, the crossing point of the curves, is taken as a reference. The curves are obtained by changing one of the variables while the two others remain constant. From the curves, the following conclusions can be drawn: Peak resolution is proportional to the square root of the plate number. Increasing the plate number by a factor of 4 ($N = 20,000$–$80,000$) doubles the peak resolution ($R_s = 1.40$–2.80). Optimization of the separation factor (α) has the greatest impact on peak resolution. Increasing α from 1.05 to 1.10 nearly doubles the peak resolution ($R_s = 1.40$–2.68), whereas a slight decrease ($\alpha = 1.03$) causes a drastic loss of resolution ($R_s = 1.40$–0.68). On the

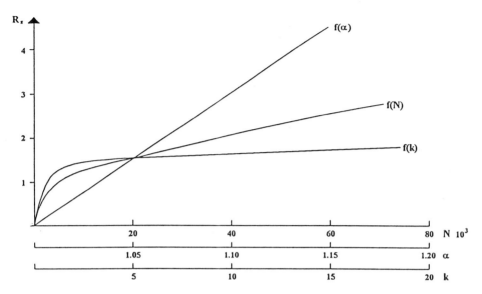

Figure 1 Effect of the plate number (*N*), the separation factor (α), and the retention factor (*k*) on resolution (*R*$_s$) [1].

other hand, low *k* values result in poor peak resolution. Increasing the *k* value from 0.25 to 5 corresponds to a resolution gain of 378% (*R*$_s$ = 0.37–1.40).

Optimization of the selectivity in the separation process is thus of utmost importance.

These theoretical considerations are definitely valid for GC and HPLC. In the case of pSFC the situation is more complicated because of the dependence of the different parameters of the resolution on the pressure/density drop. The impact of the pressure/density gradient along packed columns in pSFC on resolution has been studied by a number of research groups [2–14] and controversy statements have been advanced. Nevertheless, the above-mentioned studies indicated that "selectivity" remains the most important factor in pSFC to enhance resolution.

II. SELECTIVITY IN pSFC

It is not the aim of this chapter to discuss selectivity in pSFC in detail. The subject has been reviewed and discussed in several books [15–17] (see also Chapter 2). Only some general remarks will be given in relation to selectivity and selectivity tuning. Selectivity in GC is determined exclusively by the nature of the stationary phase because the mobile phase, an inert gas, is only a carrier of the solutes from one plate to another and has no interaction with the solutes.

Some subtle selectivity differences have, however, been noted when hydrogen, helium, or nitrogen is replaced by carbon dioxide [18]. In HPLC, adjustments in selectivity are also made possible by compositional changes in the mobile phase, although the nature of the stationary phase still predominates in the interplay with the solutes. Recently, temperature has also been exploited to introduce selectivity in HPLC [19]. In fact, in HPLC one can split selectivity into two parts: "large" adjustment by changing the stationary phase and "fine" adjustment by changing the composition of the mobile phase or the temperature. Because supercritical fluids combine many characteristics of gases and liquids, it is not surprising to find that the selectivity in pSFC lies between GC and HPLC. Both the stationary phase and the mobile phase can be varied but there are severe constraints on the mobile phase composition in pSFC because the critical parameters are a function of the mobile phase composition. The mobile phase strength is, however, also function of the density, density drop, and temperature. Selectivity adjustment in pSFC can therefore be made to some extent by selecting the appropriate density, density drop, or temperature. Such fine tuning is best performed after selection of the best stationary phase and

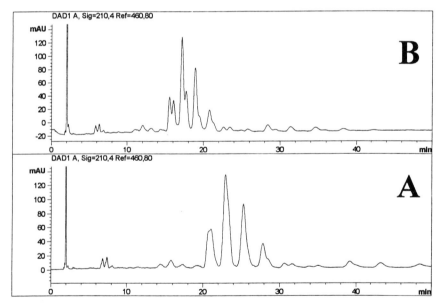

Figure 2 Influence of pressure on pSFC separation of triglycerides in peanut oil. Experimental conditions: column Shandon ODS Ultrabase C18 5 µm, 25 cm × 4.6 mm id; flow rate 1.5 mL/min; temperature 25°C; modifier methanol 5% isocratic; pressure (A) 150 bar, (B) 200 bar; detection 210 nm; sample concentration 100 mg/mL; injected volume 5 µL.

mobile phase composition for a given application. An example of how a density change at constant temperature can make an "improvement" in selectivity is shown in Fig. 2 in which the lipids of peanut oil are separated according to hydrophobicity or carbon number on an octadecyl silica column at 150 and 200 bar, respectively. A better or large adjustment is made by changing the nature of the stationary phase, as illustrated in Fig. 3 showing the analysis of the same sample on a cation exchange material impregnated with silver ions [20–22]. Separation occurs according to the degree of unsaturation and is much more complete compared to Fig. 2.

On this highly selective phase, density also plays an important role in fine tuning the resolution. Figure 4 shows the analysis of sunflower oil on a silver-loaded column at different pressure programming rates keeping all other parameters constant. The resolution at 1.5 bar/min (B) is far superior to those obtained

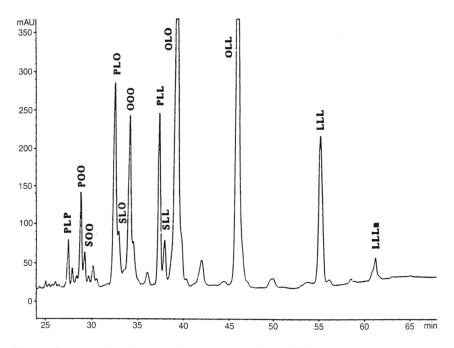

Figure 3 Separation of triglycerides in peanut oil by pSFC on a silver-loaded stationary phase. Experimental conditions: column Chromspher lipids 5 μm, 25 cm × 4.6 mm id; flow rate 1 mL/min; temperature 65°C; modifier acetonitrile/isopropanol 6:4, programmed from 1.2% (2 min) to 7.2% (28 min) at 0.3%/min, then to 12.2% at 0.5%/min; pressure from 150 bar (2 min) to 280 bar at 1.5 bar/min; detection 210 nm; sample concentration 100 mg/mL; injected volume 5 μL. Abbreviation of fatty acid moieties coupled to the glycerol backbone: P, palmitate; S, stearate; O, oleate; L, linoleate; Ln, linolenate [22].

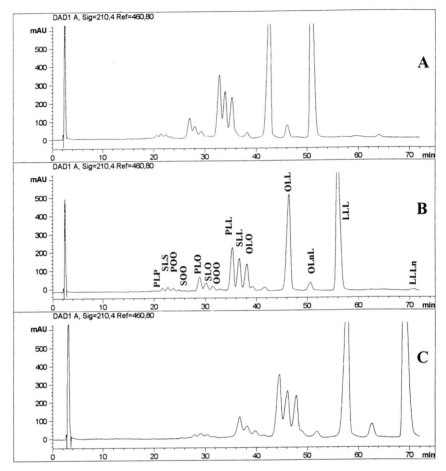

Figure 4 Influence of the pressure gradient on the pSFC separation of triglycerides in sunflower oil on a silver-loaded column. Experimental conditions and peak identification as in Fig. 3 except pressure program: from 150 bar to 280 bar at (A) 2 bar/min; (B) 1.5 bar/min; (C) 1 bar/min; detection at 210 nm; sample concentration 100 mg/mL. For identification, see Fig. 3.

at 1 bar/min (C) and 2 bar/min (A), respectively. Selectivity enhancement here was based more on trial and error than on a good understanding of the role of the density.

Although selectivity in pSFC is mainly controlled by the nature of the stationary phase, the importance of the mobile phase composition may not be underestimated. Moreover, over the years we have experienced that several

A

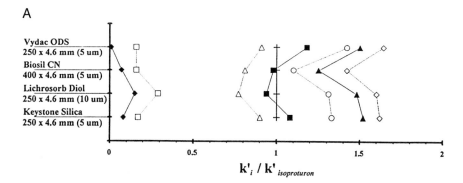

B

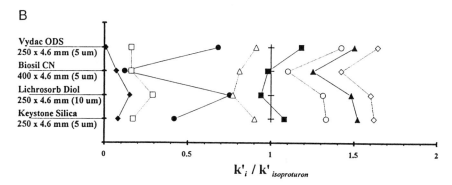

Figure 5 Relative retention of some pesticides separated in pSFC on different stationary phases. Identification of the compounds according to the following notation: ◆, metobromuron; △, fenuron; ○, diuron; □, linuron; ∓, isoproturon; ▲, chloroxuron; ●, Methabenzthiazuron; ■, chlorotoluron; ◊, metoxuron.

stationary phases provide very similar chromatograms in pSFC and this definitely when solutes with similar polarities are analyzed. In Fig. 5A, the relative retention (isoproturon is 1.00) for a number of phenylurea herbicides is plotted for several stationary phases. The elution order is very similar and, moreover, the highest similarity is observed for silica and octadecyl silica. In pSFC both phases behave as normal phases. An explanation was advanced by Berger [23] who stated that adsorbed mobile phase forms a more dense layer of the mobile phase and when modifiers are present, e.g., methanol, the concentration of methanol is higher on the surface of the stationary phase than in the mobile phase. This means that the stationary phase is always more polar than the mobile phase, i.e., a normal phase partitioning mechanism. In Fig. 5B, the relative retention for methabenzthiazuron, a phenylurea herbicide with different polarity from those shown in Fig. 5A, is included. The shifts in relative retention are much more pronounced for this solute, emphasizing the normal phase character of pSFC, i.e., separation according to polarity. Very similar data were obtained for the analysis of polynuclear aromatic hydrocarbons (PAHs).

Stationary phases in pSFC should therefore be classified into two groups: classical normal phases, e.g., silica, cyanopropyl silica, aminopropyl silica, diol silica, octadecyl silica, etc., and highly selective normal phases, e.g., silver-loaded phases, chiral phases, etc. In the latter group, selectivity rather than polarity controls the separation.

Even in the case of highly selective phases the role of the mobile phase composition can be predominant (see further).

Besides the nature of the stationary phase and the composition of the mobile phase, structural modification of the solutes by means of precolumn derivatization can be a valuable tool to increase on selectivity.

In this contribution, different ways for improving selectivity and versatility in pSFC will be discussed and illustrated by means of practical applications in various fields: environmental, natural product research, chiral separations, and polymer additives.

III. SERIAL COUPLED COLUMNS IN PSFC

Very often the efficiency and selectivity of a single packed column is insufficient for a given separation problem. Berger and Wilson [9] have shown that the pressure drop in packed columns is not a limiting factor to obtain high efficiencies. They generated more than 200,000 effective plates by coupling 11 similar 25-cm columns. This opened the way to couple columns of different polarity and to tune the selectivity between different phases in order to obtain the desired separation (selectivity tuning) or to construct separation systems with enhanced versatility. In the on-line coupling of dissimilar columns the composition of the mobile phase is constant.

A. Serial Coupled Columns to Enhance on Efficiency

Coupling of 11 columns as described by Berger and Wilson [9] is not realistic for daily use. Having the disposal of 50,000 plates by coupling two or three columns is, however, sufficient for most applications. Figure 6 shows the H vs. u plots for a 25 cm $\times$ 4.6 mm id column packed with 5-μm octadecyl silica particles operated in the HPLC and pSFC mode.

In both curves H_{min} is 12 μm at optimum linear velocities of 0.1 and 0.4 cm/s, respectively. Figure 7 compares the curves for 1 $\times$ 25 cm and 2 $\times$ 25 cm columns and the same H_{min} and u_{opt} are noted.

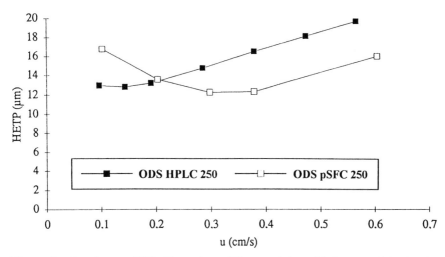

Figure 6 *H-u* plots on ODS silica column (25 cm × 4.6 mm id; 5 μm particle size) in HPLC and pSFC mode [38].

A third column 15 cm in length was coupled to the ODS tandem. Figure 8 shows the analysis of a supercritical fluid extract (SFE) of rubber on the 65 cm × 4.6 mm id 5-μm spherical ODS column. The plate number for the last eluting compound is 55,400, which corresponds to a reduced plate height of 2.34. For these column combinations the plate number increases approximately proportionally to the column length.

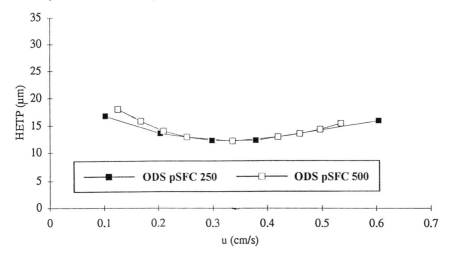

Figure 7 *H-u* plots on ODS silica 5 μm in the pSFC mode for 25- and 50-cm length columns [38].

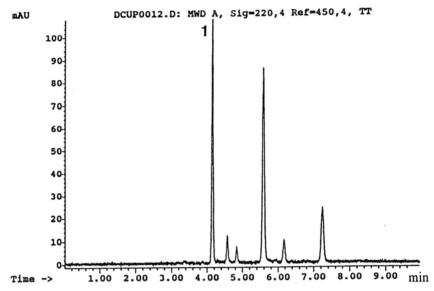

Figure 8 Analysis of an SFE extract of rubber. Experimental conditions: column BioSil ODS-HL 5 µm, 65 cm × 4.6 mm id; flow rate 2 mL/min; temperature 50°C; modifier methanol 20% isocratic, pressure 200 bar; detection 220 nm; injected volume 5 µL. Peak 1, dicumyl peroxide [38].

B. Serial Coupled Columns to Enhance on Selectivity

Selectivity tuning or combining selectivities and polarities of different stationary phases by coupling columns is an important tool for resolution optimization in GC [24] and HPLC [25,26]. The principle was applied in pSFC by A. Giorgetti et al. for the separation of a synthetic mixture of polymer additives on Spherisorb C8 and Spherisorb CN [27]; by Kot et al. [28] for the separation of PAHs on Chromspher PAH and BioSil ODS-HL; by Engelhardt et al. [29] for the separation of phytanic acid by coupling silica, aminopropyl silica; and cyanopropyl silica columns; and by M. Z. Wang et al. [30] for the separation of the enantiomers of camazepam and five metabolites on Lichrosphere CN–Chiralcel ODH. Capillary columns were serially coupled by Karlsson et al. [31] for the separation of some model lipids.

Selectivity tuning by coupled pSFC is illustrated with a standard sample and with the separation of the 16 PAHs of mixture SRM 1647a of the U.S. Environmental Protection Agency (EPA).

The standard sample contains some PAHs (anthracene/pyrene) and pesticides (propanil, aldrin, dieldrin—organochloropesticides; malathion—an organophosphorus pesticide; monolinuron and linuron—phenylurea pesticides; and carbaryl—a carbamate).

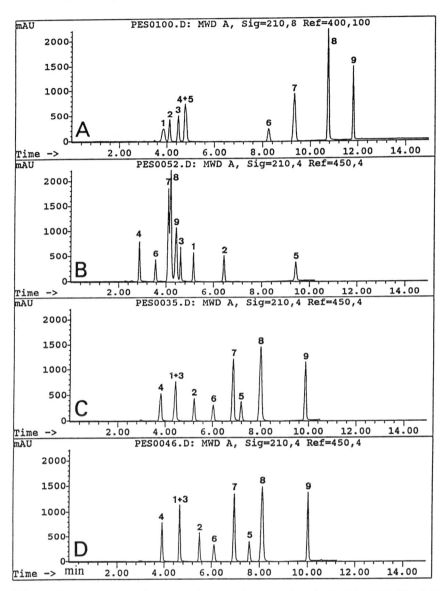

Figure 9 Standard sample separated on a column tandem A-A' (A); B-B' (B); A-B (C); and B-A (D). Experimental conditions: pressure 100 bar programmed at 30 bar/min to 300 bar; modifier methanol 5% (5 min) then programmed to 20% at 3%/min; temperature 50°C; flow rate 2 mL/min; detection UV at 210 nm; injected sample volume 5 μL. Peaks: 1, aldrin; 2, anthracene; 3, dieldrin; 4, malathion; 5, pyrene; 6, monolinuron; 7, linuron; 8, carbaryl; 9, propanil [28].

Four columns were used: 2×25 cm $\times 4.6$ mm id—spherical silica 5 μm, indicated A and A', respectively, and 2×25 cm $\times 4.6$ mm id—spherical ODS-HL silica 5 μm, indicated B and B', respectively, in the combinations 1: A-A'; 2: B-B'; 3: A-B; 4: B-A.

The chromatograms are shown in Fig. 9 and the following conclusions can be drawn: Both efficiency and inertness of the coupled columns are good. This is in agreement with Fig. 7 for columns of similar selectivity. On the tandem A-A', malathion and pyrene coelute, while on the tandem B-B', the separation of linuron and carbaryl is incomplete. On the tandems A-B and B-A, the compounds are better distributed over the chromatogram but aldrin and dieldrin coelute. Capacity factors and selectivity do not differ substantially with column order (A-B or B-A), as illustrated in Fig. 10. The small differences most probably are due to the pressure drop over the column tandem. It is important to note in Fig. 9 that focusing of solutes with small k factors is better when the hydrophobic column is placed first.

In pSFC, a challenging problem is the separation of the 16 PAHs of mixture SRM 1647a of the EPA. Complete separations on monomeric and polymeric ODS, on silica, aminopropyl and cyanopropyl silica, and on the PAH-dedicated HPLC phases Chromspher PAH and Vydac 201-TP-C18 failed. Best separations in pSFC were obtained on polymeric ODS and on the Chromspher PAH column. On the polymeric phase all compounds were separated except the pair benzo[b]fluoranthene and benzo[k]fluoranthene. These two compounds are baseline separated on the Chromspher PAH column, but the early eluting peaks

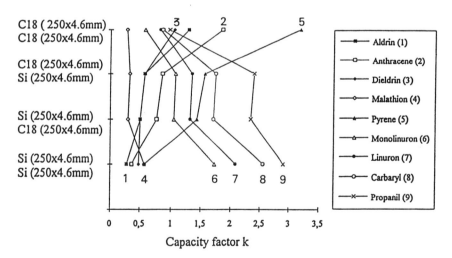

Figure 10 Influence of column tandems on retention data (k values) [28].

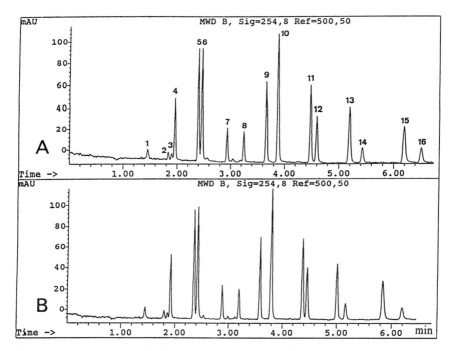

Figure 11 Separation of PAHs on a column tandem: ODS-HL / Chromspher PAH (A) and Chromspher PAH / ODS-HL (B). Optimized pSFC conditions: columns 15 cm × 4.6 mm id spherical ODS-HL silica 3 µm + 10 cm × 4.6 mm id Chromspher PAH 3 µm; pressure 80 bar programmed at 50 bar/min to 200 bar; modifier acetonitrile 1% programmed to 30% at 8%/min; temperature programmed from 40°C to 55°C at 2°C/min; flow rate 3 mL/min; UV detection at 254 nm; injected sample volume 5 µL. Peaks: 1, naphthalene; 2, acenaphtylene; 3, fluorene; 4, acenaphthene; 5, phenanthrene; 6, anthracene; 7, fluoranthene; 8, pyrene; 9, benzo[a]anthracene; 10, chrysene; 11, benzo[b]fluoranthene; 12, benzo[k]fluoranthene; 13, benzo[a]pyrene; 14, dibenzo[a,h]-anthracene; 15, indeno[1,2,3,c,d]pyrene; 16, benzo[g,h,i]perylene [28].

naphthalene, acenaphthylene, fluorene, and acenaphthene exhibited incomplete separation. Coupling the columns resulted in the separation shown in Fig. 11A for the tandem ODS-Chromspher PAH and in Fig. 11B for the tandem Chromspher PAH-ODS. As in the previous example, column order has nearly no influence on retention and resolution.

Although both columns were individually intensively used in pSFC (ODS) and HPLC (Chromspher PAH) after these experiments, the PAH separation could be repeated 15 months later on the occasion of an SFC course on the same column tandem. However, correct data on lifetime and reproducibility cannot be advanced because our laboratories are not faced with routine analysis.

C. Serial Coupled Columns to Enhance on Versatility

Enantiomer separation is one of the fields in which pSFC is recognized to have excellent characteristics. The use of pSFC for chiral separations has been pioneered by M. Caude and his group [32,33] and a large number of applications on different chiral stationary phases (CSPs) have been published in the past few years [34–38].

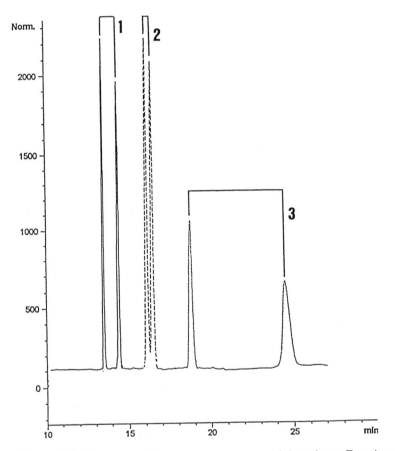

Figure 12 Separation of different racemates on a triplet column. Experimental conditions: columns Chiralpak AD (10 μm) + Chiralcel OD (10 μm) + Chirex 3022 (5 μm), 3 × 25 cm × 4.6 mm id; flow rate 2 mL/min; temperature 25°C; modifier methanol containing 0.5% triethylamine and 0.5% trifluoroacetic acid, programmed from 4% (5 min) to 30% at 5 %/min; pressure 200 bar; detection 220 nm; injected amount ~ 200 ng each compound. Peaks: 1, metoprolol; 2, tiaprofenic acid; 3, lormetazepam [44].

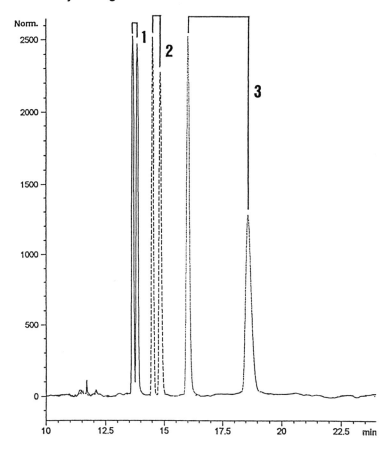

Figure 13 Separation of different racemates on a triplet column. Experimental conditions as in Fig. 12. Peaks: 1, clenbuterol; 2, mianserin; 3, medetomidine [44].

The number of CSPs has grown tremendously in the last years, and in principle all normal phase LC columns can be used with sub- and supercritical SFC. The polysaccharide phases and more especially Chiralcel OD and Chiralpak AD have shown the highest enantioselectivity in sub- and supercritical FC [37–40] (see also Chapters 3, 8, 9). The first performs better for the separation of basic compounds; the latter for acidic compounds. Pirkle- or Brush-type columns, such as Chirex 3022 with π-donor characteristics and Chirex 3005 with π-acceptor characteristics, also exhibit some enantioselectivity in SFC [37,38].

Because coupling of columns in pSFC provides good results, different CSPs have been connected in series in order to provide a versatile chiral separation system. In doing this, it is clear that enantioselectivity offered by a single chiral

DAD1 A, Sig=210,8 Ref=460,80

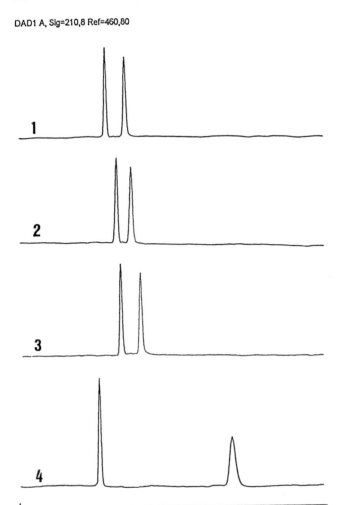

Figure 14 Separation of different barbiturates on a triplet column. Experimental conditions as in Fig. 12. Peaks: 1, pentobarbital; 2, secobarbital; 3, butabarbital; 4, hexobarbital [44].

selector operated under optimal pSFC conditions is always the highest for a particular enantioseparation [41].

The disadvantages of coupling CSPs [41,42] or mixing CSPs [43] have been described. Nevertheless, over the years, we have stated that for 90% of the enantioseparations realized on a particular CSP in pSFC, the column tandem

Chiralpak AD–Chiralcel OD or the column triplet Chiralpak AD–Chiralcel OD–Chirex 3022 performs very well [44,45]. Moreover, a single optimized pSFC program could be applied. Coupling CSPs in pSFC offers therefore a versatile tool for enantioseparation, the application of which can reduce the number of trial-and-error experiments on individual CSPs.

The optimized pSFC program is as follows: temperature 25°C, pressure 200 bar, gradient 95% carbon dioxide/methanol containing 0.5% trifluoroacetic acid (TFAA) and 0.5% triethylamine (TEA) to 70% carbon dioxide/methanol containing 0.5% TFAA and 0.5% TEA at 5%/min at a flow rate of 2 mL/min. The dual nature of the additives is required to block active sites in both Chiralpak AD (TFAA) and Chiralcel OD (TEA) and has no negative influence on the chromatographic performance of the column combinations. The column order in the tandem or triplet combination has been shown to have a neglible influence on retention and resolution [37,44,45].

Some representative chromatograms are shown in Fig. 12–14. A large number of racemates have been separated on the column triplet Chiralpak AD–Chiralcel OD–Chirex 3022 including indapamide, fenoterol, hydroxyzine, buclizine, atenolol, tropic acid, 1-(9-anthryl)-2,2,2-trifluoroethanol, 2-phenyl-cyclohexanone, and derivatized amino acids and dipeptides, to mention a few [37,44,45.]

An important aspect of sub- and supercritical fluid chromatography is the stability and lifetime of CSP columns. The functionalized cellulose and amylose coated on silica gel, and the Brush-type phases appear to tolerate the high flow rates and outlet pressures. For all experiments performed in the period 1993–1995, only one of the various CSP columns was purchased. The chromatographic performance of the different columns still is intact. Moreover, the reproducibility of several chiral separations was evaluated over a 9-month period and RSD% values in retention times and absolute peak areas are less than 0.2 and 3, respectively. Information about the stability of Chiralcel OD in pSFC in an industrial environment is given in Ref. 46.

Recently, we evaluated CSPs based on the coupling of macrocyclic antibiotics on silica [47], namely, Chirobiotic V (vancomycin) and Chirobiotic T (teicoplanin) in pSFC [45]. These CSPs embody different chiral selectors (multiselector). Fewer separations could be realized on those columns compared to the columns and column combinations discussed here.

IV. SELECTIVITY ENHANCEMENT BY ON- AND OFF-LINE MULTIDIMENSIONAL pSFC

In the previous section, the selectivity was tuned by selecting and coupling different stationary phases but the mobile phase composition was the same for the column combination. The selectivity can also be tuned by modifying both the stationary phase and the mobile phase composition. Multidimensional systems as developed for GC and HPLC can also be constructed for SFC [48]. Some schematic drawings that we are using for pSFC-pSFC are illustrated in Fig. 15. Selectivity tuning, heart cutting, backflushing, and the like can be performed by means of six-way valves. Combination C in Fig. 15 is interesting because only one SFC instrument is required. The combination was developed by Wilson for the analysis of ibuprofen racemates in urine [49]. Upon injection between the columns, the sample passes through an achiral column that separates ibuprofen from contaminants. When ibuprofen is in the 100-μL loop, as can be deduced from the UV detector signal, the valve is thrown and only the ibuprofen fraction enters the chiral column. The separated enantiomers pass through the achiral column and are detected.

In our research on natural products we are often confronted with samples of very high complexity. In those samples, major and minor components have to be elucidated and quantified. Nowadays, pSFC on a semipreparative scale plays herein an important role. An off-line multidimensional system offers more flexibility for our goals than an on-line system and, in fact, with automation of injection, fraction collection, and reinjection with the help of robotics, there is hardly any difference between the two approaches. The system that has been developed is described and two applications—namely, the separation of lipids according to hydrophobicity and unsaturation by pSFC-pSFC, and the determination of sterols in vegetable oils by pSFC-capillary gas chromatography-mass spectroscopy (CGC-MS)—have been selected to illustrate the performance of the system.

The system is based on a modular Gilson SFC instrument and composed of the following units: (1) 308, 306, and 307 high-pressure pumps used for carbon dioxide, modifier, and additional collection solvent delivery, respectively; (2) 821 automated pressure regulator; (3) 233XL valve switching system for automated injection and fraction collection; (4) 402 single-syringe low-pressure pump for automated injection; (5) 831 temperature regulator; (6) 119 UV-VIS dual-wavelength detector; (7) 506 C interface; (8) Rhodyne 7037 manual diaphragm relief valve. The pSFC system was controlled by the Gilson 715 software running under Windows. The 233XL valve configuration allows automated sample injection and fraction collection, as shown in Fig. 16. Collection of the fractions is time-controlled using a program running on the Gilson keypad controller, which is included in the 233XL configuration and is synchronized with all other units by the 715 software general contact event window (see also Chapters 3 and 12).

Delivery of an additional solvent by the standalone 307 high-pressure pump, inserted between detection and automated pressure regulation, is needed to prevent deposition of analytes on the internal walls of the tubing after carbon

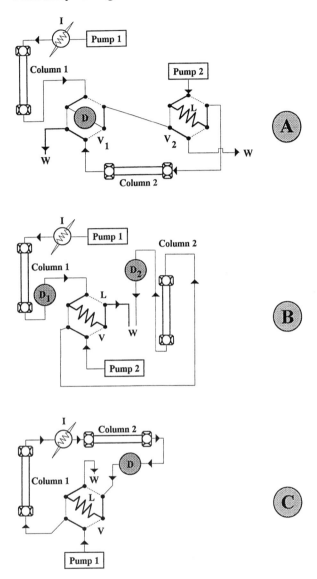

Figure 15 Experimental setup for multidimensional pSFC (I, injector; V, V$_1$, V$_2$, high-pressure six-port valves; D, D$_1$, D$_2$, detectors; L, calibrated loop; W, waste).

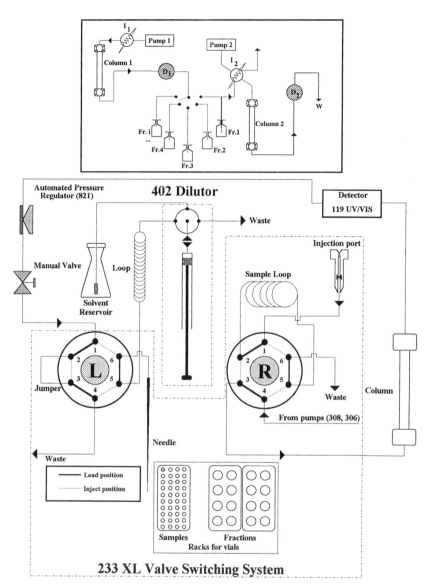

Figure 16 Setup for automated injection/fraction collection for the Gilson SF3 pSFC system. Insert: Off-line coupling pSFC with another separation technique.

dioxide expansion. In this way, collection can be performed quantitatively and the system does not give memory effects.

Triglycerides are the main components of oils and fats and a wide variety of fatty acids occur in these triacylglycerols. This results in an intricate series of compounds and it is unrealistic to except that a single separation system can unravel this complexity. In pSFC two separation mechanisms can be exploited to separate triacylglycerols. The first involves the use of octadecyl silica on which the separation is roughly based on hydrophobicity. The separation occurs in increasing order of carbon number (CN) while a double bond (DB) in the chain reduces the retention time by the equivalent of one carbon number (separation number or SN = CN − DB). On the other hand, on silver-impregnated stationary phases, separation occurs according to degree of unsaturation. Combining both mechanisms in an off-line multidimensional separation system provides a wealth of information on the triglyceride composition in oils and fats [22]. This is illustrated with the analysis of arachide oil. Figure 17 shows the analysis obtained on 2×25 cm $\times 4.6$ mm id, 5-μm ODS columns. The different

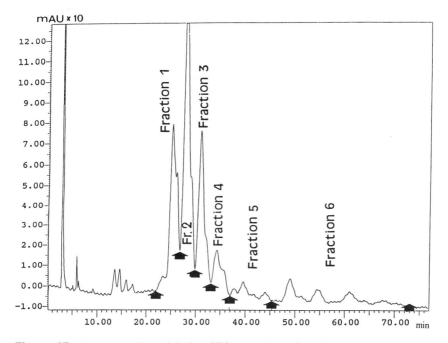

Figure 17 Fractions collected during SFC separation of triglicerides in arachide oil. The column tandem was operated in the subcritical mode at 25°C and 150 bar with carbon dioxide containing 10% methanol and at a flow rate of 2 mL/min. The sample loop was 5 μL and a 10% solution of arachide oil in chloroform was injected. Detection at 210 nm [22].

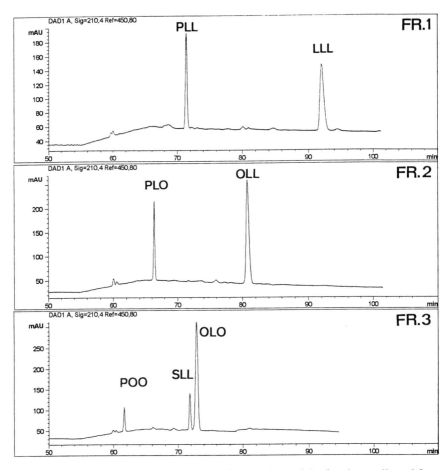

Figure 18 Analysis on a silver-loaded stationary phase of the fractions collected from arachide oil during SFC semipreparative separation. Experimental conditions: temperature 65°C; flow rate 1 mL/min; pressure programmed from 150 bar (2 min) to 300 bar at 1.5 bar/min; modifier acetonitrile/isopropanol in ratio 6:4 was programmed from 1.2% (2 min) to 7.2% (28 min) at 0.3%/min and the to 12.2% at 0.54%/min; detection 210 nm [22].

fractions were collected five times with 0.8 mL/min chloroform as make-up solvent. The solvent was evaporated to 100 μL and the fractions were reinjected on a silver-impregnated column. Figure 18 shows the chromatograms for fractions 1–3 analyzed on a home silver-impregnated strong cation exchanger 25 cm × 4.6 mm id, 5 μm (Nucleosil 100-5SA, Machery and Nagel).

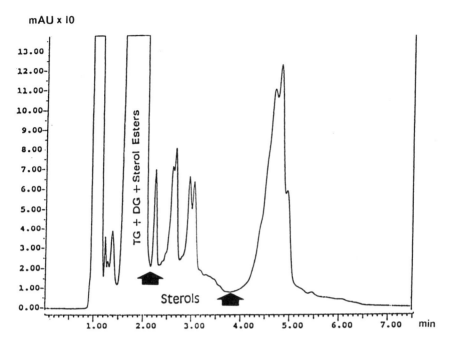

Figure 19 Isolation of the sterolic fraction in sunflower oil by semipreparative SFC. Experimental conditions: pressure 150 bar, column temperature 70°C, flow rate 2 mL/min, 10% methanol as modifier, sample injection volume 5 µL, sample concentration 10%; detection at 210 nm. The additional solvent was composed of methanol/chloroform 1:1 at a flow rate of 0.8 mL/min. TG, triglycerides; DG, diglycerides [50].

As can been seen from the chromatograms, each fraction from the ODS separation corresponds to a specific SN number: fraction 1–48, fraction 2–49, and fraction 3–50. Note that P is palmitic acid CN 16, DB 0; S is stearic acid CN 18, DB 0; O is oleic acid CN 18, DB 1 and L is linoleic acid CN 18, DB 2. On the silver-impregnated column the triglycerides are separated according to degree of unsaturation (PLL < LLL in fraction 1; PLO < OLL in fraction 2; and POO and SLL/OLO in fraction 3). For the separation of SLL and OLO, the higher the number of unsaturated fatty acids, the higher is the retention.

For sterol fractionation of vegetable oils a 20 cm × 4.6 mm id, 5-µm aminopropyl silica gel (APSG) column was applied [50]. Sunflower oil purchased from the local supermarket was diluted in chloroform (1 g/10 mL) and 50 µL of an internal standard solution (10 mg cholesterol/mL chloroform) was added before fractionation. Figure 19 shows the pSFC chromatogram. The free sterols elute between between 2.15 and 3.8 min and this fraction was 10 times collected in a fully automated way. The solvent was then evaporated and the residue

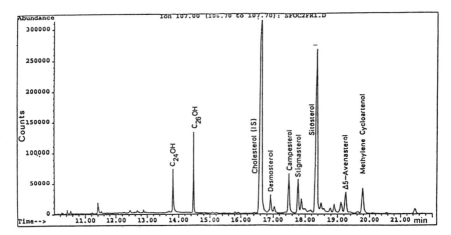

Figure 20 CGC-MS analysis of the sterolic fraction in sunflower oil isolated by semipreparative SFC. Experimental conditions: injector temperature 280°C, splitless time 50 s, column temperature from 50°C (1 min) to 290°C at 20°C/min, then to 330°C (5 min) at 2°C/min. The MS interface temperature was 300°C. The GC column was a WCOT-HP5MS, 30 m × 0.25 mm id × 0.25 μm film thickness. Helium was used as carrier gas at a flow rate of 0.8 mL/min (constant flow mode, linear velocity 33.2 cm/s) [50].

taken up in 100 μL dichloromethane. The sterolic fraction was then analyzed on an HP 5890 series II gas chromatograph equipped with electronic pressure control and split/splitless injection. The instrument was coupled to a 5972 mass selective detector and 1 μL was injected in the splitless mode. The chromatograms was recorded in the full-scan mode applying electron impact ionization. Quantitation of the sterols was made from the extracted ion chromatogram at m/z 107 (Fig. 20). The response factors at m/z 107 compared to cholesterol were calculated for campesterol, stigmasterol, and β-sitosterol. The reproducibility of the complete method was evaluated and the sample was fractionated and analyzed five times over a 1-month period. The concentrations with RSD % were 249 ppm (3.7%) for campesterol, 127 ppm (4.8%) for stigmasterol, and 833 ppm (2.1%) for β-sitosterol.

V. SELECTIVITY ENHANCEMENT BY FINE-TUNING THE MOBILE PHASE IN pSFC

For some separation problems fine tuning of the mobile phase composition in pSFC is of utmost importance. A typical case concerns the separation of D,L- and *meso-N,N'*-di(carbobenzyloxy)-2,6-diaminopimelic acid [51]. Previous investigations [38,39] indicated that Chiralpak AD is the best chiral stationary phase for the analysis of acidic enantiomeric pairs by pSFC. Chiralcel OD and Brush-type CSPs with π-donor and π-acceptor characteristics were evaluated as well in the study [51], but the selectivity on Chiralpak AD for both the D,L separation and for the D,L/meso separation were far superior.

Experimental parameters that can be varied to optimize the resolution between the D and L form, indicated by R_s^E, and between the last eluting enantiomer and the mesoform, indicated by R_s^M, are the inlet pressure, the flow rate, the temperature, the modifier and its concentration, and the nature of the acidic additive.

The pressure was varied in the interval 73–300 bar keeping the other conditions constant, i.e., flow rate 1 mL/min, 40°C, methanol concentration programmed from 10% (2 min) to 18% at 0.4%/min and 0.5% TFAA as additive. The best value for R_s^E was at 73 bar (1.17) and for R_s^M at 100 bar (1.54). 100 bar, at which R_s^E was 1.07, was selected as compromise for further optimization. Recording H vs. mobile phase velocity plots under isoconfertic conditions ($P = 100$ bar, $T = 40$°C), but with the modifier programmed as described above, indicated that the highest efficiency was obtained at 1 mL/min. The apparent plate number measured for the meso form was 1.8 times higher

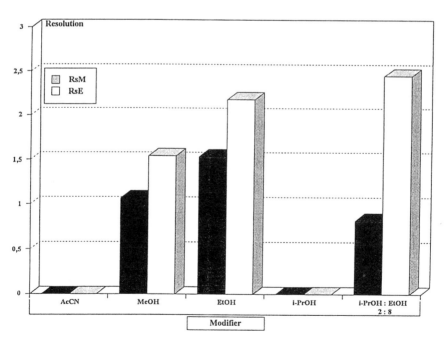

Figure 21 Influence of the nature of the modifier on R_s^E and R_s^M. For abbreviation, see text [51].

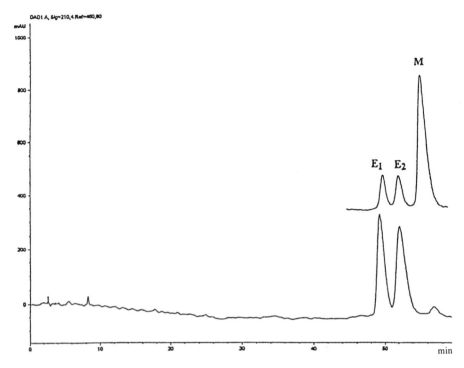

Figure 22 Optimized separation of two samples of D,L- and *meso-N,N'*-di(carbo-benzyloxy)-2,6-diaminopimelic acid. Experimental conditions: pressure 100 bar, temperature 40°C, carbon dioxide at 1 mL/min with ethanol as modifier containing 0.5% trifluoroacetic acid and programed from 10% (2 min) to 18% at 0.4%/min; detection at 210 nm; injected amount ~200 ng. E1, enantiomer 1; E2, enantiomer 2; M, mesoform [51].

compared to the plate number for the first eluting enantiomer and 2.2 times higher compared to the plate number for the last eluting enantiomer. The starting concentration of the modifier methanol was varied between 8% and 16%, with a 2% interval. Ten percent methanol yielded the highest enantioselectivity whereas R_s^M was unaffected by the starting modifier concentration. The most important parameter to the resolution optimization was the nature of the modifier. For some racemates similar results were observed in HPLC [52]. Figure 21 shows the resolution values for acetonitrile, methanol, ethanol, isopropanol, and the mixture isopropanol/ethanol (2:8), keeping all other conditions the same.

For acetonitrile and isopropanol, the three solutes were not resolved at all, whereas the highest selectivity was obtained for ethanol as modifier. No general rules that correlate selectivity on the nature and concentration of the modifier can be advanced in chiral SFC [37], and this is a clear illustration. We have no explanation for the superior behavior of ethanol as modifier and the data are purely based on trial-and-error experiments. The role of the nature of the acidic additive with ethanol as modifier was also investigated and TFAA exhibited far better selectivities compared to formic acid (FA) and acetic acid (AA). No selectivity was observed for the enantiomers applying FA and AA ($R_s^E = 0$) and R_s^M dropped to 0.5 and 1.2 for FA and AA, respectively. As far as the temperature is concerned, measurements were carried out in the range 30–45°C with a 5°C interval and, as expected, the lower the temperature, the better was the enantioseparation but the worse was the separation of the enantiomeric pair from the meso form. The best compromise was 40°C. Figure 22 shows the optimized chromatograms for the separation of D,L- and *meso-N,N'*-di(carbo-benzyloxy)-2,6-diaminopimelic acid (samples 1 and 2).

VI. STRUCTURAL MODIFICATION OF SOLUTES TO ENHANCE ON SELECTIVITY IN pSFC

Structural modification of solutes or derivatization is a well-established technique in chromatography. The main reasons to derivatize a sample in GC are to volatilize the solutes, to block the active functionalities, and/or to enhance on detectability, and in HPLC to offer opportunities to analyze the sample with reversed phase HPLC or to enhance detectability. In pSFC, the most important reason is to allow solubilization of the solutes in the supercritical medium. However, the selection of a derivatization reaction for pSFC application is also important to enhance resolution and, more especially, selectivity. This is illustrated with the enantioseparation by pSFC of amino acids.

Amino acids are very polar and it is difficult to solubilize them in a supercritical medium based on carbon dioxide, and this even at very high modifier concentrations. Camel et al. [53] described elution of underivatized amino acids under subcritical conditions in subcritical fluid chromatography (subFC) but modifiers containing up to 7% water were needed. Recently, we were able to separate the enantiomers of free amino acids by pSFC on Chirobiotic T applying very high concentration of water-rich modifiers [45]. More common is to block at least one of the two functionalities (amino or carboxylic function). For the enantioseparation of amino acids we selected the Chiralpak AD column, which means that, in the first instance, the amino function had to be derivatized. Derivatization into the tosylated, dansylated, and trifluoroacetylated amino acids was not successful. The first two could not be solubilized in a supercritical fluid whereas the latter didn't yield quantitative reactions especially for the basic amino acids.

For chiral recognition of amino acids, benzoylation of the amino group provides the highest enantioselectivity on different CSPs. Dependent on the

Table 1 Resolution Values Obtained for Some Derivatives of Alanine, Aminobutyric Acid and Leucine During Chiral pSFC Separation of Chiralpak AD Column. Experimental Conditions as in Fig. 24

Compound	Derivative (*) structure	Resolution between enantiomers							
		Bz	Bz,OMe	mNBz	mNBz,OMe	pNBz	pNBz,OMe	DNBz	DNBz,OMe
Alanine	$CH_3CH(NHR)COOR_1$	2.2	2.0	1.4	6.6	1.9	8.5	2.8	10.0
Aminobutyric acid	$CH_3CH_2CH(NHR)COOR_1$	2.4	1.7	1.0	2.9	2.6	11.1	0.0	6.1
Leucine	$(CH_3)_2CHCH_2CH(NHR)COOR_1$	4.7	2.2	4.7	7.8	6.5	8.6	5.2	6.5

(*)Derivative Structure

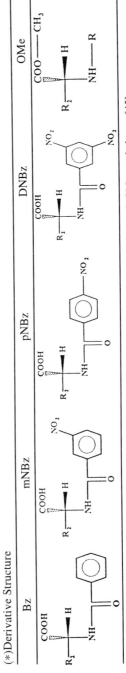

Notations: Bz–benzoyl; mNBz–meta nitrobenzoyl; pNBz–para nitrobenzoyl; DNBz–3, 5-dinitrobenzoyl; OMe–methyl ester [45].

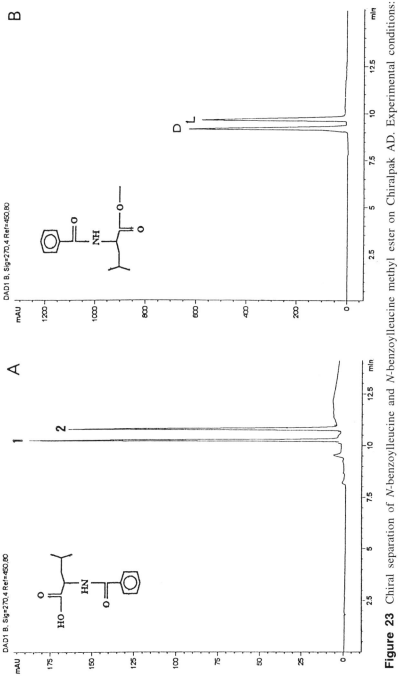

Figure 23 Chiral separation of *N*-benzoylleucine and *N*-benzoylleucine methyl ester on Chiralpak AD. Experimental conditions: column Chiralpak AD 10 μm, 25 cm × 4.6 mm id; flow rate 2 mL/min; temperature 30°C; modifier methanol containing 0.1% triethylamine and 0.1% trifluoroacetic acid, programmed from 5%(5 min) to 30% at 5%/min; pressure 200 bar; detection 270 nm; injected amount ~ 200 ng [45].

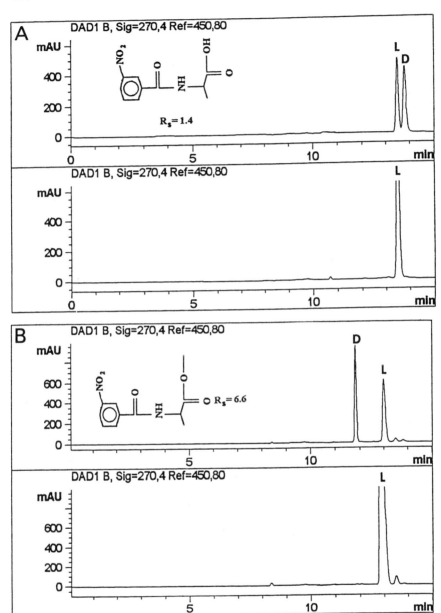

Figure 24 Chiral separation of *N-(para)*nitrobenzoyl alanine and *N-(para)*nitro-benzoyl alanine methyl ester on Chiralpak AD. Experimental conditions as in Fig. 23 [45].

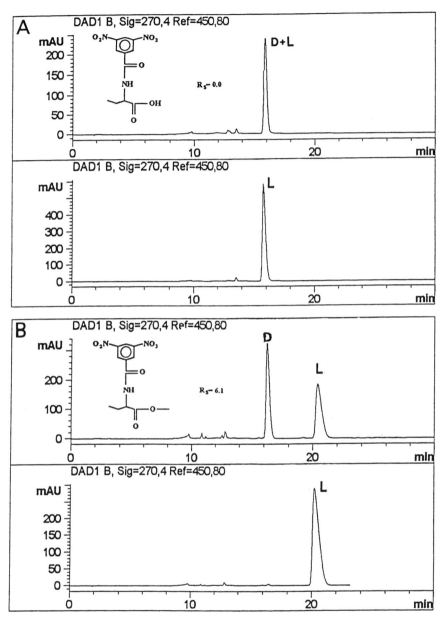

Figure 25 Chiral separation of *N*-dinitrobenzoylaminobutyric acid and *N*-dinitroben-zoylaminobutyric acid methyl ester on Chiralpak AD. Experimental conditions as in Fig. 23 [45].

chiral selector a derivative with π-donor or π-acceptor characteristics is prepared. In this framework, different benzoyl groups were introduced into the amino acids, and the benzoylated amino acids were analyzed as such and after esterification into the methyl ester derivatives. Although no general rules can be advanced on which derivative gives the highest enantioseparation, the differences in selectivities were remarkable [45]. This will be illustrated with alanine (ALA), aminobutyric acid (ABA), and leucine (LEU) as examples (Table 1).

The formation of nonsubstituted benzoyl derivatives (π-donor) gives good enantioseparation for ALA, ABA, and LEU. Methylation decreases resolution as illustrated in Fig. 23 for LEU. When a mononitro-substituted benzoylation (π-acceptor) is performed, methylation will always increase resolution. This is illustrated in Fig. 24A and B for ALA. In this particular case the elution order L/D is even reversed, which can be very useful for the determination of enantiomeric excess. Para substitution gives for both nonesterified and esterified amino acids better results than meta substitution. Dinitro-substituted benzoylation can destroy enantioselectivity as for ABA but will give enhanced separation upon forming the methyl esters (Fig. 25A and B).

The above examples illustrate once more how difficult it is to understand and predict retention and separation mechanisms in enantioseparations. Nevertheless, it shows how important the structural modification of solutes can be to enhance resolution in pSFC.

VII. CONCLUSIONS

Selectivity is the key parameter in pSFC for resolution optimization. Besides appropriate selection of the stationary phase, fine tuning of the mobile phase is a very important tool. By the low viscosity of the mobile phase coupling columns is not a problem in pSFC. This can result in improved efficiency, enhanced selectivity, and greater versatility. Last but not least, appropriate solute derivatization can enhance selective interactions.

REFERENCES

1. P. Sandra, *J. High Resolut. Chromatogr.*, *12*:82 (1989).
2. S. M. Fields and M. L. Lee, *J. Chromatogr.*, *349*:305 (1989).
3. P. J. Schoenmakers and F. C. C. J. G. Verhoeven, *J. Chromatogr.*, *352*:315 (1986).
4. P. J. Schoenmakers and L. G. M. Uunk, *Chromatographia*, *24*:51 (1987).
5. P. A. Mourier, M. H. Caude, and R. H. Rosset, *Chromatographia*, *23*:21 (1987).
6. K. D. Bartle, T. Boddington, A. A. Clifford, and G. F. Shilstone, *J. Chromatogr.*, *471*:347 (1989).
7. D. P. Poe and D. E. Matire, *J. Chromatogr.*, *517*:3 (1990).
8. T. A. Berger and J. F. Deye, *Chromatographia*, *30*:57 (1990).
9. T. A. Berger and W. H. Wilson, *Anal. Chem.*, *65*:1451 (1993).

10. H.-G. Janssen, H. M. J. Snijders, J. A. Rijks, C. A. Cramers, and P. J. Schoenmakers, *J. High Resolut. Chromatogr.*, *14*:438 (1991).

11. H.-G. Janssen, H. Snijders, C. A. Cramers, and P. J. Schoenmakers, *J. High Resolut. Chromatogr.*, *15*:458 (1992).

12. U. Koehler, P. Biermanns, and E. Klesper, *J. Chromatogr. Sci.*, *32*:461 (1994).

13. U. Koehler and E. Klesper, *J. Chromatogr. Sci.*, 32:525 (1994).

14. X. Low, H.-G. Janssen, H. Snijders, and C. Cramers, *J. High Resolut. Chromatogr.*, 19:449 (1996).

15. R. M. Smith (ed.), *Supercritical Fluid Chromatography*, Royal Society of Chemistry, Burlington House, Piccadilly, London (1988).

16. M. L. Lee and K. E. Markides (eds.), *Analytical Supercritical Fluid Chromatography and Extraction*, Chromatography Conferences, Inc., Brigham Young University, Provo, Utah (1990).

17. T. A. Berger, *Packed Column SFC* (R. M. Smith, ed.), Royal Society of Chemistry, Thomas Graham House, The Science Park, Cambridge, U. K. (1995).

18. V. G. Berezkin, I. V. Malynkova, V. R. Alishoev, and J. de Zeeuw, *J. High Resolut. Chromatogr.*, *19*:272 (1996).

19. R. Trones, A. Iveland, and T. Greibrokk, "Recent advances of High Temperature LC in Packed Capillary Columns," Proceedings to the XVIIIth Int. Symposium on Capillary Chromatogr. (P. Sandra and G. Devos, eds.), Huethig Verlag, Heidelberg, pp. 135–139 (1996).

20. W. W. Christie, *J. Chromatogr.*, *454*:273 (1988).

21. M. Demirbüker and L. G. Blomberg, *J. Chromatogr.*, *550*:765 (1991).

22. A. Medvedovici, F. David, and P. Sandra, in preparation.

23. T. A. Berger, *Packed Column SFC* (R. M. Smith, ed.), Royal Society of Chemistry, Thomas Graham House, The Science Park, Cambridge, U. K., p. 145 (1995).

24. P. Sandra and F. David, *Multidimensional Chromatography: Techniques and Applications* (H. J. Cortes, ed.), Marcel Dekker, New York, p. 145 (1990).

25. D. Repka, P. Sandra, and J. Krupcik, *Analusis*, *20*:555 (1992).

26. Th. Welsch, V. Dornberger, and D. Lerche, *J. High Resolut. Chromatogr.*, *16*:18 (1993).

27. A. Giorgetti, N. Periclès, H. M. Widmer, K. Anton, and P. Dätwyler, *J. Chromatogr. Sci.*, *27*:318 (1989).

28. A. Kot, F. David, and P. Sandra, *J. High Resolut. Chromatogr.*, *17*:272 (1994).

29. P. Lembke and H. Engelhardt, *J. High Resolut. Chromatogr.*, *16*:700 (1993).

30. M. Z. Wang, M. Klee, and S. K. Yang, *J. Chromatogr. B. Biomed. Appl.*, *665*:139 (1995).

31. L. Karlsson, M. Jeremo, M. Emilsson, L. Mathiasson, and J. A. Jönsson, *Chromatographia*, *37*:402 (1993).

32. P. A. Mourier, E. Eliot, M. H. Caude, R. H. Rosset, and A. Tambuté, *Anal. Chem.*, *57*:2819 (1985).

33. P. Macaudière, A. Tambuté, M. H. Caude, R. Rosset, M. A. Alembik, and I. W. Wainer, *J. Chromatogr.*, *317*:177 (1986).

34. P. Macaudière, M. H. Caude R. Rosset, and A. Tambuté, *J. Chromatogr. Sci.*, *27*:383 (1989).

35. R. Rosset, M. H. Caude, and A. Jardy, *Chromatographies en Phase Liquide et Supercritique*, Publ. Masson, Paris, France, p. 555. (1991).

36. X. Lou, X. Liu, Y. Sheng, and L. Zhou, *J. Chromatogr.*, *605*:103 (1992).

37. A. Kot, P. Sandra, and A. Venema, *J. Chromatogr. Sci.*, *32*:439 (1994).
38. A. Kot, The role of supercritical fluids in separation sciences, Ph. dissertation, University of Ghent (1995).
39. K. E. Garcia, A. Medvedovici, V. Ferraz, and P. Sandra, *J. High Resolut. Chromatogr.*, *19*:569 (1996).
40. A. van Overbeke, P. Sandra, A. Medvedovici, W. Baeyens, and H. Y. Aboul-Enein, submitted.
41. W. H. Pirkle and C. J. Welch, *J. Chromatogr. A*, *731*:322 (1996).
42. W. H. Pirkle and C. J. Welch, *J. Chromatogr.*, *589*:45 (1992).
43. T. Zhang and E. Francotte, *Chirality*, *7*:425 (1995).
44. P. Sandra, A. Kot and F. David, *Chemistry Today*, *9*:33 (1994).
45. A. Medvedovici, L. Toribio, F. David, and P. Sandra, submitted.
46. K. Anton, J. Eppinger, L. Frederiksen, E. Francotte, T. Berger, and A. Wilson, *J. Chromatogr.*, *666*:395 (1994).
47. D. W. Armstrong, Y. Tang, S. Chen, Y. Zhou, C. Bagwill, and J.-R. Chen, *Anal. Chem.*, *66*:1473 (1994).
48. H. J. Cortes (ed.), *Multidimensional Chromatography: Techniques and Applications*, Marcel Dekker, New York, p. 301 (1990).
49. W. H. Wilson, *Chirality*, *6*:216 (1994).
50. A. Medvedovici, F. David, and P. Sandra, *Chromatographia*, *44*:37 (1997).
51. A. Medvedovici, P. Sandra, A. Kot, and A. Kolodziejczyk, *J. High Resolut. Chromatogr.*, *19*:227 (1996).
52. K. M. Kirkland, *J. Chromatogr. A*, *718*:9 (1995).
53. V. Camel, D. Thiébaut, M. Caude, and M. Dreux, *J. Chromatogr.*, *605*:95 (1992).

7

Subcritical Fluid Chromatography with Organic Modifiers on Octadecyl Packed Columns: Recent Developments for the Analysis of High Molecular Organic Compounds

Eric Lesellier and Alain Tchapla

University Institute of Technology, Orsay, France

I. INTRODUCTION

High molecular weight compounds are of great interest in the food, pharmaceutical, and environmental industries. Some biological compounds such as carotenoids, steroids, triacylglycerols, or vitamins are involved in the protection of human health. Others are of importance for the plastics (antioxidants, polymers) or cosmetics (waxes) industries, or in soil and water preservation (polyaromatic hydrocarbons, surfactants).

The analysis of such compounds is generally difficult and requires techniques having a high elution power. High-temperature gas chromatography (HT-GC) or nonaqueous reversed phase liquid chromatography (NARP-LC) are generally used to separate complex mixtures containing high molecular weight compounds such as unsaponified fractions of plants extracts, vegetable oils, organic extracts of plastics, soils, etc. Packed column subcritical fluid chromatography (SubFC) is an interesting alternative that combines low temperatures and short retention times. Although the most exciting and rapidly growing application area of SubFC turned to be chiral separations as described in Chapters 3, 8, and 9, SubFC is also particularly well suited for the analysis of thermolabile high molecular weight compounds such as carotenoids pigments that can be carried out only at low temperature.

Compared to packed column supercritical fluid chromatography (pSFC), SubFC is also characterized by the use of high modifier contents that may induce a change in the fluid state, the critical temperatures and pressures of mixed mobile phases being usually higher than those of pure carbon dioxide [1]. Therefore understanding phase transitions in fluid mixtures is necessary to operate the mobile phase under conditions where phase separation does not occur. Methods of fluid behavior and physicochemical measurements, which will help the chromatographer optimizing his separation more efficiently, are discussed in this chapter.

Furthermore, fundamentals of retention and selectivity in SubFC will be described, in particular how modifiers affect the special organization of the stationary phase, the mobile phase density and polarity, and how they influence the retention behavior of high molecular weight compounds. An original feature of SubFC with modifiers having a high dielectric constant where selectivity can be easily tuned without influencing the retention, simply by adjusting the modifier content, is demonstrated. In that particular case the logic of method development commonly used in reversed phase liquid chromatography (RPLC), i.e., a retention diagram, applies for SubFC too.

As the solutes studied in this chapter are mostly hydrophobic compounds, the separation were carried out on apolar octadecyl-bonded silica (ODS) stationary phases that provides a good selectivity for compounds of the same family showing only slight molecular differences. This choice allows also a direct comparison with NARP-LC. In the last section, we present a simple characterization test for ODS phases that estimates the retention power of the stationary phase, i.e., its ability to shape discrimination and to avoid hydrophobic parasite interactions. The stationary phases can then be classified according to the two selectivity criteria described in the test. This classification, which reflects characteristics of the phases such as functionality, loading density, and residual silanols, is an original tool to select the best ODS phase for a specific separation.

II. CHARACTERIZATION OF CARBON DIOXIDE/MODIFIER MIXTURES

In the presence of high level of modifiers, the number of phases (mono- or biphasic), the occurrence of phase transitions (from the supercritical to the subcritical state), and the density of the mixed mobile phase are all relevant to the chromatographer. Direct and indirect methods used to determine these physicochemical characteristics are described here. Many organic solvents, such as methanol, acetonitrile, methylene chloride, or hexane, are totally soluble in carbon dioxide at 25°C, below its critical temperature of 32°C and can therefore be used as modifiers in SubFC [2].

A. Determination of the Fluid State and of Phase Transitions

Most phase behavior studies are done using view cells. These transparent, high-pressure-resistant, sapphire cells have allowed direct visual observation of the fluid state [2–5]. This type of cell can also be used with a laser light. The scattered or the transmitted light was measured, and demixing or phase transitions due to changes in pressure, temperature, or modifier content could be deduced from the changes in signal intensity [4,5].

pSFC can alternatively be used in these studies. The baseline stability, recorded at 210 nm with a UV-visible (UV-vis) detector, gave information on the monophasic or biphasic nature of a mobile phase on changing the operating conditions [6,7]. In case of demixing, the previously quiet baseline became erratic and noisy. A similar phenomenon has been observed when decreasing the pressure from 80 to 70 bar while maintaining the temperature at 40°C and the mobile phase composition at methanol-carbon dioxide 2:98 (v/v) [7].

This effect can also be used to investigate phase transitions. According to liquid-vapor equilibrium measurement at 0.76 g/cm^3, the mobile phase should be considered supercritical when it contains 3% methanol and subcritical with 5% methanol. A quiet baseline indicated that no phase transition took place when the methanol content was increased from 3% to 5%. This conclusion was also supported by the constant chromatographic efficiency observed for the two sets of conditions. Moreover, the peak shape remained unchanged, when the temperature of a mobile phase containing 5% methanol was increased from 28°C (subcritical state) to 80°C (supercritical state) at 182 bar outlet pressure, which also indicated that no phase transition occurred during the temperature change [6].

Beside the observation of the baseline, the observation of the peak shape of a solvent injected in a capillary tube connected to a flame ionization detector (FID) can be used to determine if phase transitions are possible, depending on the mixtures' compositions [8].

B. Density Measurement

It is well known in SFC that the elution power of neat carbon dioxide, which can be varied from that of perfluoroalkane to that of chloroform, depends on density [9]. Addition of modifiers as required in SubFC induces changes both in density of the mobile phase and in solute solubility. Density measurements are then needed to understand the effect of modifiers in mixed mobile phases.

Density variations have been measured at different temperatures, pressures, and mobile phase compositions by either a static method [10] or a dynamic method [6,11]. In the latter case, a densitometer fitted with a high-pressure-resistant U tube was connected on-line to the chromatographic system. The density of neat carbon dioxide lies between 0.700 g/cm^3 (80 bar, 30°C) and 0.953 g/cm^3 (150 bar, 10°C) for temperatures ranging from 10°C to 30°C and pressures from 80 to 150 bar [11]. Density decreases continuously with an increase of temperature and a decrease of pressure. However, the density variations are less pronounced at low temperatures and high pressures.

These characteristics can be quite different for mixed mobile phases. At high temperature (30°C) and low pressure (80 bar), density increases quickly for a methanol content up to 40%, whereas at low temperature (10°C) and high pressure (150 bar), density rises to a maximum and then decreases rapidly when the modifier is in the 0–40% range. Thus the addition of a modifier shifts density independently of temperature and pressure toward that of the pure organic modifier. This effect has been verified for acetonitrile, methanol, ethanol, nitromethane, heptane, and methylene chloride. For higher temperatures ranging between 40°C and 60°C, the density of the mobile phase increased with the addition of methanol until the fluid became biphasic at a methanol content 28%. The lower pressure limit, at which the fluid stayed monophasic, was between 80 and 120 bar when the temperature was in the 40–60°C range [60].

These density measurements have been used to prove that the retention behavior of high molecular weight compounds in SubFC is generally not influenced by variations of the mobile phase density at the opposite of what can be observed in pSFC with neat carbon dioxide.

III. INFLUENCE OF CHROMATOGRAPHIC PARAMETERS ON SELECTIVITY AND RETENTION IN SUB-FC

Before describing the effects of organic modifier on retention and selectivity, we will discuss whether their addition to carbon dioxide modifies the efficiency of separations carried out in the subcritical state.

Changes in efficiency can be due either to a variation in the diffusion coefficient or to the presence of a density gradient along the column. Investigations with a methanol-carbon dioxide (10:90 v/v) mixture showed no significant change in the diffusion coefficient [9]. Thus, although the addition of modifiers causes an increase of the density gradient in the column, no loss of efficiency was reported [7,12]. This is also the case when moving from a super- to a subcritical state because of the addition of a modifier [6] or on decreasing the temperature [13]. When peaks broaden, the apparent loss of efficiency is due mainly to solubility limitations [12]. Sometimes the addition of methanol to carbon dioxide enhances the efficiency, this phenomenon was explained as either the capping of the residual silanol groups of bonded silica by the modifier, which prevents the adsorption of solutes onto these active sites [14], or an increase in the solubility of high molecular weight solutes in the mobile phase [15] (see Chapter 2).

Taking into account that the pressure drop across the column is lower in SubFC and in pSFC than in HPLC, it is possible to get up to 200,000 plates for an analysis simply by coupling eight columns in series [16]. This technique is not possible in conventional HPLC and shows once more the unique potential of SubFC and pSFC.

A. Influence of Modifiers on Retention

1. Classification of Modifiers According to the Retention Behavior of High Molecular Weight Solutes

In general, chromatographic retention can be expressed by Eq. (1), where x represents the molar fraction of the organic modifier in reversed phase HPLC or the density of neat carbon dioxide in pSFC [17]:

$$k' = Ax^2 + Bx + C \qquad (1)$$

Both in HPLC and in pSFC, as the first constant A is small, its contribution to retention can be neglected. Retention thus decreases linearly as the molar fraction of the organic modifier or the density of the mobile phase is increased.

However, the A term cannot be neglected in describing the retention of numerous high molecular weight solutes, such as carotenes, in SubFC. Figure 1 shows the variation of log k' vs. modifier content for α-carotene. The analysis was carried out with different concentrations of different types of organic modifiers in carbon dioxide on a high-density monofunctional ODS column [18].

Depending on the retention behavior of high molecular weight solutes, the organic modifiers used in SubFC can be classified in three groups. The first group, which includes tetrahydrofuran, methylene chloride, and heptane, is characterized by a rapid decrease of retention and a loss of resolution for carotenoid pigments as the modifier content is increased in the mobile phase. Modifiers, such as acetone, ethanol, propionitrile, and propanol, belong to the second group. With these solvents, retention decreases more gradually than with the solvents in the first group. The third group comprises methanol, acetonitrile, and nitromethane. In this case retention decreases to a minimum. In the case of methanol, this minimum is displaced toward higher modifier contents. The same kind of behavior has been observed for other high molecular weight compounds, such as squalane [19], various antioxidants [20], and triacylglycerols [21]. However, this behavior seems to be more pronounced for solutes with a hydrophobic part as illustrated by the fact that the minimum of retention is reached earlier for triacylglycerols than for carotenoids or antioxidants.

In conclusion, Eq. (1) is valid to describe the retention of numerous high molecular weight compounds in SubFC and the A term cannot be neglected for

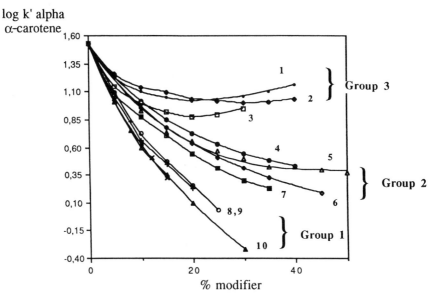

Figure 1 Variation of the log k' for α-carotene vs. the percentage of modifier in carbon dioxide. $T = 25°C$; P_{out} = 150 bar; flow rate = 3 mL/min; UV-vis detection: λ = 450 nm; column: UB 225 (250 × 4.6 mm; 5 μm). 1, Acetonitrile; 2, methanol; 3, nitromethane; 4, ethanol; 5, proprionitrile; 6, acetone; 7, 1-propanol; 8, hexane; 9, tetrahydrofuran; 10, methylene chloride. (From Ref. 18.)

solvents of the third group. However, the physicochemical reasons for the particular behavior of the solvents in the third group are not well understood.

2. Influence of Modifiers on Solute-Mobile Phase Interactions

In liquid chromatography, changes in retention are related to modifications of the interactions between the solute and stationary phase, between the solute and mobile phase, and between the mobile and stationary phases. In the particular case of SubFC and pSFC, the influence of solute–mobile phase interactions are due either to a change in density or to the particular solvating properties of modifier.

Figure 2 shows the density profiles obtained as the organic modifier content in carbon dioxide is increase [18]. Three types of behavior are observed for different solvents: a steady decrease for heptane, a steady increase for methylene chloride and nitromethane, or an initial increase followed by a decrease in density for acetonitrile, methanol, acetone, and ethanol. If the retention would correlate with density, the three groups of solvents should comprise alkanes in the first group, nitromethane and methylene chloride in the second group, and

volumic mass
(g/l)

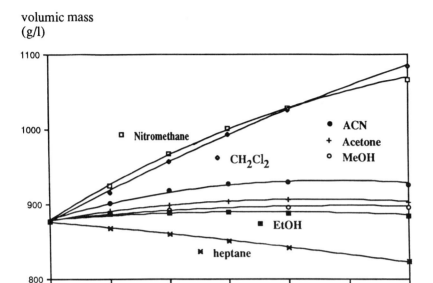

Figure 2 Variation of the volumic mass vs. the percentage of modifier in carbon dioxide. (From Ref. 12.)

methanol, ethanol, acetone, and acetonitrile in the third group. As this is not the case, mobile phase density variations alone cannot describe the retention behavior of carotene (Fig. 1).

As an alternative, the effect of the modifiers on the solute solubility needs to be investigated (see Chapter 1). This approach can be tested by estimating the retention factor k' as a function of the Hildebrand solubility parameter: δ_s for the stationary phase, δ_m for the mobile phase, and δ_i for the solute. The retention variation is represented by Eq. (2) [22].

$$\ln k' = \frac{V_i}{RT} \cdot ((\delta_m + \delta_s - 2\,\delta_i)\,(\delta_m - \delta_s)) + \ln\left(\frac{n_s}{n_m}\right) \qquad (2)$$

Where V_i is the molar volume of the solute; n_s the total number of moles of the stationary phase; and n_m the total number of moles of the mobile phase.

This relationship describes a decrease of retention with increasing modifier content if $\delta_m < \delta_i$, a minimum of retention when δ_m equals δ_i, and an increase when $\delta_m > \delta_i$. Methanol, acetonitrile, nitromethane, and ethanol have very similar solubility parameters (Table 1) and would therefore be expected to generate the same type of retention profile. As nitromethane and ethanol, which have the closest solubility parameters, cause different retention behavior, this

Table 1 Classification of the Modifiers and Values of Dielectric Constant (ε), Superficial Tension (γ), and Hildebrand Solubility Parameters (δ)

Group	Solvents	ε	γ	δ
1	Heptane	1.92	20.8	7.4
	Methylene chloride	8.9	28.1	9.6
	Tetrahydrofuran	7.6	27.6	9.1
2	Ethanol	24.6	22.3	11.2
	Propionitrile	27.2	27.2	10.8
	Acetone	21.4	23.3	9.4
	1-Propanol	20.1	23.7	10.2
3	Methanol	32.7	22.5	12.9
	Acetonitrile	37.5	29.1	11.8
	Nitromethane	36	37	11.0

equation also cannot describe the retention phenomenon of high molecular weight solutes in SubFC.

However, a correlation can be found between the dielectric constant (ε) and the chromatographic influence of the modifiers. All solvents in the first group have a dielectric constant lower than 10, those in the second group one between 20 and 30, and those in the third group one higher than 30. Therefore, the energy of solute transfer between the stationary and mobile phases will be modified differently depending on the dielectric constant of the pure modifier. As the dielectric constant is high for mobile phases that can create strong polar interaction, solvents in the third group enhance the solubility of solutes in the mobile phase at low concentration and have the opposite effect when the modifier concentration increases. For these modifiers, the shift to the higher level of modifier of the retention minimum (Fig. 1) increases with the surface tension (γ) of pure modifier (Table 1). The same retention behavior can be generalized for large rigid hydrocarbons (carotenoids, antioxidants) or for high molecular weight hydrocarbons having mobile chains, such as triacylglycerols (Fig. 3).

The modifier content at the retention minimum depends on the nature of the solute. In a comparative study of xanthophylls (β-cryptoxanthin, zeaxanthin) and β-carotene, the minimum was found to be related to the number of hydroxyl groups in the molecule [23]. The more hydrophobic the solute, the smaller the proportion of organic modifier needed to achieve the minimum retention. This behavior was independent of the nature of the ODS-bonded phase (monofunctional or polyfunctional) [20]. Although the reasons for this phenomenon are not well understood, it is interesting to use this idea to tune selectivity of high molecular weight solutes in SubFC, as will be described in Sec. III.B.

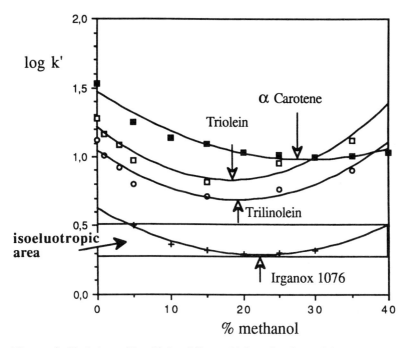

Figure 3 Variation of log k' for different high molecular weight compounds vs. the percentage of methanol in carbon dioxide. $T = 25°C$; $P_{out} = 150$ bar; flow rate = 3 mL/min; column: UB 225 (250×4.6 mm; 5 μm).

3. Influence of Modifiers on Solute–Stationary Phase Interactions

In SubFC, the solute–stationary phase interactions can be influenced by the presence of modifiers in the mobile phase as reported for the separation of rigid molecules such as polyaromatic hydrocarbon (PAH) or cis/trans isomers of β-carotene [18] on a high-density monofunctional ODS stationary phase. The elution order and selectivity factors for benzo[a]pyrene (BaP), a planar PAH, and tetrabenzonaphthalene (TBN), a nonplanar PAH, have been used to demonstrate the modification of the molecular organization of the bonded phase according to the nature of the organic modifier [24]. The TBN/BaP selectivity decreases with addition of modifier because of an increase of the shape discrimination of the stationary phase as explained in more detail in Sec. IV [25]. This can be interpreted either as an increase of the elongation of the bonded chains or as an enhancement of the rigidity and of the order of the bonded chains when the modifier content is increased. However, as the decrease of TBN/BaP selectivity is less important for high-dielectric-constant modifiers, the solvation ability of the octadecyl chain is apparently weaker in these modifiers.

Other rigid solutes do not penetrate the stationary phase but stick to the outer surface of the bonded chain [25,26]. This behavior is illustrated by the separation of cis/trans isomers of carotene, which are, respectively, a bent and a linear molecule. Thus, solvents could be classified in three categories according to the observed effects:

1. No modification of selectivity was observed for methanol, ethanol, and nitromethane.
2. A small decrease of selectivity was observed with acetonitrile, acetone, and propionitrile.
3. A strong decrease was noted for heptane, tetrahydrofuran, and methylene chloride.

This classification is almost identical to the one previously established for the retention behavior of carotenes (Sec. III.A.1.6) or for the selectivity of the PAH [25], but the subtle differences observed for carotene and PAH indicate that the two separation mechanisms are not the same, even if they are both due to shape selectivity. These differences are probably related to the chemical structures of compounds that are used as molecular probes. The overall results prove that modifier addition to carbon dioxide is able to modify the surface state of the ODS phases but cannot explain the change of retention described previously for the modifiers of the third group (Sec. III.A.1.6).

B. Selectivity Tuning with High Dielectric Constant Modifiers

From the analytical point of view, the solvents of the first and second groups (Table 1) are capable of decreasing the retention because their presence in the mobile phase improves the solubility of the solutes. Unfortunately, this decrease of retention is coupled with a decrease of selectivity. Therefore, addition of these modifiers is of interest only if the resolution was originally satisfying.

As previously demonstrated, the influence of solvents in the third group is different because they do not cause a constant decrease in retention. For a modifier content between 0 and 50%, it is possible to consider that the retention of solutes does not change greatly. Thus, an isoelutropic area can be defined (Fig. 3) in which the retention factor of a solute is nearly constant. According to the Purnell equation [Eq. (3)], the chromatographic resolution depends in this case on both the selectivity α and the efficiency N_2 of the system:

$$R = \frac{1}{4} \ \sqrt{N_2} \ \frac{\alpha - 1}{\alpha} \ \frac{k'}{k_2' + 1} \tag{3}$$

In this equation, subscript 2 designates the most retained compound.

As efficiency does not vary greatly with addition of modifier (see Sec. III), the variation of the selectivity mainly governs the separation in this isoeluent

log k'

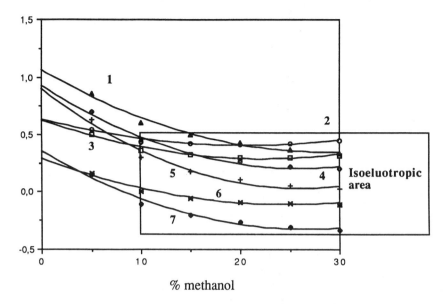

% methanol

Figure 4 Variation of log k' vs. the percentage of methanol in carbon dioxide for different antioxidants. $T = 25°C$; $P_{out} = 150$ bar; flow rate = 3 mL/min; UV - vis detection: $\lambda = 220$ nm; column: UB 225 (250×4.6 mm; 5 μm). 1, Irganox 1330; 2, Irgafos 168; 3, Irganox 1076; 4, Irganox 565; 5, Irganox 1010; 6, Chimassorb 81; 7, Irganox 1035.

area. Figure 4 shows the changes in retention of seven commercial antioxidants vs. the percentage of methanol in carbon dioxide. This isoelutropic region covered a range of log k' between 0.4 and 0.7. In this region, some of the compounds coeluted when the retention was increased by changing the proportion of methanol.

By drawing a window diagram of selectivity, it was possible to determined that 25% methanol in carbon dioxide was the only mobile phase composition that allows the separation of the seven compounds (Fig. 5) within 4 min.

The separation of carotenoids was optimized in a similar way. The selectivity curves were plotted for the dihydroxylated carotenoids (lutein, zeaxanthin) and showed a degradation of the separation with increasing methanol content in the mobile phase [23]. However, the addition of methanol enhanced the separation of pigments, which have a very different chemical structures. Taking into account these two antagonistic effects, the best compromise was found to be 35% of methanol in the mobile phase.

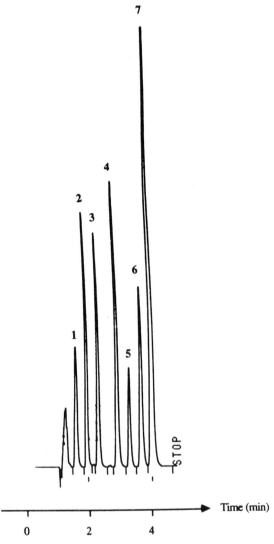

Figure 5 Separation of seven commercial phenolic antioxidants in SubFC. $T = 25°C$; $P_{out} = 150$ bar; flow rate = 3 mL/min; mobile phase: methanol-carbon dioxide (25:75 v/v); UV-vis detection: $\lambda = 220$ nm; column: UB 225 (250 × 4.6 mm; 5 μm). Peak identification, see Fig. 4.

C. Extension of the Isoeluent Concept to Temperature and Pressure

Temperature and pressure play an important role in modulating pSFC through their influence on eluent density. Thus, density increases with increasing pressure; consequently, retention decreases with increasing pressure. The effect of temperature is more complex because temperature also influences the volatility of solutes. For this reason, retention may decrease even when density decreases due to changes in temperature, but this effect is not observed in SubFC with high molecular weight compounds.

Figure 6 shows the changes in the retention of triacylglycerol vs. temperature for different acetonitrile-carbon dioxide compositions in the subcritical state. For modifier contents between 5% and 10%, retention first decreased and then increased as the temperature was raised. This profile is the opposite of the one observed in pSFC. At low temperatures, this phenomenon can be explained as in HPLC by a decrease of solute–mobile phase interactions when the temperature is raised. At higher temperatures, the diminution of the mobile phase density predominates, as confirmed by experimental measurement [11]. This effect, usually observed at high temperatures, disappears at higher levels of organic modifier in the mobile phase (from 15% to 30%).

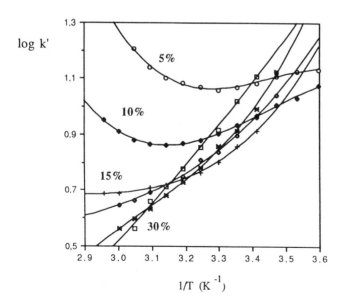

Figure 6 Variation of log k' for triolein vs. $1/T$ for different percentage of acetonitrile in carbon dioxide. $P_{out} = 150$ bar; flow rate = 3 mL/min.; UV-vis detection: $\lambda = 210$ nm; column: UB 225 (250×4.6 mm; 5 µm). (From Ref. 21.)

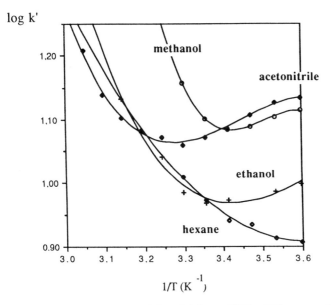

Figure 7 Variation of log k' for triolein vs. $1/T$ for different modifiers at 5% content in carbon dioxide. P_{out} = 150 bar; flow rate = 3 mL/min.; UV-vis detection: λ = 210 nm; column: UB 225 (250×4.6 mm; 5 μm).

The influence of the solvent nature was also investigated at low modifier content. The two protic solvents (methanol and ethanol) had curves similar to that of acetonitrile (Fig. 7). However, the retention minimum, which underlines the change in the governing retention effect (from solute–mobile phase interactions to mobile phase density), was shifted to a higher temperature with acetonitrile.

Such a shift probably due to a better cohesion of the acetonitrile-carbon dioxide mixture than that of protic modifiers-carbon dioxide. This reflects stronger dipole–dipole interactions between the nitrile part of acetonitrile and carbon dioxide. Surprisingly, retention increases along with temperature for a hexane-carbon dioxide mixture, what can be explained by a steady decrease in density. This behavior reflects the lack of strong interactions between these two solvents, whose elution powers are often considered to be very close.

As in pSFC, an increase of pressure leads to an increase of density for any type of organic modifier. However, a decrease in retention, presumably due to a modification of density, occurs only at low levels of modifier in carbon dioxide (less than 15%). At higher modifier concentrations, pressure has no effect on retention [11].

In conclusion, it can be said that for high molecular weight and hydrophobic compounds:

If the proportion of modifier is lower than 15%, the variation in retention due to temperature or pressure approximately follows the variations in mobile phase density as in pSFC.

When the modifier content is higher than 15%, the effect of temperature or pressure on retention is similar to that observed in RPLC.

In the same manner that the isoelutropic region was defined previously with respect to the organic modifier content in the mobile phase, it is possible to determine two new isoelutropic regions: one at low modifier content that is independent of temperature, the other that is independent of pressure at a high percentage of modifier. Thus in these areas, the influence of the chromatographic parameters (modifier content, temperature, and pressure) on retention is governed by slight differences in the chemical nature of the solutes.

Therefore, it is possible to tune the selectivity of a separation between two unsatisfactorily resolved compounds while maintaining the analysis time. This methodology was successfully applied to optimize the separation of triacylglycerols on a packed ODS column [21]. This was possible because the solubility of saturated hydrocarbonaceous chains was enhanced to a greater extent than that of unsaturated hydrocarbonaceous chains on increasing temperature.

D. Application of the Isoeluent Concept to the Separation of Selected High Molecular Weight Solutes

The introduction of an organic modifier changes the global polarity of the mobile phase. It may also modify the solubility of a particular solute by modifying specific interactions between the mobile phase and solute. Modifiers may also be able to change the state of the stationary phase and modify the separation based on a shape selectivity, in particular if the modifier is adsorbed as a thick layer on the stationary phase as described in Chapter 2.

Regarding the interactions modified by the solvent, it is easy to differentiate the solvents from each other by examining the Hildebrand partial solubility parameter (Table 1) or, more simply, by examining their chemical structure. One can distinguish between protic solvents (methanol, ethanol, propanol, butanol, etc.) and dipolar aprotic solvents (acetonitrile, propionitrile, nitromethane). Their addition to the mobile phase, as in HPLC, leads to subtle modifications in the separation either for molecules possessing hydroxyl groups or for those differentiated by the location of double bonds, e.g., carotenoids, triacylglycerols, or PAH [27]. In the following section, we examine the influence of the mobile phase composition on the elution behavior of these compounds.

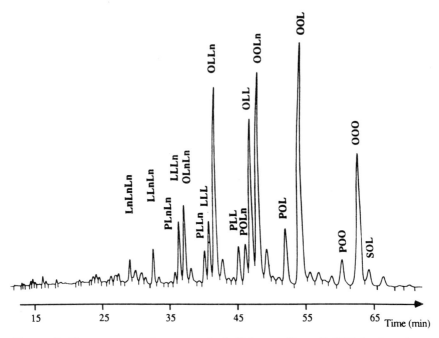

Figure 8 Chromatogram of a rapeseed oil. Columns: Hypersil ODS (series of seven 120 × 4.6 mm columns; 5 μm;), mobile phase: acetonitrile-methanol-carbon dioxide (5.4:0.6:94 v/v/v), T = 16°C; P_{out} = 100 bar; flow rate = 3 mL/min.; UV-vis detection: λ = 210 nm. Ln, linolenate; L, linoleate; P, palmitate; O, oleate; S, stearate. (From Ref. 21.)

A triacylglycerol molecule consists of trihydric alcohol glycerol, each position of which is esterified by a more or less unsaturated long chain fatty acid. Using ODS columns, the best separation (Fig. 8) is achieved with acetonitrile as a modifier, which gave the best selectivity between several weakly resolved pairs of triacylglycerols (PLL/OLL; POL/OOL; POO/OOO) [21]. The analysis may need a small quantity of methanol to allow the simultaneous separation of other, less hydrophobic solutes present in fats (sterols and tocopherols).

Carotenoids are polyunsaturated compounds that may contain a terminal six-membered cyclic group functionalized by a hydroxyl group. An increase in the proportion of protic solvent used as a modifier decreases the retention and, consequently, the selectivity between lutein and zeaxanthin, which are both hydroxylated carotenoids [23]. On the other hand, these organic modifiers have no effect on the separation of apolar carotenoids such as the α- and γ-carotene pair. However, the nature of nonprotic solvents plays an important role in the separation of these two carotenes. Selectivity is increased when the modifier is chosen in the following order: acetone < methylene < chloride < propionitrile

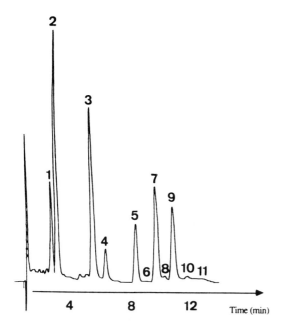

Figure 9 Chromatogram of a mixture of carotenes. $T = 25°C$; $P_{out} = 150$ bar; flow rate = 3 mL/min; UV-vis detection: $\lambda = 450$ nm; mobile phase: acetonitrile-methanol-carbon dioxide (33.25:1.75:65 v/v/v); column: UB 225 (250 × 4.6 mm; 5 μm). 1, lutein; 2, zeaxanthin; 3, β-cryptoxanthin; 4, lycopene; 5, all-*trans*-γ-carotene; 6, *cis*-γ-carotene; 7, all-*trans*-α-carotene; 8, *cis*-α-carotene; 9, all-*trans*-β-carotene; 10,11, *cis*-β-carotene. (From Ref. 26.)

< acetonitrile < nitromethane. This order reflects the dipole-dipole partial solubility parameter of Hildebrand (Table 1). Figure 9 shows the separation of seven common carotenoids and some of their cis isomers as present in many natural extracts. The search for optimal conditions was governed by the need to obtain sufficient separation for both xanthophylls and carotenes. As for the analysis of fats, a small amount of methanol (1.75% content) favored the elution of the more polar components, in this case lutein and zeaxanthin, but a too-high methanol content leads to coelution of these compounds. For carotenes, the optimal acetonitrile content, determined mainly by the resolution of α and γ isomers, was between 30% and 35%.

E. Application of the Isoeluent Concept to the Prefractionation of Complex Mixtures

In the previous sections, the different effects of organic modifiers on retention and selectivity were demonstrated. These effects can also be used to fractionate

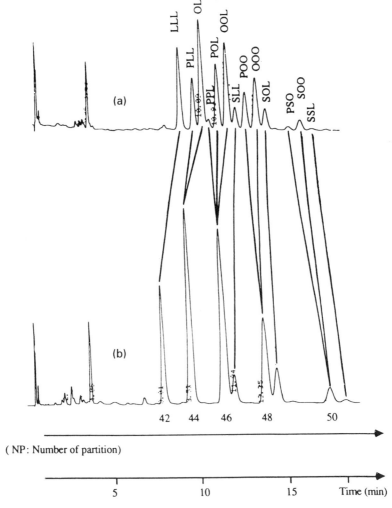

Figure 10 Chromatograms of an argan oil. $T = 25°C$; $P_{out} = 150$ bars; mobile phase: (a) 5% of acetonitrile in carbon dioxide; (b) 30% of acetonitrile in carbon dioxide; column: UB 225 (250 × 4.6 mm; 5 μm). Peak identification, see Fig. 8. (From Ref. 21.)

complex mixtures, such as fats [21]. Different fractions can be collected for identification purpose or for reanalysis under other conditions as discussed more extensively in Chapters 3 and 12.

A close examination of the retention order of triacylglycerols shows that these compounds elute according to their partition number (PN), as in NARP-

LC (Fig. 8). For a given compound, PN is equal to the total number of carbon atoms in the fatty acid chains minus twice the number of double-bond fatty acids, e.g., LLL, PN = 42; PLL and OLL, PN = 44 (L = linoleate chain, R = $C_{18}H_{34}$; P = palmitate chain, R = $C_{16}H_{34}$; O = oleate chain, R = $C_{18}H_{36}$) [28]. Thus, PLL and OLL have close retention times, owing to their common partition number. The elution order of these two compounds and all other pairs differing only by a palmitate or an oleate chain is inverted compared to the one in NARP-LC.

Obviously, the more compounds possessing the same partition number, the more difficult the separation. Because the separation of triacylglycerols in SubFC is partly related to the acetonitrile content in the mobile phase, the separation is quite different for low and high modifier content. Indeed, Fig. 10 compares the separations of argan oil using two different levels of acetonitrile in the mobile phase.

With 5% acetonitrile in the mobile phase the separation obtained at 25°C is quite good for the main triacylglycerols of argan oil. However, an increase in acetonitrile content (30%) results in a loss of resolution for some compounds and in group separation of compounds of a given partition number. An oleate chain (O) becomes chromatographically equivalent to a palmitate chain (P). Such behavior is also observed at low temperature with methanol as a modifier.

In addition, selectivity can be differently modulated by changing temperature for a given mobile phase composition. Thus, with low modifier content, less than 10% acetonitrile or methanol, and high temperature (35°C for methanol and 50°C for acetonitrile), a palmitate (P) residue becomes chromatographically equivalent to a linoleate (L) residue.

The overall effects are related to solubility changes in triacylglycerols induced by variations of the mobile phase composition and temperature. Thus, by choosing appropriate experimental conditions, it is possible to perform fractionations of fats with the objective of the isolation or collection of a particular triacylglycerol of interest.

IV. CHARACTERIZATION AND CHOICE OF ODS STATIONARY PHASES

Most of the published separations of nonpolar high molecular weight solutes are achieved with ODS phases. Unfortunately, even if the bonded chains are the same, the performance of the resulting stationary phases can vary greatly, and transposition of given analytical conditions from one commercial support to another can produce very disappointing chromatograms.

Many different parameters account for this phenomenon: diversity of silicas employed (spherical or irregular, pore diameter from 60 Å to 300 Å, particle size from 1.5 μm to 10 μm, pellicular or porous, etc.), silica pretreatment (acidic, basic, thermal), bonding treatment (reactivity of silanizing agent: dimethylamino-, chloro-, or alkoxy-; mono-, di-, or trifunctional reagents; coated, bonded, or sandwich nature), end-capping (trimethylsilyl-) or steric protection, or the presence of water during the reaction steps. All these points lead to columns that have very different performances and properties, including:

The hydrophobicity of the stationary phase, which is commonly related to its carbon content (classically from 5% to 30%).

The steric discrimination due to the mono- or polyfunctionality of the bonding.

The secondary interaction mechanism due to accessible residual hydroxyl (or siloxane) groups, which is mainly noticeable with acid/basic solutes or hydroxylated ones.

The bonding density (up to 3 or 4 μmol/m^2 for high-density supports), which is related to the carbon content of the monofunctional stationary phases and to the specific surface of the silica. This parameter can also modulate the last two properties. For a successful separation, it is therefore necessary to have a better understanding of the adsorbent employed.

A. Description of a Simple Characterization Test

To characterize a chromatographic support two types of methods are in use [29]: (1) static methods, either nondestructive (diffuse reflectance, Fourier transform-infrared, spectrofluorometry, mass spectrometry, microscopy, thermal analysis, thermal neutron diffusion, ^{29}S or ^{13}C CP-MAS NMR spectroscopy) or destructive (elemental analysis, chemical degradation by HF or alkaline reaction, followed by GC analysis); (2) dynamic methods that measure the chromatographic properties of the support.

Attempts have been made to establish generally accepted procedures involving standardized test solutes and conditions [30]. Of seven tests validated for HPLC, the three following tests have been confirmed as acceptably measuring the properties listed above in SubFC: The selectivity of a theophylline/caffeine mixture gives information on the activity of residual silanols. A mixture of toluene/naphthalene/anthracene can be used to determine the hydrophobic character of the stationary phases. Finally, a mixture of benzo[a]pyrene (BaP), tetrabenzonaphthalene (TBN), and phenanthrylphenan-threne (PhPh) has been used to determine the mono- or polyfunctionality of stationary phases (NIST Test 869) [24]. Although these tests may be easier to perform than static methods, they deliver only fragmented information that has to be compiled for interpretation.

Based on our recent studies on carotenoids separations we propose to use these compounds as probes to characterize commercial ODS phases [25]. The separation of cis/trans isomers of β-carotene as well as the selectivity of β-carotene/zeaxanthin are namely dependent on the nature of the stationary phase.

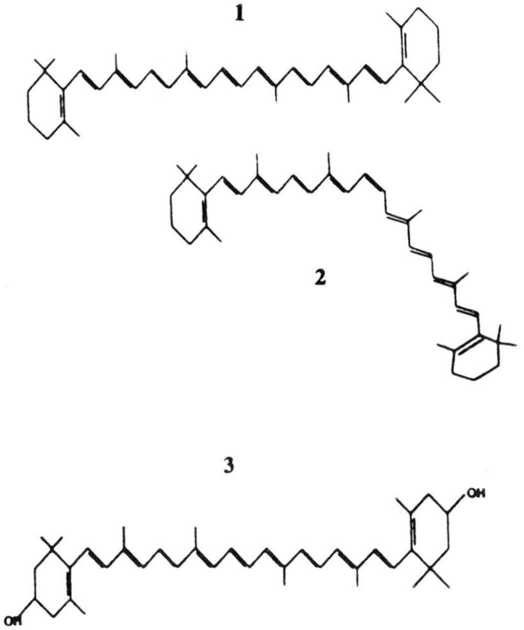

Figure 11 Chemical structures of the carotenes used for the SubFC characterization test. 1, all-*trans*-β-carotene; 2, 15-*cis*-β-carotene; 3, zeaxanthin.

Because cis/trans isomers are either bent rigid (cis) or linear rigid (trans) molecules (Fig. 11), their chromatographic differentiation is due to a shape recognition mechanism at the surface of the stationary phase. Thus, bonded silicas that present irregular surface topography are particularly efficient at separating these types of compounds. This is why using a mixture of *cis/trans* β-carotene is useful as a probe to characterize the macroscopic state of the reverse bonded stationary phase.

Moreover, for identical analytical conditions, the elution order of the β-carotene/zeaxanthin pair is sometimes inverted on different ODS phases. Because the only structural difference between these two solutes is the presence of two hydroxyl groups in zeaxanthin and none in β-carotene (Fig. 11), it is assumed that this selectivity depends on whether residual silanols are present. Thus, such a pair of compounds can be used as a probe to measure siloxane or residual silanol activity.

Finally, the absolute value of the retention time of β-carotene, which is directly related to the hydrophobic nature of the analyte, allows the comparison of solvophobic power for different stationary phases.

B. Validation of the Results

Before introducing this new test mixture for the characterization of chromatographic reversed phases, it is worthwhile to compare it with those previously described.

1. Comparison with NIST Test 869

The most useful criterion in this test is the selectivity factor α-TBN/BaP. In HPLC, the value of this selectivity is lower than 1 for polyfunctional phases and greater than 1.7 for monofunctional phases [24]. For intermediate values, the type of material is not readily characterized.

Results obtained in SubFC with the same solutes were compared to those gained from HPLC. Plots of α-TBN/BaP in HPLC vs. α-TBN/BaP in SubFC are linear [25]. The good correlation coefficient (0.975) indicates a good agreement between the two sets of results. Instead of 1.0 and 1.7, the subcritical values were, respectively, 0.8 and 1.5. Such a correlation underlines the analogy of the shape recognition mechanism in HPLC and SubFC, using both ODS columns.

Another study that compares the selectivity of cis/trans-β-carotene and NIST test results in SubFC proved that the cis/trans-carotene selectivity is well suited for the evaluation of reversed phase functionality [25].

2. Validation of the Test Based on a Mixture of β-Carotene and Zeaxanthin

Experimental information suggests the use of zeaxanthin as a probe to get information on the activity of residual silanols of the bonded reversed phases. This conclusion was drawn from a study on selectivity between β-carotene and zeaxanthin carried out on both end-capped and uncapped basic silica [25]. Table 2 shows that an end-capping treatment enhances the selectivity of these compounds by a factor of approximately 3 [31]. Because the retention times of β-carotene are similar on both treated and untreated silicas, the observed enhancement is mainly caused not by an increase in retention of β-carotene but by a decrease in the retention of zeaxanthin.

The values listed in Table 3 are in good agreement with this hypothesis [32]. Indeed, the higher the density of residual silanols, the lower the selectivity.

Table 2 Comparison of the β-Carotene/Zeaxanthin Selectivity for Different Encapped or Non-End-Capped Commercial Supports

Columns	Selectivity β-carotene/zeaxanthin
Lichrospher 100 RP-18	2.78
Lichrospher 100 RP-18e	8.84
Superspher 100 RP-18	3.59
Superspher 100 RP-18e	10.30

Table 3 Relationship Between the Residual Silanophobic groups and the β-Carotene/Zeaxanthin Selectivity for Partisil Supports

Column	Number of OH/nm^2	Selectivity β carotene/zeaxanthin
Partisil ODS-1	2.73	0.178
Partisil ODS-2	1.75	0.511
Partisil ODS-3	1.05	2.27

Source: Ref. 31.

Because one could not eliminate the ability of xanthophylls to interact with the siloxane oxygen atoms, the previous result shows that measurement of the selectivity of this pair of solutes is a useful criterion by which to evaluate the presence of residual silanols [25].

In conclusion, a test based on the analysis of a mixture of partially light isomerized β-carotene and zeaxanthin provides three important pieces of information in a single chromatographic run [32]. The proposed experimental conditions are as follows: mobile phase, methanol-carbon dioxide (15:85 v/v); temperature, 25°C; flow rate, 3 mL/min; and outlet pressure 150 bar. UV-vis detection was carried out at 450 nm. The retention factors of all *trans*-β-carotenes (major compound of the isomer peaks), 13-*cis*-β-carotene (more intense *cis* peak isomer), and zeaxanthin, the selectivity of *cis/trans*-β-carotene and that of β-carotene/zeaxanthin, are calculated and used to characterize ODS phases.

C. Classification of the Commercial Supports

Our purpose is to use the results obtained using the carotenoid probes to classify commercial supports and as a guide to optimize the choice of a support for a particular analytical problem. For instance, low-retentive supports may be particularly useful for very hydrophobic compounds.

The solvophobic power of many commercial columns is given by the retention factor of β-carotene, as shown in Table 4. The higher the carbon content of the bonded phase, the higher the retention. Bonded silicas, for which the values of the retention factor are low, are either low-density monofunctional bonded silicas (Partisil ODS-1) or polyfunctional wide-pore silicas (Vydac 201 TP).

Both the residual silanols and steric discrimination power can be evaluated from the diagram shown in Fig. 12, which correlates the selectivity of β-carotene/zeaxanthin with the selectivity of *cis/trans*-β-carotene. An initial classification can be deduced from each criterion. The cis/trans selectivity allows differentiation between four types of support: coated, monofunctional, intermediate, and polyfunctional (Table 5). Gammabond C_{18} was the only commercial support tested that had a selectivity lower than 1. The mechanism causing inversion of the retention order of *cis/trans* isomers of β-carotene,

Table 4 Retention Factor k' for all-*trans*-β-Carotene
for Different ODS Phases

Columns	k'
Gammabond C18 (ES Industries)	18.65
Alltima (Alltech)	12.6
Adsorbosil C18 (Alltech)	12.3
Ultrapak C18 (Phenomenex)	11.66
Ultrabase UB 225 (Shandon-SFCC)	11.2
Econosil C18 (Alltech)	10.99
Kromasil C18 (Shandon)	10.99
Superspher 100 RP-18 (Merck)	10.99
Adsorbosphere HS C18 (Alltech)	10.58
Supelco LC18DB (Supelco)	9.83
Superspher 100 RP-18e (Merck)	9.73
Lichrospher 100 RP-18 (Merck)	9.64
Lichrospher 100 RP-18e (Merck)	9.36
Capsell pak C18 (Interchim)	8.75
Inertsil ODS-2 (Interchim)	7.88
Partisil ODS-2 (Whatmann)	7.75
Spherisorb ODB (Phase Sep)	7.46
Hypersil BDS (Shandon)	7.4
Spherisorb ODS 2 (Phase Sep)	7.32
Nucleosil C18 100 (Nucleasil)	7.23
Hypersil ODS (Shandon)	7.11
Purospher RP18 (Merck)	7.09
Partisil ODS 3 (Whatmann)	6.98
Brownlee Spheri 5 (Brownlee Labs)	6.91
Econosphere (Alltech)	6.28
Lichrospher PAH (Merck)	3.95
Spherisorb ODS 1 (Phase Sep)	3.63
Vydac 201 HS (Hesperia group)	3.42
Bondasorb C18 (Shandon-SFCC)	2.30
Vydac 218 TP (Hesperia group)	1.62
Vydac 201 TP (Hesperia group)	1.41
Partisil ODS 1 (Whatmann)	1.37
Baker WP C18 (Baker)	1.21

which occurs only on this support and not on the other tested columns, is still unknown. The second class, with selectivity parameter α ranging from 1 to 1.1, corresponds to low-density monofunctional bonded silicas (Partisil ODS-3 or Hypersil ODS, for instance). In this case, the interfunctional distance between two bonded chains is large enough to allow carotene to diffuse into the stationary phase, independent of their configuration. The third group consists of high-density monofunctional bonded silica (Ultrabase UB 225, Kromasil) as well as bifunctional bonded silica (Lichrospher). The higher the density of alkyl bonded chains, the more rigid the stationary phase, preventing the carotenes from pene-

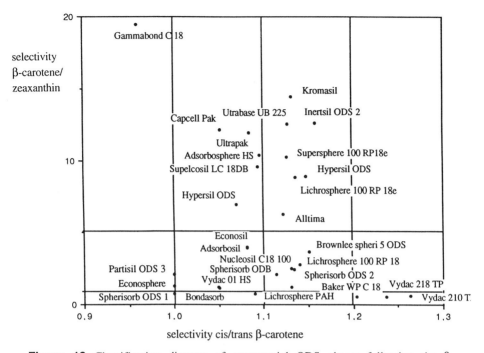

Figure 12 Classification diagram of commercial ODS phases following the β-carotene/zeaxanthin and the *cis/trans*-β-carotene selectivities.

trating between the octadecyl chains. Thus, the interactions developed between the solute and the surface of the stationary phase allow a better separation between rigid linear and rigid bent isomers of β-carotene. Finally, the fourth group is characteristic of the polyfunctional bonded silicas (Vydac 201 TP, Baker WP 300), whose stereochemistry is well adapted to separate compounds with spatial differences, such as the PAHs.

Based on the selectivity of β-carotene/zeaxanthin, three groups of silicas could be determined, according to the following values: $\alpha < 1$, $1 < \alpha < 5$, and $\alpha > 5$. An increase of this selectivity is representative for a minor role of residual silanols. It must be noted that characteristic polyfunctional bonded silicas are

Table 5 Discrimination of the Stationary Phases Based on the *cis/trans*-β-Carotene Selectivity

Kind of support	Selectivity *cis/trans*-β-carotene
Coated	$\alpha < 1.0$
Monofunctional	$1.0 < \alpha < 1.1$
Intermediate	$1.1 < \alpha < 1.2$
Polyfunctional	$1.2 < \alpha$

not supports that avoid silanophilic interactions. In fact, the polymerization reaction should lead to the presence of siloxane groups on the silica surface, which are easily accessible for solutes.

A global examination of the two criteria allows a better characterization of tested columns. For example, it is now possible to distinguish between two low-bonded monofunctional silicas (similar cis/trans selectivity) with similar hydrophobic character (indicated by k' value in Table 4), such as Hypersil ODS and Partisil ODS-3. Indeed, their silanophilic characteristics are very different. Thus, it is easier to choose a support with regard to the required separation based on the structure of analyzed solutes. The efficiency of such a characterization test is interesting in SubFC of nonpolar high molecular weight solutes because the solvophobic and silanophilic power as well as the mono- or polyfunctional nature of the supports can be determined with a single chromatographic injection.

V. CONCLUSION

SubFC using packed columns and a mixed mobile phase is a method of choice for the analysis of complex mixtures of high molecular weight solutes. It has very interesting properties, particularly for thermolabile or very low-water-soluble solutes. This is due to the particular retention behavior of these solutes in the presence of modifiers having a high dielectric constant.

The isoeluent concept, which was first developed with respect to the mobile phase composition, can be extended to temperature and pressure.

The transition from a supercritical type of behavior, in which density is the most important parameter, to a liquid type of behavior depends on the chemical nature of the modifier, as underlined in this chapter.

In addition, a very simple test allowing characterization of different reversed phase supports in a single injection has been described.

Finally, SubFC/ pSFC provides an opportunity to collect nonvolatile fractions after the back pressure regulator. The solutes can be reconcentrated through evaporation of carbon dioxide and are ready for use in molecular characterization or as a standard for other separation methods.

ACKNOWLEDGMENTS

The authors thank Dr. C. Berger and Dr. R. M. Smith for clarifying the text and Dr. C. Berger for helpful suggestions on the presentation of this chapter.

REFERENCES

1. J. B. Crowther and J. D. Hennion, *Anal. Chem.*, *57*:2711 (1987).
2. S. H. Page, H. Yun, M. L. Lee, and S. R. Goates, *Anal. Chem.*, *65*:1493 (1993).
3. T. L. Chester, J. D. Pinkston, and D. E. Raynie, *Anal. Chem.*, *68*:487R (1996).

4. E. Lesellier and A. Tchapla, "Utilisation, caractérisation et compréhension du rôle des mélanges carbon dioxide/modificateurs en chromatographie subcritique sur colonnes remplies," oral communication, SFC 94, Villeurbanne, France (1994).

5. S. H. Page, J. F. Morrison, R. G. Chrisensen, and S. J. Choquette, *Anal. Chem.*, *66*:3553 (1994).

6. T. A. Berger, *J. High Resolut. Chromatogr.*, *14*:312 (1991).

7. T. A. Berger and J. F. Deye, *Chromatographia, 31*:529 (1991).

8. J. W. Ziegler, J. G. Dorsey, T. L. Chester, and D. P. Innis, *Anal. Chem.*, *67*:456 (1995).

9. P. Mourier, M. Caude, and M. Rosset, *Analusis, 13*:299 (1985).

10. J. J. Langerfeld, S. B. Hawthorne, D. J. Miller, and J. Tahrani, *Anal. Chem.*, *64*:2263 (1992).

11. E. Lesellier, A. Tchapla, P. Subra, and R. Tufeu, "Mesure de la masse volumique des mélanges carbon dioxide/modificateurs et relation avec la rétention de molécules biologiques en chromatographie subcritique," Proceedings du 3éme colloque sur les fluides supercritiques, Grasse, France, pp. 257–264 (1996).

12. T. A. Berger and J. F. Deye, *Chromatographia, 30*:57 (1990).

13. M. Ashraf-Khorassani, S. Shah, and L. T. Taylor, *Anal. Chem.*, *62*:1173 (1990).

14. J. G. M. Janssen, P. J. Schoenmakers, and C. A. Cramers, *J. High Resolut. Chromatogr.*, *12*:645 (1989).

15. E. Lesellier, unpublished results.

16. T. A. Berger and W. H. Wilson, *Anal. Chem.*, *63*:1451 (1993).

17. D. E. Martire and R. D. Boehm, *J. Phys. Chem.*, *91*:2433 (1988).

18. E. Lesellier, A. M. Krstulovic and A. Tchapla, *Chromatographia, 36*:275 (1993).

19. D. Upmoor and G. Brunner, *Chromatographia, 33*:261 (1992).

20. E. Lesellier and A. Tchapla, "La chromatographie subcritique à haute teneur en modificateur: une démarche inhabituelle pour de multiples applications," oral communication, Sept. 95, Lyon France (1995).

21. E. Lesellier and A. Tchapla, "Mise au point de l'analyse des triglycerides en chromatographie subcritique sur colonnes remplies," Proceedings du 3éme colloque sur les fluides supercritiques, Grasse, France, pp. 115–126 (1996).

22. P. J. Schoenmakers, *Optimization of Chromatographic Selectivity* (P. J. Schoenmakers, ed.), Elsevier, Amsterdam, Chap. 3, (1986).

23. E. Lesellier, A. M. Krstulovic, and A. Tchapla, *J. Chromatogr.*, *641*:137 (1993).

24. L. C. Sander and S. A. Wise, *LC-GC, 5*:377 (1990).

25. E. Lesellier, A. Tchapla, and A. M. Krstulovic, *J. Chromatogr.*, *645*:29 (1993).

26. E. Lesellier, M. R. Pechard, A. Tchapla, C. R. Lee, and A. M. Krstulovic, *J. Chromatogr.*, *557*:59 (1991).

27. A. Tchapla, S. Heron, E. Lesellier, and H. Colin, *J. Chromatogr.*, *656*:81 (1993).

28. J. P. Goiffon, C. Reminiac, and M. Olle, *Rev. Fr. Corps Gras, 28*:327 (1981).

29. S. Heron and A. Tchapla, *Analusis, 21*:327 (1993).

30. K. K. Unger, *Packings and Stationary Phases in Chromatographic Techniques* (K. K. Unger), Marcel Dekker, New York, Chap 6 (1990).

31. D. Y. Shang, J. L. Grandmaison, and S. Kadiaguine, *J. Chromatogr.*, *672*:185 (1994).

32. E. Lesellier and A. Tchapla, "Description d'un test simple en chromatographie Subcritique pour une multicaractérisation des supports greffés Octadecyl (C 18)," poster A6, Sept. 95, Lyon (1995).

8

Chiral Chromatography Using Sub- and Supercritical Fluids

Roger M. Smith

Loughborough University, Loughborough, Leicestershire, United Kingdom

I. INTRODUCTION

Chiral chromatography is an area where sub- and supercritical fluid chromatography (subFC and SFC) have frequently demonstrated real advantages over high-performance liquid chromatography (HPLC). It has been a principal area of success for SFC and probably the one that has been most widely adopted on the analytical and preparative scale in industrial laboratories. This interest has been fueled by the requirements of regulatory bodies in the pharmaceutical industry for evidence of chiral purity, by the need to examine the rates of degradation and metabolism of individual enantiomers, and by the need to develop rapid methods for chiral separations to yield individual enantiomers on a small scale for biological and chemical testing [1]. Because enantiomeric resolution comes primarily from a close interaction between the analyte and the stationary phase, the normal phase-like properties of most supercritical fluids prompt high interactions and high enantioselectivity. Most of the studies have employed commercial chiral stationary phases (CSP) designed for normal phase HPLC and these usually work equally well or better in packed column SFC (pSFC). However, because of the sensitivity and commercial importance of many of the applications, the great majority have probably gone unreported.

Chiral separations can also be achieved in SFC by using an achiral column and the formation of transient diastereoisomers by the addition of a chiral selector to the mobile phase, such as *N*-benzoxycarbonylglycyl-L-proline (ZGP) for the separation of chiral 1,2-diamines [2] (see also chap. 10). Alternatively, diastereoisomeric precolumn derivatives can be prepared for achiral SFC, including the reactions of chiral alkanols with (*S*)-Trolox methyl ether [3] (*R*)-(-)-1-(1-naphthyl)ethyl isocyanate [4]. Supercritical solvents have also been employed in a number of other areas of chiral studies. There are many cases in which supercritical fluid extraction (SFE) has been used as a mild achiral solvent to isolate chiral analytes before separation by SFC, gas liquid chromatography (GLC) or HPLC. Fogassy and coworkers (5) have also examined the use of supercritical fluids to enhance chiral recognition in the resolution of enantiomers. Enantioselective biological reactions can also be carried out in supercritical solvents and Cernia [6], Glowacz [7], and their coworkers have reported that lipases can selectively hydrolyse esters and lipids in supercritical carbon dioxide.

The methods and applications of chiral SFC on packed columns have been extensively reviewed over the last 7 years, initially by Macaudière and coworkers [8,9] and then by Bargmann and coworkers [10]. Chiral separations on packed and open-tubular columns were compared by Petersson and Markides [11] and Perrut [12] and the application of chiral SFC in the pharmaceutical industry was reported by Anton and coworkers [13] and, more recently, by Berger [14]. Chiral separations have also been included in Chirbase, a database of chiral separation methods [15,16].

Column temperatures of 25°C have been frequently selected in these SFC studies as providing the optimum resolution. Under these conditions, a carbon dioxide–based mobile phase will be below its critical temperature of 31.3°C and will thus be subcritical. However, there are no significant discontinuities in chromatographic properties between sub- and supercritical fluids. In many cases in this review, the term SFC will often be applied to both conditions and subcritical fluid chromatography (subFC) will be used for specific applications as appropriate.

II. STATIONARY PHASES FOR PACKED COLUMN CHIRAL SFC

Since the first studies of chiral SFC were reported in 1985 by Mourier and coworkers using a classic Pirkle phase for the separation of phosphine oxides [17], a wide range of stationary phases has been examined for pSFC. The majority of these phases were either originally developed for HPLC or belong to related structural groups. The details of the materials will only be considered briefly as the chemistry, structures, and liquid phase chromatographic properties of these columns have been extensively compared and reviewed, for example in Refs. 18 and 19. A few studies have examined phases developed specifically for packed column SFC but the general success of the commercial stationary phases has usually meant that few additional advantages could be demonstrated. This contrasts with the situation in open-tubular SFC, where the comparisons have usually been with GLC. Because supercritical fluids have a greater solubilizing power than gases and the columns are usually used at higher temperature than the packed columns (typically 70°C), considerable work has been needed to design specific stationary phases for SFC open-tubular columns, including the immobilization of the chiral selectors on the capillary surface [11].

One difficulty facing the user is that, as with HPLC chiral separations, there are few clear guidelines for the choice of stationary phase for a particular analyte and often more than one phase may need to be examined in order to obtain a suitable resolution. A comparison of different stationary phases employed in pSFC has been reported by Kot et al. [20] and the properties of open-tubular and packed columns have been reviewed by Petersson and Markides [11].

The conditions usually employed in pSFC separations are often very mild with column temperatures typically 40°C or less. As a result, the columns can be very stable and have prolonged lifetimes. Anton and coworkers [13] found reproducible results on a Chiralcel OD column after 690 h of use even when operated at four times the manufacturer's recommended flow rates. In another study, Kot and coworkers [20] found that only one column was needed for a study lasting 9 months and over that period they obtained highly reproducible separations.

The different phases reported for chiral pSFC separation can be divided into three main groups based on polysaccharides, cyclodextrin inclusion complexes, and amino acid and amide derivatives.

A. Polysaccharide-Based Phases

Polysaccharide stationary phases based on cellulose or amylose have proved to be one of the more successful and widely applicable groups of stationary phases for chiral SFC. The polysaccharides are usually derivatized by esterification or carbamate formation to generate a wide range of phases [21] many of which are available commercially. Generally with these phases the addition of alkanol modifiers to the carbon dioxide has a similar effect to their addition to hexane in normal phase HPLC. SubFC generally provided higher efficiency and shorter separation times than the corresponding HPLC separations using the same stationary phase. Most of the published work has concentrated on a small number of selected phases of which the most widely employed has been cellulose tri(3,5-dimethylphenylcarbamate) coated on silica gel (Chiralcel OD) (Fig. 1), which has also been very successful in HPLC separations.

In a typical study, the application of this phase has usually been compared to normal phase HPLC, using hexane-2-propanol as the eluent. For example, Lyman and Nicolas [22] examined the applications of the Chiralcel OD column. They separated *trans*-stilbene oxide, carbobenzyloxy (CBZ) phenylalaninol, and CBZ alanine using carbon dioxide plus methanol, ethanol, or 2-propanol as modifier. The SFC separations yielded a higher resolution than the HPLC separations and had faster equilibration times. The stationary phase was found to be

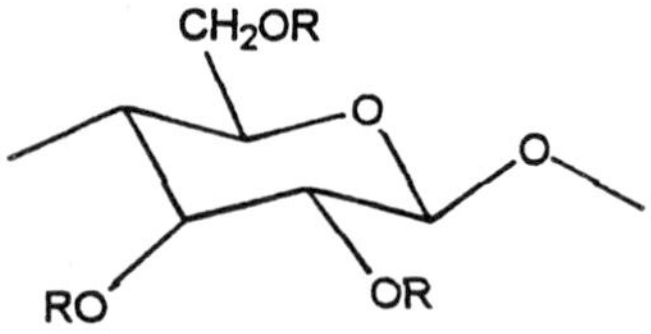

Chiracel OB R = benzoate(COPh)
Chiracel OD R = 3,5-dimethylcarbamate(CONHC$_6$H$_3$(CH$_3$)$_2$)
Chiracel OJ R = 4-methylbenzoate (COC$_6$H$_4$CH$_3$)

Figure 1 Structure of Chiralcel OB, OD, and OJ phases.

stable in both separation methods. Chiralcel OD has also been considered in detail by Anton and coworkers [13]. They examined the separation of 2,2,2-trifluoro-1-(9-anthryl)-ethanol and N-[2-dihydrobenzofuran-3-yl)ethyl]-β-alanine methyl ester. They demonstrated that with subFC sufficient resolution could be obtained for an analytical separation in a much shorter time than in conventional HPLC (Fig. 2). These results contrasted with an earlier study by Lee and coworkers [23], who found little difference between the HPLC and SFC separation of β-blockers on the same stationary phase.

In many separations, the optimization of separation conditions is an important time factor and Stringham and colleagues [24] have shown how subFC on Chiralcel OD could be used for the rapid development of a method for the precursors of an antiviral compound DMP323. The retentions and selectivity on subFC at 25°C and on HPLC were similar but the subFC separation had higher efficiency providing a higher resolution (Fig. 3). They also showed that SFC could be used to also identify a minor enantiomer at the 0.5% level in a partially racemic mixture.

The second most popular phase has been cellulose tribenzoate (Chiralcel OB, Fig. 1), which was also examined by Macaudière [9,25]. They were concerned that as the CSP is only coated on the surface of the silica, it might be eluted from the column but no loss of activity was observed, even up to 200 bar column pressure. The 2-naphthoyl amides of a series of secondary amines were resolved with lower selectivities than in HPLC but the corresponding *para*-substituted benzamide derivatives gave higher selectivities. There was no clear preference whether HPLC or subFC gave better resolutions for a series of lactone and lactams. Different alkanols were compared as eluent modifiers. Branched alkanols generally gave better resolution than *n*-alkanols [9]. They concluded that there were significant differences in the interaction of the alkanol modifiers in pSFC with carbon dioxide and in HPLC using hexane. The selectivity increased with the proportion of alkanol up to a maximum of about

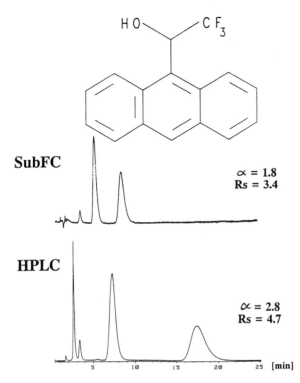

Figure 2 Comparison of HPLC and subFC separation of 2,2,2-trifluoro-1-(9-anthryl)-ethanol on a Chiralcel OD column. SubFC conditions, carbon dioxide-ethanol (90:10), 225 bar at room temperature. HPLC conditions, hexane-2-propanol (90:10). (Reprinted from Ref. 13 with permission of Elsevier Science NL, Amsterdam.)

0.25×10^{-3} mol of alkanol per gram of mobile phase. It decreased at higher concentrations suggesting that by then the alkanol was saturating the chiral sites on the column diminishing the stereoselective interactions. At very low alkanol concentrations, the selectivity again decreased suggesting a different mechanism of modification to liquid chromatography. The principal effect of SFC was much more rapid separations, for example, an α-methylene-γ-lactone could be resolved by SFC in 4 min with an improved resolution ($Rs = 1.2$) compared to 40 min in HPLC ($Rs = 0.9$).

Smith and Ma [26] found that increasing the proportion of modifier reduced the retention factors of a number of model and drug compounds but had little effect on selectivity for both supercritical and subcritical separations on Chiralcel OD or Chiralcel OB columns. In other studies, Kaida and Okamoto [27] found poorer separations using SFC for a number of enantiomers on three cellulose phenylcarbamate phases and on an amylose phenylcarbamate phase

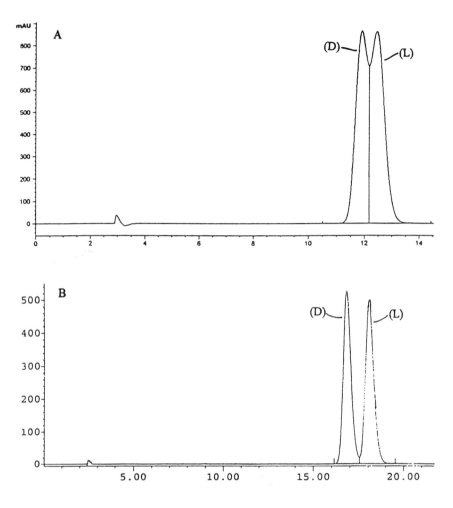

Figure 3 Comparisons of separation of a racemic mixture of Q8266 by (A) HPLC with hexane-ethanol (85:15) and (B) subFC with carbon dioxide-ethanol (90:10) at 40°C on a Chiralcel OD column. (Reprinted with permission from Ref. 24 © 1994, American Chemical Society.)

than with HPLC using hexane-2-propanol. However, the SFC separations were slightly better on a cellulose 4-methylbenzoate phase. More recently, Whatley [28] also compared the Chiralcel OD column and the corresponding amylose tri(3,5-dimethylphenylcarbamate) (Chiralcel AD) column for the resolution of glibenclamide derivatives. Medvedovici and coworkers [29] have compared these columns for the separation of D,L- and meso-N,N'-(CBZ)-2,6-diamino-

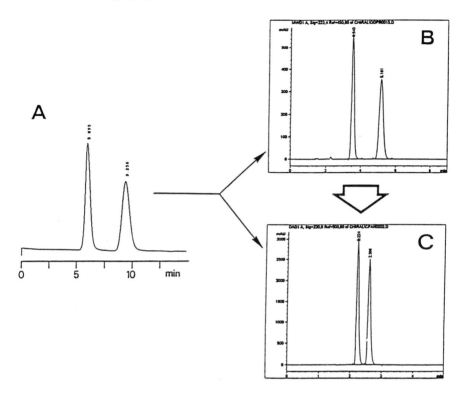

Figure 4 Comparison of separations of propranolol on (A), HPLC on Chiralcel OD column with hexane-2-propanol (75:25) plus 0.5% diethylamine; (B), on Chiralcel OD; and (C), Chiralpak AD columns both with carbon dioxide-methanol (70:30) with 0.5% diethylamine [20]. (Reproduced from *J. Chromatograph Sci.* by permission of Preston Publications, a division of Preston Industries Inc.)

pimelic acids. The application of the amylose phases in SFC has also been compared with the cellulose phases by Bargmann-Leyder and coworkers [30]. Kot and coworkers [20] found a faster separation for propranolol on both Chiralcel OD and AD using subFC compared to HPLC (Fig. 4). They attributed the increased resolution to a higher selectivity for the analytes with carbon dioxide compared to hexane rather than changes in the physical state of the eluent.

Another derivative cellulose tri(4-methylbenzoate) (Chiralcel OJ, Fig. 1) was used by Siret and coworkers [31] for the separation of dihydropyridines used as calcium channel blockers using binary and ternary eluents (carbon dioxide-ethanol-isobutanol 84:8:8). A noncommercial tri(phenylcarbamate)cellulose coated on silica gel was used to separate *trans*-stilbene oxide by Nitta and coworkers [32] at 0, 25, and 40°C and they obtained a better separation than with normal phase chromatography using hexane as the eluent.

B. Cyclodextrin-Based Inclusion Phases

The second major group of chiral stationary phases, that have been also widely used in HPLC, are α-, β-, or γ-cyclodextrin (CD) inclusion columns. One attraction in HPLC was that they could be used within reversed phase eluents, broadening their range of applications. Their selectivity is based on the presence of a chiral hydrophobic cavity within the cyclodextrin, which can trap the analyte or a part of it from a relatively polar eluent, and a ring of chiral secondary hydroxyl groups around the rim of the cavity (Fig. 5) [18,19].

The first studies with these materials in pSFC were carried out by Macaudière and coworkers [9,33], who examined the β-cyclodextrin (Cyclobond I) columns. They observed that in normal phase HPLC nonpolar mobile phases, such as hexane, can often occupy the internal cavity and block analyte interactions. There was therefore concern that supercritical carbon dioxide, being relatively nonpolar, would also occupy this site and block the entry of analytes. However, they obtained SFC chiral resolution for phosphine oxides and 2-naphthoyl amides with a better separation than in normal phase HPLC [33]. They concluded that although the carbon dioxide could enter the cavity, its molecules were small and could be easily displaced from the cavity. Subsequently they separated 3-thienylcyclohexylglycolic acid [25] on the cyclodextrin column but could not resolve the corresponding methyl ester, although this could be resolved on a Chiralcel OB column. They suggested [33] that the chiral recognition required two stages, an interaction with

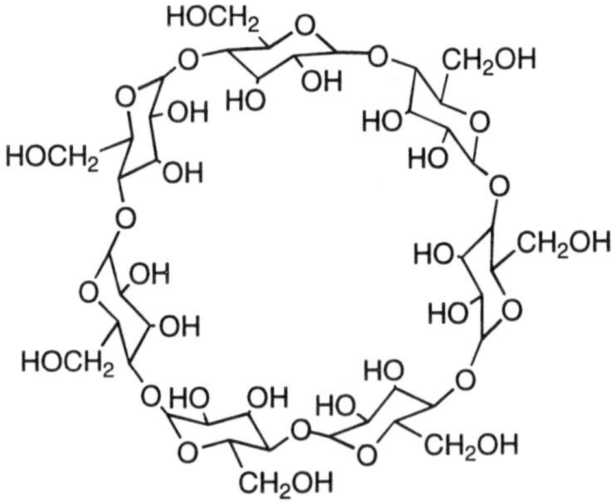

Figure 5 β-Cyclodextrin stationary phase material.

the secondary hydroxyl groups on upper edge of the CD cavity followed by inclusion complex formation. In a comparison [9] of the separation of phosphine oxides and β-cyclodextrin (Cyclobond I), α-cyclodextrin (Cyclobond III), γ-cyclodextrin (Cyclobond II), and an acetylated cyclodextrin, the Cyclobond I column always provided the highest enantioselectivity. The acetylated material showed a much reduced interaction suggesting that the analytes needed to "see the secondary hydroxyl groups on the edge of the cavity." However, Kot and coworkers [20] found that for a series of pharmaceuticals, no separation could be obtained on the β-CD phase for nonsteroidal antiinflammatory agents (NSAIDs) or benzodiazepines, with only a poor resolution for β-blockers.

A specially synthesized DMP (dimethylphenyl carbamylated) cyclodextrin phase has been designed for open tubular chiral separations but was also examined as a coating on packed columns [34]. However, when the column was compared using reversed phase and normal phase HPLC and SFC conditions, the latter performed poorly and the best separations were obtained under normal phase conditions. Smith and Li have found that a β-RN CD [(R)-(-)-1-naphthyl ethyl isocyanate–derivatized cyclodextrin] phase could be used to separate ibuprofen and 2,2,2-trifluoro-1-(9-anthryl)ethanol but a untreated α-cyclodextrin column would only separate benzoin (with a high resolution, Rs up to 6.05) [26].

Many of these alkylated or acylated cyclodextrins were first developed for chiral gas chromatography. Their first use in capillary SFC was by Schurig and coworkers who immobilized permethylcyclodextrin (Chirasil-Dex) [35]. Subsequently they considerably extended these studies to a wide range of open-tubular chromatography. Shen and Lee have also examined the use of a number of bonded CD phases encapsulated on silica in packed capillary columns for SFC (see Chapter 5).

C. Stationary Phases Based on Amides and Amino Acids

These materials can be divided into three groups: (1) those specifically designed to provide multipoint interactions, the π-acid and π-base phases primarily based on the work of Pirkle; (2) the amino acid-based columns, and finally (3) diamide phases, whose chirality is derived from a enantiomeric 1,2-aminocyclohexane ring. These materials are also widely known in HPLC [36] where an abundance of commercial columns is available. Although they can work well and provide high resolution, they often require a fairly specific structural match between the stationary phases and analyte to obtain a strong interaction. When this interaction is present the enantioselectivity can be high but column selection can be more difficult than with the more generally applicable cellulose-based columns.

1. π-Acid and π-Base Materials

These stationary phase materials usually incorporate ligand groups, which possess either electron donor or acceptor groups, coupled with hydrogen-bonding groups arranged so that multiple interactions can occur. Originally considered to require three or more specific interactions for chiral analytes, some two-point interactions can also yield resolution. The active groups include dipoles, such as the carbonyl group, hydrogen-bonding groups including amino groups, and typically a π-π complexation between an electron-rich naphthalene group and an electron-withdrawing dinitrophenyl group. Commonly known as Pirkle columns after their originator or brush columns because the active groups stand out from the silica surface, they were among the first chiral stationary phases to be examined by SFC.

The initial studies in chiral SFC by Mourier and coworkers [17] were based on a Pirkle-type column containing (R)-N-(3,5-dinitrobenzoyl)phenylglycine bonded to an amino-bonded silica gel (Fig. 6). The best separations were found at subcritical temperatures in the presence of a low polarity alcohol (2-propanol). They found a minimum in the plate height h vs. density curve; the density had to be greater than 0.7 or significant losses occurred. This work was rapidly followed by further studies on the effect of eluent and modifiers on the separation of naphthoyl amides of 2-aminoalkanes [33,37]. The separations in SFC were generally similar to those found in normal phase HPLC with hexane-alcohol mobile phases. Because of better mass transfer properties in subcritical carbon dioxide mobile phases, SFC usually gave higher column efficiencies and more plates in unit time [37]. The resolution of enantiomers usually improved as the temperature decreased.

There were, however, some differences. For example, it was noted that increasing the chain length of analytes caused a systematic reduction in retention in HPLC but the higher homologs were more highly retained in subFC [37]. In the latter case, the effect is presumably a function of the decrease in the volatility of the analytes with increasing molecular weight. This study was

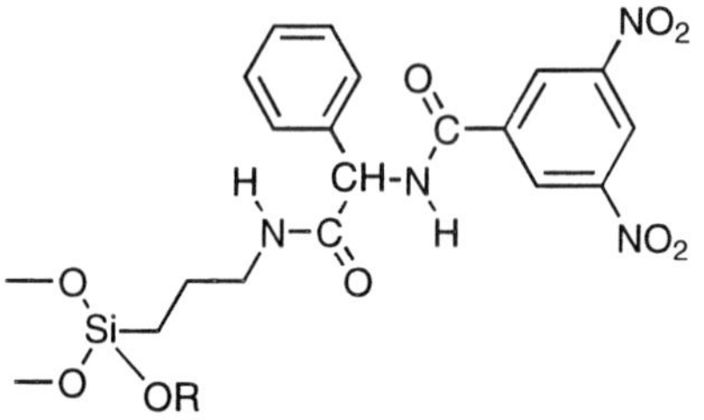

Figure 6 (R)-N-(3,5-dinitrobenzoyl)phenylglycine bonded to an amino-bonded silica gel—a typical π-acceptor stationary phase.

followed by the examination of dinitrobenzoyl (DNB) derivatives of amino acids, amino alcohols, and amides [38]. The effect of different spacer chains between the silica surface and the enantioselectivity groups for phases based on phenylalanine and phenylglycine [39] have been studied by Bargmann-Leyder and coworkers.

This type of column can be readily prepared synthetically and a number of related columns have been examined for use in SFC. However, many show few advantages except for specific analytes and as a consequence have not been applied to real samples. One group of these specialized stationary phases is the 3,5-dinitrobenzoyltyrosine based columns (ChyRoSine-A) materials, which were developed and studied by Siret and coworkers [40]. The influence of the substitution on the system and of the chain length of the spacer on these columns was determined [10,39,41]. The use of these tyrosine based stationary phases in HPLC and SFC has been reviewed [42].

In other studies [43] β-lactones have been used to prepare 1,1'-butylnaphthol amide derivatives for the analysis of drug compounds using carbon dioxide-2-propanol. A π-donor phase based on 3,5-dinitropivaloylnaphthylethylamide coated onto silica has been used to separate dinitrobenzoyl amides [44].

More recently, a new series of stationary phases has been developed based on a combined π-acid and π-base chiral selector [45,46]. They have been used to separate a range of analytes on both preparative and analytical scales, and are now available commercially as the Whelk-O 1 phase. Their development as rationally designed CSPs and their properties and structures are discussed in detail in Chapter 9.

The mechanism of the chiral separation in SFC with the Pirkle types of chiral stationary phase has been examined in detail by Edge and coworkers using molecular modeling [47]. They compared the energies of full range of possible configurations of the interactions between the enantiomers of 2,2,2-trifluoro-1-(9-anthryl)-ethanol on a (R)-3,5-dinitrobenzoylphenylglycine CSP and of (R/S)-N-(3,5-dinitrobenzoyl)-α-methylbenzylamine on a (R)-phenylurea CSP. In a similar study by Bargmann-Leyder and coworkers [41] using NMR spectroscopy and limited molecular modeling the interactions between ChyRoSine-A and (R)-propranolol have been examined and it was determined that the presence of carbon dioxide altered the configuration of the analyte.

Other workers have used columns based on ureas to separate DNB derivatives of amines. They found that the selectivity was higher if the silica was end-capped [48]. A further group of phases has been the (S)-naphthylureas, which have been used by Peytavin and coworkers for the assay of antimalarials by SFC [49].

2. Amino Acid-Based Materials

Because the amino acid groups can be readily used to impart a chiral center to a stationary phase, many researchers have used substituted amino acids as amides or ureas to prepare synthetic stationary phases for chromatography. However, as with many of the synthetic Pirkle phases few have gained wide usage. Dobashi and coworkers separated derivatized amino acids [50] by SFC using a column prepared from a valine diamide linked to silica gel with a decamethylene spacer chain. They compared the differences between methanol and propanol as additives. The latter gave better resolution but was less efficient. This material was further developed by Hara and coworkers [51,52] as (N-formyl-L-valylamino) propylsilica phases and the effects of different modifiers and different spacer chains were examined.

3. Diamide-Based Materials

Rather than employing an amino acid to impart chirality to a stationary phase, Gasparrini and coworkers formed diamides from chiral (R,R)-1,2-diaminocyclohexanes (DACH) to give DACH-N-N'-3,5-dinitrobenzoyl (DNB) phase [53–55] (for structure see Fig. 11) and DACH-acryloyl (ACRYL) phases, which were examined under super- and subcritical conditions, including using the ternary eluent, carbon dioxide-methanol-dioxane (80:10:10), to optimize the separation.

D. Other Stationary Phase Materials

1. Helical Polymers

Chiral groups can also be created by restricted movement, such as in helical polymers. These materials have been used to develop the (+)-polytriphenyl methyl methacrylate coated stationary phases (Chiralpak OT(+)) [9]. This column was designed to resolve enantiomers containing aromatic groups or atropoisomers such as bi-β-naphthols. Because neat methanol or acetonitrile is needed as the mobile phase in HPLC, Macaudière found that it was difficult to directly compare the SFC and HPLC properties because of the marked differences in the eluents required. In SFC α-methylene-γ-lactones and bi-β-naphthols were examined and they concluded that, as found by Okamoto and Hatada [56] in HPLC, two different chiral selective sites were present, firstly the polymeric chains and secondly the cavities formed between the chains.

2. Porous Graphitic Carbon-Based Phases

Wilkins and coworkers [57] have shown that chiral selectors, such as (R,R)-(+)-tartarate derivatives of anthrylamines, can be physically adsorbed on the surface of a porous graphitic carbon (Hypercarb) to generate a stable chiral stationary phase that can be used in SFC for the separation of ibuprofen and flurbiprofen. The separations were better than those obtained for the same selector on a silica surface.

III. SEPARATION CONDITIONS

A. Mobile Phases and Specific Interactions

The great majority of SFC separations and virtually all the published chiral SFC work has used carbon dioxide as the primary mobile phase component, often with alkanol modifiers for more polar analytes. Typically between 5% and 15% of methanol, ethanol, or 2-propanol has been employed (for example, Fig. 7). However, different researchers have found that the choice and proportion of alkanol are sample and stationary phase specific, and no general rules have been developed [20].

Acids and bases can be also added to the mobile phases to control stationary phase and analyte ionization. For example, isopropylamine was added by Anton

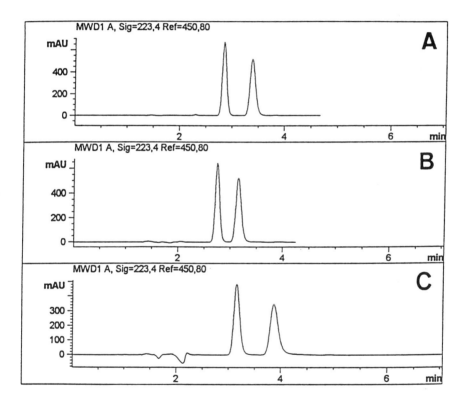

Figure 7 Effect of different alkanol modifiers on the separation of mianserin on a Chiralcel OD column. Conditions: carbon dioxide-modifier plus 0.5% isopropylamine (70:30) at 30°C and 200 bar. Modifier A, methanol; B, ethanol; C, 2-propanol [20]. (Reproduced from *J. Chromatograph Sci.* by permission of Preston Publications, a division of Preston Industries Inc.)

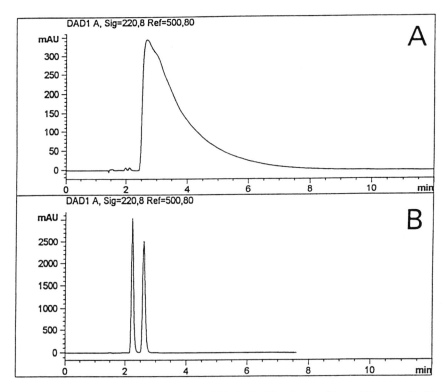

Figure 8 Separation of propranolol on Chiralpak AD column without (A) and with (B) 0.5% diethylamine in the carbon dioxide-methanol (70:30) eluent at 30°C and 200 bar [20]. (Reproduced from *J. Chromatograph. Sci.* by permission of Preston Publications, a division of Preston Industries Inc.)

and coworkers [13] to suppress ionization and improve the peak shape of the piperazine drug, hydroxyzine, on a Chiralcel OD column. A dramatic effect of a pH modifier was demonstrated by Kot and coworkers [20] for propranolol, particularly with the Chiralcel AD column, where the peaks had poor shapes and were unresolved until 0.5% diethylamine was included in the carbon dioxide eluent (Fig. 8). For acids, trifluoroacetic acid could be added, and it greatly improved the peak shape of β-agonists on Chirex 3022 columns.

It has been suggested that for specific analytes the chiral selectivity has been enhanced by an interaction or complexation between the analyte and carbon dioxide, which enhances the chiral selectivity [40,59]. This was confirmed by NMR studies by Bargmann-Leyden for selected β-blockers [41] and the authors suggested that the same effect could be used to explain unusual separations observed by previous workers [23]. Anton [13] also noted effects with phenylalaninol where changes in the NMR spectrum were observed in carbon dioxide, especially for the benzylic protons. However, unlike the β-blockers, the interaction resulted in a loss of resolution for phenylalaninol and a basic drug compound [(2R,4S)-2-benzyl-1-(3,5-dimethylbenzoyl)-N-(4-quinoline-methyl)-4-piperidineamine] CGP 49823 on a Chiralcel OD column. In contrast, these compounds were well resolved by HPLC. They suggested that the shifts in the ^{1}H-NMR spectra might be caused by salt formation between the carbon dioxide and primary amines. They concluded that the effects of these interactions were unpredictable.

B. Temperature Effects

Temperature has an important role in chiral separations of any kind but is particularly important in SFC because of the effect on the eluent density and elution strength. However, altering the flow rate or pressure of the separation had little effect on resolution [13]. The principal effect as the temperature is increased is usually to reduce retention times and resolution. This primarily results from a reduction in the interaction of the enantiomers with the stationary phase. Although there is normally an increase in column efficiency with temperature, the loss of enantioselectivity is more significant. As a consequence, many of the pSFC studies have suggested the use of the minimum reasonable temperature, usually 25°C, resulting in subcritical conditions.

In a typical example, the effect of temperature was examined by Nishikawa in his studies of synthetic pyrethroids on a Pirkle-type (Sumichiral OA-4000) column [59]. He found that the enantiomeric resolution increased as the temperature was reduced and that there was a direct relationship between ln α and $1/T$ for fenpropathrin (Fig. 9), which agreed with theory. He analyzed the differences in terms of the changes in the enthalpy and entropy of the interactions with the column. The order of elution was the same as on LC suggesting a similar mechanism of chiral recognition. The best separation was obtained using subcritical FC at 25°C using carbon dioxide-hexane (0.7%).

Kot [20] found that the lower temperatures had no effect on selectivity and resolution but were needed with 3-hydroxybenzodiazepines (such as oxazepam and lorazepam) to avoid thermal racemization. Interconversion occurred on Chiralcel OD columns at room temperature using carbon dioxide-methanol as the modifier. The corresponding 3-acetates were thermally stable. Because racemization was promoted by protic solvents, better results were obtained with carbon dioxide plus ethanol-acetonitrile as modifier (Fig. 10) [20]. Further improvement could be obtained by reducing the temperature and the proportion of ethanol in the modifier. Partial replacement of the ethanol by acetonitrile, an aprotic solvent, gave optimum conditions of complete baseline resolution.

The use of carbon dioxide as the mobile phase also means that it is possible to carry out assays at subambient temperatures, down to −50°C, which can be

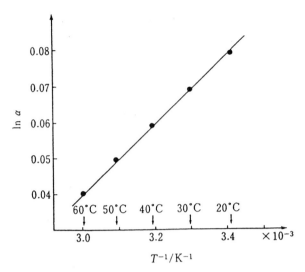

Figure 9 Temperature dependence of the separation factor (α) of fenpropathrin on a Sumichiral OA-4000 column with carbon dioxide at 170 kg cm^{-2} as the eluent [59]. (Reproduced with permission of the publisher.)

valuable for enantiolabile analytes. Gasparrini and coworkers used a DACH-DNB column to examine thermally enantiolabile 2-methyl-1-(2,2-dimethyl-propanoyl)naphthalene, which can undergo rotation around the CO-C$_{Ar}$ bond. No resolution was obtained at room temperature; when the temperature was reduced to −50°C, resolution of the enantiomers was obtained (Fig. 11) [54]. Reducing the temperature resulted in negligible degradation in column performance. This ability of SFC to separate thermally labile compounds is often claimed as one of its advantages over GLC or HPLC. However, apart from the above there are few practical examples.

Although most chiral separations improved as the temperature was reduced, this does not occur in all cases. Macaudière observed with an Chiralpak OT column that the resolution of bi-β-naphthol decreased on increasing the temperature but that of α-methylene-γ-lactone went to a maximum and then decreased [9]. Contrasting experiences were observed by Smith, et al. [60] in a detailed study of temperature effects. They found that the resolution of a benzamide derivative of a cyanoamine increased but that of the pentanoyl amide derivative decreased with increasing temperature (Fig. 12). They suggested that there are differences in the nature of the interaction of the enantiomers with the stationary phase. They compared separations by HPLC and SFC and observed the presence of an isoenantioselective temperature (T_{iso}) at which $Rs = 0$. This temperature

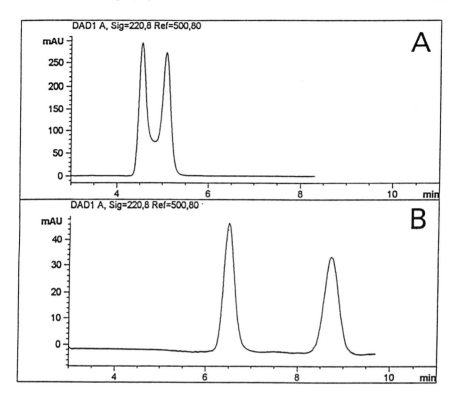

Figure 10 Effect of an aprotic eluent on enantiostability of oxazepam on a Chiralcel OD column. Conditions: 40°C and 200 bar. (A), Eluent carbon dioxide-methanol (70:30) with 0.5% diethylamine showing racemisation. (B), Eluent carbon dioxide-ethanol-acetonitrile (70:10:20) with 0.5% diethylamine [20]. (Reproduced from *J. Chromatograph. Sci.* by permission of Preston Publications, a division of Preston Industries Inc.)

differed between the compounds in the study so that at room temperature one was above and one below this value. As a consequence, while in one case resolution increased with temperature, for the second compound the reverse was found. These differences were ascribed to differences in the $\Delta\Delta H°$ values for the interaction of the analytes with the stationary phase.

The effects of temperature were recently discussed in detail in a paper by Stringham and Blackwell [61] who have studied "entropically driven" separations. They reported a number of examples, including chlorophenylamide whose enantiomers reversed their elution order between −10 °C, 70°C (T_{iso}), and 190°C (see Fig. 4 in Chapter 9).

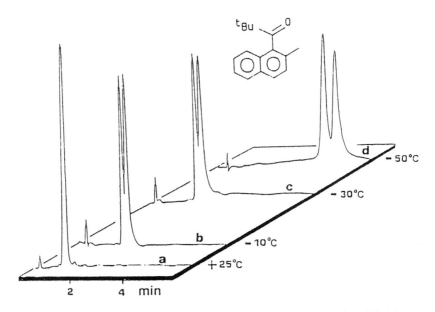

Figure 11 Dynamic chromatography of an enantiolabile-hindered naphthyl ketone (I) at subambient temperatures. Eluent carbon dioxide-2-propanol (95:5): column DACH-DBN [54]. (Reproduced with permission of the publisher.)

IV. APPLICATIONS

Many of the published papers on packed column chiral SFC compare the resolutions with those obtained in HPLC. Usually a separation can be obtained in both cases. However, there have been a limited number of samples for which resolution can only be obtained by one of the techniques. The SFC method will

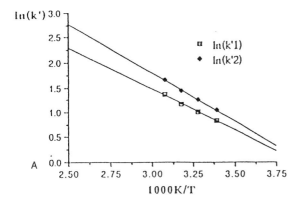

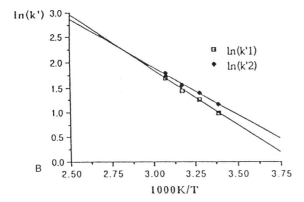

Figure 12 Effect on changes in retention factors with temperature for (A), benzamide and (B), pentanoylamide derivatives substituted compounds on Chiralcel OB column depending on isoenantioselective temperature [60]. (Reprinted from Ref. 60 with permission of Elsevier Science NL Amsterdam.)

usually provide a separation with high resolution, in less time or under milder conditions of lower temperatures.

Although authors have frequently claimed significant advantages for SFC over HPLC methods, a cautionary note has been expressed by Massey and Tandy [62], who commented that they had found no examples where SFC worked but HPLC did not. They felt that overall SFC offered little advantage over HPLC in the analytical laboratory. Anton and coworkers also noted that "the superior characteristics of these fluids do not guarantee superior performance to HPLC. In several instances apparent interactions . . . can produce less resolution of the chiral compounds by SFC or subFC than by HPLC" [13].

A. Pharmaceuticals

As might be expected, by far the majority of real samples separated by chiral SFC have come from the pharmaceutical field. In most cases, these have been of physiologically active drugs but it is difficult to tell from the references whether many of these methods have been adopted for routine quality control measurement or were primarily examined as an alternative to HPLC methods. Many of the methods were examples used to determine if resolution could be obtained and if SFC offered any advantages, such as speed, over HPLC separations. Some studies have examined the application of SFC for the determination of trace impurities in chiral analytes including enantiomeric impurities but little work has been carried out on metabolites, primarily because the aqueous sample medium is unsuitable for direct injection onto SFC. The final area of importance is preparative separation (see also Chapters 3 and 14).

The separation of β-blockers has been of particular interest and it has been suggested that specific solvation or salt formation by the carbon dioxide can lead to enhanced separations [10,13,40,58,63]. For example, β-blockers were separated on ChyRoSine-A [40] and subsequently propranolol and pindolol were resolved on an improved version of the phase [58]. This work also demonstrated the ability of SFC to separate even minor proportions of minor enantiomers during studies of enantiomeric purity (Fig. 13). A comparison of the different stationary phases was carried out by Kot and coworkers [20], who successfully separated β-blockers, including propranolol and metoprolol, on both Chiralcel OD and AD columns (Fig. 4). The former column provided the best resolution but the latter the faster elution times. Modifiers played an important part and separation of the β-blockers was possible on Chiralcel OD with methanol alone but on the Chiralcel AD column the peaks tailed unless 0.5% base was added (Fig. 8). Oxazolidine-2-one derivatives of β-receptor blockers have also been resolved [55].

Analgesics, such as ibuprofen and flurbiprofen, have been separated on Chiralpak AD using 5% methanol in carbon dioxide [64], on a DACH-DNB column [53], or on a Whelk-O 1 column [45]. Other drugs that have been examined, include warfarin [46,65], alprenolol [53], antimalarials including chloroquine, primaquine, mefloquine, quinine, and quinidine [49], abscissic acids [45,46], a calcium channel blocker [31], a series of analogs of the potassium channel activator Cromakalim [60], a potassium channel blocker [28], oxazepam [9], the benzodiazepine drugs lorazepam, lormetazepam, and temazepam [26,66], hexobarbitone [26]; and, recently, a drug targeted for cardiac arrhythmia [67]. The enantiomeric forms of camazepan and a number of its metabolites and derivatives have been separated by Wang and coworkers by both achiral and chiral SFC [68]. They reported that separations by SFC on a Chiralcel OD-H column were better resolved and required less time than the corresponding HPLC separation. Recently, Alasandro has reported the separa-

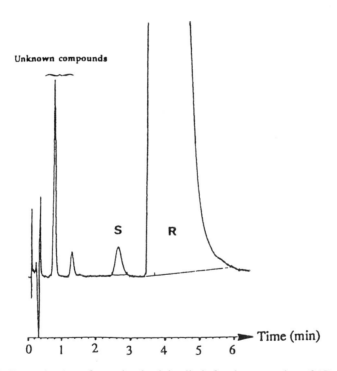

Figure 13 Determination of enantioselectivity limit for the separation of (S)-propranolol (0.05%) in the presence of (R)-propranolol (99.95%) on a butyl-substituted tyrosine-based column (structure shown R = *tert*-butyl) with carbon dioxide-ethanol (containing 1% *n*-propylamine) (90:10) at 0°C and 200 bar [58]. (Reproduced with permission of the publisher.)

tion of diastereoisomers of DuP 105, an oxazolidone antibacterial agent, also on a Chiralcel OD column [69].

A comprehensive study was carried out by Kot et al. [20], who examined a range of drugs including β-blockers, β-agonists, NSAIDs, and benzodiazepines. They found that none of the compounds was fully resolved on a cyclodextrin column and only the β-blockers showed any separation. The β-agonists showed no separation on the Chiralcel OD, Chiralpak AD, or on an (R)-naphthylglycine 3,5-dinitrobenzamide column (Chirex 3005). On a urea brush-type π-donor phase (Chirex 3022), clenbuterol, salbutamol, and Org-B showed resolutions better than 2. The basic drugs, oxazepam, lorazepam, mianserin, and Org-B, were baseline-separated on Chiralcel OD. The NSAIDs (ibuprofen, ketoprofen, flurbiprofen, and fenoprofen) could be resolved on the Chirex 3005 column but the resolution on the Chiralpak AD column was much better and the retention times were shorter. They concluded that the best results for basic drugs were obtained on the Chiralcel OD column, for the acidic NSAIDs on Chiralpak AD, and for the acidic β-agonists on Chirex 3022. They examined the experimental parameters in detail and compared the effect of flow rate, modifier, and temperature and of acidic and basic additives. However, no general rules could be concluded and the choice of alkanol modifier depended very much on the analyte and stationary phases. Usually these columns are used separately but Kot [20] also examined the linking of three columns of different types— Chiralpak AD, Chiralcel OD, and Chirex 3022—which is possible because the low viscosity of supercritical fluid results in only a low pressure drop across each column. This combination enabled a number of different types of analytes to be resolved in a single run, each of which was only separated on a single phase (Fig. 14; for further examples of separations see Chapter 6, Figs. 12–14).

Further examples of the use of chiral and achiral pSFC in the development of drugs are described in Chapter 10.

B. Other Analytes

Chiral separations have also been an important area in the development of agrochemicals, where the chirality of an analyte can play an important role in its biological activity. As well as formulation and product quality studies, this area also provides the need for trace and environmental analysis. In one of the few studies on agrochemicals, Nishikawa [59] examined the separation of synthetic pyrethroids. He was able to separate both enantiomers and diastereoisomers of fenvalerate and enantiomers of fenpropathrin.

A widely studied group of analytes has been the enantiomeric phosphine oxides [8,9,17,33]. They were among the first compounds to be separated by chiral SFC and this work has been followed by a series of related studies including the separation of 1,2,5-triphenylphospholanes [70]. The alkyl phenyl sulfoxides have also been separated [9,45,53,55].

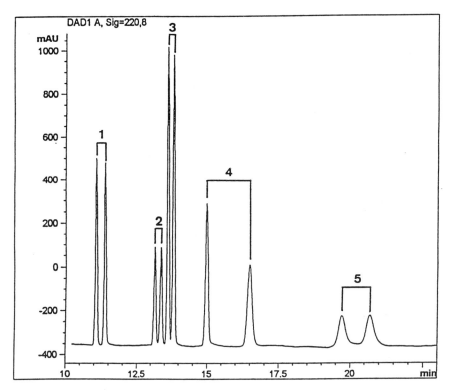

Figure 14 Chromatogram obtained by the use of three columns in series to achieve the separation of drugs of different structural types in a single separation (Chiralpak AD—Chiralcel OD—Chirex 3022). Eluent carbon dioxide-methanol (containing 0.5% triethylamine and 0.5% trifluoroacetic acid), gradient elution from 4% to 5%, 25°C, and 200 bar. Peaks: 1, ibuprofen; 2 fenoprofen; 3, clenbuterol; 4, propranolol; and 5, lorazepam [20]. (Reproduced from *J. Chromatograph. Sci.* by permission of Preston Publications, a division of Preston Industries Inc.)

Many authors have also examined simple test compounds, particularly when new stationary phases are being examined. Typically these include *trans*-stilbene oxide [22,32], benzoin [13,26,27,61], 1,1-bi-2-naphthols [8,9,43,55], and numerous examples of 2,2,2-trifluoro-1-(9-anthryl)-ethanol [for example, 12,13,27]. In a study to determine the mode of interaction of analytes with the stationary phases, Smith and Ma examined a series of substituted phenylalkanols and related compounds [26]. They found (Table 1) that on Chiralcel OD and Chiralcel OB columns, the separation was enhanced when the chiral center carried both the hydroxyl group and phenyl groups.

One group of chiral analytes of particular interest is the amino acids. Because of their ionization and polarity, they have usually been examined as

Table 1 Comparison of Separation of Phenylpropanols and Related Analytes on Chiralcel OD and OB Columns [26]

	Column			
	Chiralcel OB		Chiralcel OD	
Compound	k	α	k	α
1-Phenyl-1-propanol	0.65	1.12	1.47	1.20
1-Phenyl-2-propanol	0.60	1.00	1.39	1.00
2-Phenyl-1-propanol	1.05	1.00	1.69	1.00
1,2,3,4-Tetrahydro-1-naphthol	1.22	1.24	2.44	1.10
1-Phenyl-1,2-ethanediol	1.54	1.13	5.88	1.00
Styrene oxide	0.68	1.00	0.72	1.00
Benzoin	1.27	1.11	—	—

Conditions: eluent, carbon dioxide-methanol (93:7); temperature, 40°C; pressure, 125 bar.

derivatives by SFC. For example, the group of Hara and Dobashi have separated N-acetyl amino acid tert-butyl esters [52] and N-4-nitrobenzoyl amino acid isopropyl esters [50]. CBZ derivatives of amino acids [22,45] and of diaminopimelic acid [29] have been separated. N-3,5-Dinitrobenzoyl and N-acetyl methyl esters were examined by Brügger and coworkers [44]. Amines have also required derivatization before examination, for example, as 3,5-dinitrobenzoyl amides [44] or 2-naphthoyl amides [37]. Phenylalaninol [13] and Z-phenylalaninol [61] have been resolved.

C. Preparative Separations

Because of the ease with which the mobile phase can be removed after a separation and the mild separation conditions, there has been considerable interest generally in the use of SFC for preparative separations [70] (see also Chapters 3, 14, and 15. Preparative chiral chromatography has been used to separate the 1,2,5-triphenylphospholanes [12,45,72]. Fuchs and coworkers [69] separated phosphine oxides, using a Pirkle phase and carbon dioxide-ethanol. They obtained resolutions of $\alpha = 1.15–1.35$ and up to 100 mg an hour could be separated.

This is a field of particular interest in the pharmaceutical laboratory as the separated product can be virtually solvent-free and a small amount of the enantiomers can be obtained quite rapidly in good yields. They could scale up SFC to 200 mg warfarin [46]. Verapramil isomers were separated using a mixture of acetic acid and 2-propanol in carbon dioxide [45]. The effects of scaling up a separation were examined by Kot and coworkers [20] for the separation of mianserin on a Chiralcel OD column and for flurbiprofen on a Chiralpak AD column. Up to 250 µg of the former could be separated on each run with 100 µg collection. In the latter case, better resolution ($Rs = 1.8$) enabled 1 mg to be separated per run even though the efficiency was reduced ($N = 300$). No peak overlap occurred and the fractions were obtained in 99.8% enantiomeric purity.

V. CONCLUSION

Supercritical and subcritical fluids have been found to demonstrate valuable properties for the separation of enantiomers on a wide range of analytical columns. They frequently provide a higher resolution of enantiomers in shorter times than for the corresponding HPLC separations. This has lead to their application to a wide range of pharmaceutical compounds and other analytes. The high volatility of the carbon dioxide eluent has prompted their application for the preparative separations of enantiomers.

REFERENCES

1. W. H. de Camp, *Chirality*, *1*:2 (1989).
2. W. Steuer, M. Schindler, G. Schill, and F. Erni, *J. Chromatogr.*, *447*:287 (1988).
3. W. Walther, W. Vetter, and T. Netscher, *J. Microcol. Sep.*, *4*:45 (1992).
4. K. Sakaki and H. Hirata, *J. Chromatogr.*, *585*:117 (1991).
5. E. Fogassy, M. Ács, T. Szili, B. Simándi, and J. Sawinsky, *Tetrahedron Lett.*, *35*:257 (1994).
6. E. Cernia, C. Palocci, F. Gasparrini, D. Misiti, and N. Fagnano, *J. Mol. Catalysis*, *89*:L11 (1994).
7. G. Glowacz, M. Bariszlovich, M, Linke, P. Richter, C. Fuchs, and J. T. Morsel, *Chem. Phy. Lipids*, *79*:101 (1996).
8. P. Macaudière, M. Caude, R. Rosset, and A. Tambuté, *J. Chromatogr. Sci.*, *27*:383 (1989).
9. P. Macaudière, M. Caude, R. Rosset, and A. Tambuté, *J. Chromatogr. Sci.*, *27*:583 (1989).
10. N. Bargmann, A. Tambuté, and M. Caude, *Analusis*, *20*:189 (1992).
11. P. Petersson and K. E. Markides, *J. Chromatogr. A.*, *666*:381 (1994).
12. M. Perrut, *J. Chromatogr. A*, *658*:293 (1994).
13. K. Anton, J. Eppinger, L. Frederiksen, E. Francotte, T. A. Berger, and W. H. Wilson, *J. Chromatogr.*, *666*:395 (1994).
14. T. A. Berger, Packed Column SFC, Royal Society of Chemistry, Cambridge (1995).
15. B. Koppenhoefer, R. Graf, H. Holzschuh, A. Nothdurft, U. Trettin, P. Piras, and C. Roussel, *J. Chromatogr. A*, *666*:557 (1994).
16. B. Koppenhoefer, A. Nothdurft, J. Pierrotsanders, P. Piras, C. Popescu, C. Roussel, M. Stiebler, and U. Trettin, *Chirality*, *5*:213 (1993).
17. P. A. Mourier, E. Eliot, M. H. Caude, R. H. Rosset, and A. G. Tambuté, *Anal. Chem.*, *57*:2819 (1985).
18. A. M. Krstulovic, *Chiral Separations by HPLC*, Ellis Horwood, Chichester (1989).
19. S. G. Allenmark, *Chromatographic Enantioseparations*, 2nd ed., Ellis Horwood, Chichester (1991).
20. A. Kot, P. Sandra, and A. Venema, *J. Chromatogr. Sci.*, *32*:439 (1994).

21. Y. Okamoto and Y. Kaida, *J. Chromatogr. A.*, *666*:403 (1994).

22. K. G. Lynam and E. C. Nicolas, *J. Pharm. Biomed. Anal.*, *11*:1197 (1993).

23. C. R. Lee, J. P. Porziemsky, M.-C. Aubert, and A. M. Krstulovic, *J. Chromatogr.*, *539*:55 (1991).

24. R. W. Stringham, K. G. Lynam, and C. C. Grasso, *Anal. Chem.*, *66*:1949 (1994).

25. P. Macaudière, M. Caude, R. Rosset, and A. Tambuté, *J. Chromatogr.*, *450*:255 (1989).

26. R. M. Smith and L. Ma, *J. Chromatogr.* (submitted).

27. Y. Kaida and Y. Okamoto, *Bull. Chem. Soc. Jpn.*, *65*:2286 (1992).

28. J. Whatley, *J. Chromatogr. A*, *697*:251 (1995).

29. A. Medvedovici, P. Sandra, A. Kot, and A. Kolodziejczyk, *J. High Resolut. Chromatogr.*, *19*:227.

30. N. Bargmann-Leyder, A. Tambuté, and M. Caude, *Chirality*, *7*:311 (1995).

31. L. Siret, P. Macaudière, N. Bargmann-Leyder, A. Tambuté, M, Caude, and E. Gougeon, *Chirality*, *6*:440 (1994).

32. T. Nitta, Y. Yakushijin, T. Kametani, and T. Katayama, *Bull. Chem. Soc. Jpn.*, *63*:1365 (1990).

33. P. Macaudiè M. Caude, R. Rosset, and A. Tambuté, *J Chromatogr.*, *405*:135 (1989).

34. T. Hargitai and Y. Okamoto, *J. Liq. Chromatogr.*, *16*:843 (1993).

35. V. Schurig, Z. Juvancz, G. J. Nicholson, and D. Schmalzing. *J. High Resolut. Chromatogr.*, *14*:58 (1991).

36. W. H. Pirkle and T. C. Pochapsky, *Chem. Rev.*, *89*:347 (1989).

37. P. Macaudière, A. Tambuté, M. Caude, R. Rosset, M. A. Alembik, and I. W. Wainer, *J. Chromatogr.*, *371*:177 (1986).

38. P. Macaudière, M. Lienne, M. Caude, R. Rosset, and A. Tambuté, *J. Chromatogr.*, *467*:357 (1989).

39. N. Bargmann-Leyder, J. C. Truffert, A. Tambuté, and M. Caude, *J. Chromatogr. A*, *666*:27 (1994).

40. L. Siret, N. Bargmann, A. Tambuté, and M. Caude, *Chirality*, *4*:252 (1992).

41. N. Bargmann-Leyder, C. Sella, D. Bauer, A. Tambuté, and M. Caude, *Anal. Chem.*, *67*:952 (1995).

42. M. Caude, A. Tambuté, and L. Siret, *J. Chromatogr*, *550*:357 (1991).

43. D. Lohmann and R. Däppen, *Chirality*, *5*:168 (1993).

44. R. Brügger, P. Krähenbühl, A. Marti, R. Straub, and H. Arm, *J. Chromatogr.*, *557*:163 (1991).

45. A. M. Blum, K. G. Lynam, and E. C. Nicolas, *Chirality*, *6*:302 (1994).

46. G. J. Terfloth, W. H. Pirkle, K. G. Lynam, and E. C. Nicolas, *J. Chromatogr. A*, *705*:185 (1995).

47. A. M. Edge, D. M. Heaton, K. D. Bartle, A. A. Clifford, and P. Myers, *Chromatographia*, *41*:161 (1995).

48. R. Brügger and H. Arm, *J. Chromatogr.*, *592*:309 (1992).

49. G. Peytavin, F. Gimenez, B. Genissel, C. Gillotin, A. Bailett, I. W. Wainer, and R. Farinotti, *Chilarity*, *5*:173 (1993).

50. A. Dobashi, Y. Dobashi, T. Ono, S. Hara, M. Saito, S. Higashidate, and Y. Yamauchi, *J. Chromatogr.*, *461*:121 (1989).

51. S. Hara, A. Dobashi, T. Hondo, M. Saito, and M. Senda, *J. High Resolut. Chromatogr. Commun.*, *9*:249 (1986).

52. S. Hara, A. Dobashi, K. Kinoshita, T. Hondo, M. Saito, and M. Senda, *J. Chromatogr.*, *371*:153 (1986).

53. F. Gasparrini, D. Misiti, and C. Villani, *J. High Resolut. Chromatogr.*, *13*:182 (1990).

54. F. Gasparrini, F. Maggio, D. Misiti, C. Villani, F. Andreolini, and G. P. Mapelli, *J. High Resolut. Chromatogr.*, *17*:43 (1994).

55. F. Gasparrini, D. Misiti, and C. Villani, *Trends Anal. Chem.*, *12*:137 (1993).

56. K. Okamoto and K. Hatada, *J. Liq. Chromatogr.*, *9*:369 (1986).

57. S. M. Wilkins, D. R. Taylor, and R. J. Smith, *J. Chromatogr. A*, *697*:587 (1995).

58. N. Bargmann-Leyder, D. Thiébaut, F. Vergne, A. Bégos, A. Tambuté, and M. Caude, *Chromatographia*, *39*:673 (1994).

59. Y. Nishikawa, *Anal. Sci.*, *9*:33 (1993).

60. R. J. Smith, D. R. Taylor, and S. M. Wilkins, *J. Chromatogr. A*, *697*:591 (1995).

61. R. W. Stringham and J. A. Blackwell, *Anal. Chem.*, *68*:2179 (1996).

62. P. R. Massey and M. J. Tandy, *Chirality*, *6*:63 (1994).

63. P. Biermanns, C. Miller, V. Lyon, and W. Wilson, *LC-GC*, *11*:711 (1993).

64. W. H. Wilson, *Chirality*, *6*:216 (1994).

65. N. Bargmann-Leyder, A. Tambuté, A. Bégos, and M. Caude, *Chromatographia*, *37*:433 (1993).

66. R. M. Smith and L. Ma, Proceedings of the 2nd European Symposium on Analytical Supercritical Fluid Chromatography and Extraction, Riva del Garda, 27–28th May 1993.

67. T. Loughlin, R. Thompson, G. Bicker, P. Tway, and N. Grinberg, *Chirality*, *8*:157 (1996).

68. M. Z. Wang, M. S. Klee, and S. K. Yang, *J. Chromatogr. B, Biomed. Appl.*, *665*:139 (1995).

69. M. Alasandro, *J. Pharm. Biomed. Anal.*, *14*:807 (1996).

70. G. Fuchs, L. Doguet, D. Barth, and M. Perrut, *J. Chromatogr.*, *623*:329 (1992).

71. E. Francotte, *J. Chromatogr. A*, *666*:565 (1994).

72. J. C. Fiaud and J. Y. Legros, *Tetrahedron Lett.*, *35*:5089 (1991).

9

Enantiomer Separation by Sub- and Supercritical Fluid Chromatography on Rationally Designed Chiral Stationary Phases

Christian Wolf and William H. Pirkle

University of Illinois, Urbana, Illinois

I. INTRODUCTION

Preparative separations of enantiomers and determinations of enantiomeric purity are of great interest, e.g., in asymmetrical synthesis, pharmaceutical and toxicological studies of chiral bioactive compounds, and stereochemical analysis of biomolecules. Especially in the pharmaceutical industry, a detection limit of 0.1% is required. In the precise determination of minute enantiomeric impurities, enantioselective chromatography offers many advantages relative to chiroptical methods and to nuclear magnetic resonance (NMR) spectroscopy using chiral shift reagents. Moreover, neither of the latter techniques provide a means of separating enantiomers.

Accordingly, a variety of chiral stationary phases (CSPs) have been developed for enantioselective high-performance liquid chromatography (HPLC) or gas chromatography (GC). Compared to HPLC, subcritical fluid chromatography (subFC) and supercritical fluid chromatography (SFC) on packed columns can provide higher efficiency, shorter analysis time, faster method development, and reduced usage of organic combustible solvents. In contrast to GC, the separation of thermally labile and nonvolatile analytes is also possible. Mourier et al. were the first to separate some phosphine oxides by packed column SFC (pSFC) in 1985 [1]. They used the commercially available (*R*)-*N*-(3,5-dinitrobenzoyl)phenylglycine CSP covalently bonded to aminopropyl silica and developed by Pirkle et al. for HPLC [2]. Consequently, a broad variety of CSPs originally developed for HPLC purposes have successfully been introduced to packed column subFC and pSFC (Chapters 3, 6, 8, 10, and 14).

Most CSPs have been empirically developed and optimized using the chiral pool of natural products, e.g., cyclodextrins and polysaccharides. Some of these CSPs are very useful and broadly applicable in chiral pSFC but do not allow a comprehensive understanding of the enantiodiscrimination process or any prediction of enantioseparations, i.e., the user has to carry out troublesome trial-and-error procedures to find a suitable CSP.

II. RATIONALLY DESIGNED CSPs

In a more systematic approach, small synthetic molecules can be used to study the chiral interactions between a selector and one enantiomer of the analyte to develop rationally designed CSPs. Several methods, such as ultraviolet absorption spectroscopy, NMR spectroscopy, structure–activity relationships, and X-ray structure determination of selector analyte complexes, are useful for studying chiral recognition mechanisms. Additionally, Pirkle and coworkers introduced the concept of reciprocity to screen effective selectors for a CSP [2]. This concept is based on the assumption that a single enantiomer of a given racemate that is well separated on a CSP can be used as an efficacious selector in a new CSP for the resolution of the enantiomers of compounds that are similar in their structure to the selector of the original CSP (Fig. 1).

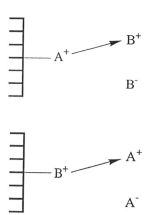

Figure 1 Concept of reciprocity. An immobilized chiral selector A^+ is used to separate the enantiomers B^+ and B^- to screen selectors of interest. It is presumed that immobilization of one well-separated enantiomer, B^+, will provide an efficacious CSP for the separation of the enantiomers of A.

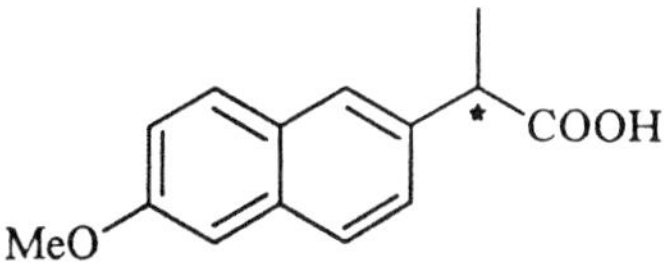

Figure 2 Structure of naproxen.

A. Enantioseparation on a Broadly Applicable Brush-Type CSP in pSFC

Systematic studies of the chiral recognition mechanism and the concept of reciprocity have been used by Pirkle and coworkers to develop and optimize dozens of CSPs [3]. Accordingly, a CSP derived from the (S) enantiomer of the nonsteroidal antiinflammatory drug (NSAID) naproxen (Fig. 2) as well as NMR studies have been used to design the commercially available Whelk-O 1 [4–6]. The Whelk-O 1 has been shown to be broadly applicable in the resolution of the enantiomers of a variety of different compound classes [4,7–10]. This selector contains a π-basic aromatic group and a π-acidic 3, 5-dinitrobenzoylamide and thus represents a combination of π-basic and π-acidic brush-type CSPs (Fig. 3).

The π-acidic and π-basic aromatic systems are held perpendicular to each other and form a flexible cleft in which the analyte is held by simultaneous face-to-face and face-to-edge interactions. Additionally, hydrogen bonding to the amide N-H and steric interactions occur near the stereogenic center. In general, compounds with an aromatic group and a hydrogen bond acceptor near a chiral center, axis, or plane can be separated into the enantiomers. Since the Whelk-O 1 is not only available in the (3R, 4S) form but also in the enantiomeric (3S, 4R) form, the elution order of the enantiomers can easily be controlled. This is a great advantage for trace enantiomer analysis with respect to sensitivity and

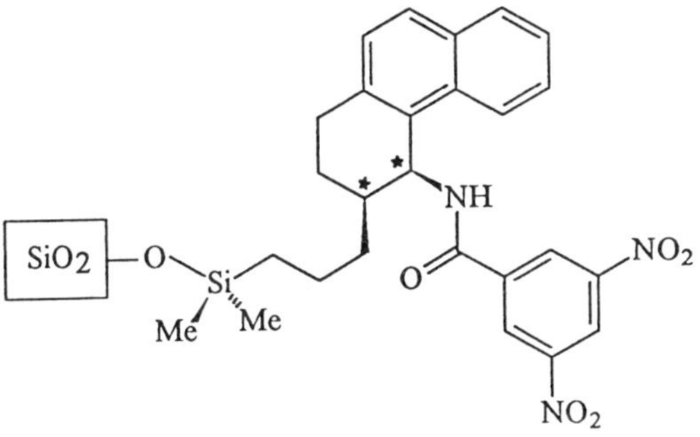

Figure 3 Structure of (3R, 4S)-Whelk-O 1.

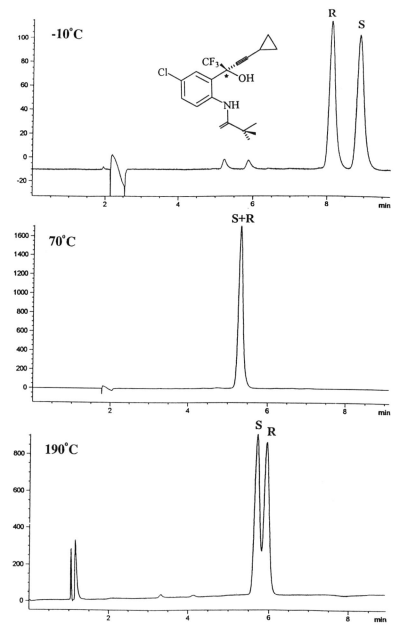

Figure 4 pSFC separation of an aromatic amide at −10, 70, and 190°C on (S,S)-Whelk-O 1 (250 × 4.6 mm). Operating conditions: Mobile phase: carbon dioxide containing 10% ethanol; flow rate: 1.5 mL/min; back pressure: 300 bar; UV detection at 210 nm. (From Ref. 12.)

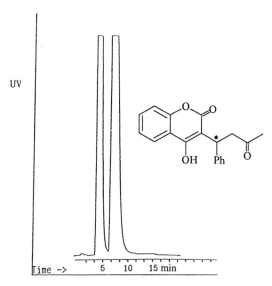

Figure 5 Preparative pSFC separation of warfarin on Whelk-O 1 (250 × 25.4 mm). Operating conditions: Mobile phase: carbon dioxide containing 25% isopropanol and 0.5% acetic acid; loading: 200 mg; flow rate: 100 mL/min; back pressure: 253 bar; temperature: 25°C; UV detection at 260 nm. (From Ref. 13.)

precision as it is especially required in the pharmaceutical and agrochemical industry; see also Chapter 10 [5,6,11]. Stringham et al. determined the isoenan tioselective temperature and reversal of the elution order of several test solutes in pSFC by using the brush-type Whelk-O 1 at temperatures up to 190°C, thus proving the thermal stability of this CSP (Fig. 4) [12].

Blum et al. noted a long column life under subFC conditions and showed that Whelk-O 1 is of great utility for the separation of enantiomeric compounds with different structures, such as aromatics, carboxylic acids, alcohols, ketones, epoxides, esters, ethers, carbamates, amines, sulfoxides, and sulfonamides, in analytical as well as in preparative scales [13]. Comparison with HPLC revealed a general superiority of subFC in terms of higher speed, efficiency, and faster method development.

pSFC is also suitable for enantioseparations on a preparative scale. Compared to preparative HPLC, the use of subFC or SFC on packed columns results in easier product and solvent recovery, reduced solvent waste and cost, and higher output per unit time. Due to reduced sample capacity, pSFC usually allows the separation of 10–100 mg per run. The usefulness of preparative pSFC is discussed in detail in Chapters 3 and 15. Nevertheless, Blum et al. reported on a preparative separation of the enantiomers of warfarin with a loading of 200 mg on the Whelk-O 1 in 10 min (Fig. 5) [13].

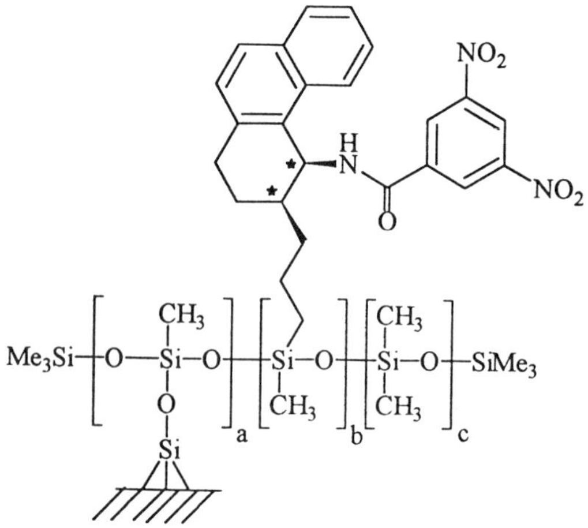

Figure 6 Structure of (3R, 4S)-PolyWhelk-O.

B. A Broad-Spectrum Polysiloxane-Based CSP for pSFC Separation of Enantiomers

Recently, Terfloth et al. reported on the incorporation of the Whelk-O selector and two of its homologs into polymethylsiloxanes and immobilization on silica gel. These polysiloxane side chain–modified CSPs (Fig. 6) display the physical and chemical robustness to withstand high pressures, temperatures, and different mobile phase additives [14].

Among the CSPs evaluated, the commercially available PolyWhelk-O proved to give the best results for the enantiomer separations of profens and several test compounds, i.e., an increase of the tether length or the introduction of methyl substituents in the π-basic moiety of the chiral selector are detrimental to enantioselectivity under subFC conditions. This is partly in contrast with

Table 1 Enantioseparation of Arylpropionic Acids and Several Test Solutes on PolyWhelk-O (250 × 4.6 mm)

Sample	Condition	k'_1	α	Sample	Condition	k'_1	α
Naproxen	A	1.25	2.28	Etodolac	A	0.49	1.20
Ibuprofen	A	0.09	1.44	*trans*-Stilbene oxide	B	0.28	3.57
Fenoprofen	A	0.14	1.71	Styrene oxide	B	0.07	1.7
Flurbiprofen	A	0.18	1.39	Methyl phenyl sulfoxide	B	0.92	1.41
Cicloprofen	A	1.29	1.75	Warfarin	B	3.29	2.04
Pirprofen	A	0.83	1.40	Benzoin	B	0.29	3.04
Carprofen	A	3.09	1.79	Abscissic acid	B	0.40	1.45

Operating conditions: Mobile phase: carbon dioxide containing 10% (0.2% acetic acid) isopropanol (A) and 10% isopropanol (B); flow rate: 2 mL/min; back pressure: 200 bar; temperature: 25°C; UV detection at 220 mm. (From Ref. 14.)

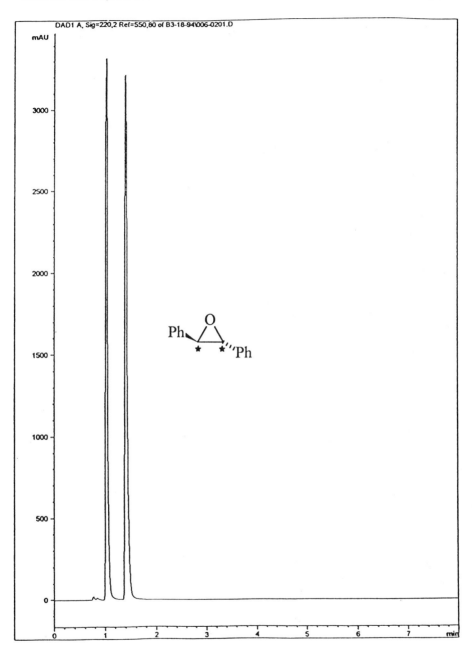

Figure 7 Enantioseparation of *trans*-stilbene oxide on PolyWhelk-O (250 × 4.6 mm) in SubFC. Operating conditions: Mobile phase: carbon dioxide containing 10% isopropanol; flow rate: 2 mL/min; back pressure: 200 bar; temperature: 25°C; UV detection at 254 nm. (From Ref. 14.)

Table 2 HPLC and subFC Separation of the Enantiomers of Several α-Naphthyl-1-ethylamine Carbamates on PolyWhelk-O (250 × 4.6 mm)

R	HPLC			SubFC		
	k'_1	α	Rs	k'_1	α	Rs
n-propyl	0.18	4.17	9.61	0.72	6.06	24.9
i-propyl	0.17	4.47	9.56	0.66	6.44	25.5
neopentyl	0.19	4.63	9.58	0.63	5.52	19.0
n-octyl	0.25	4.44	11.5	0.87	5.98	25.0

Operating conditions: HPLC: Mobile phase: 100% methanol; flow rate: 2 mL/min; temperature: 27°C; UV detection 254 nm. SubFC: Mobile phase: carbon dioxide containing 20% methanol; flow rate: 2 mL/min; back pressure: 200 bar; temperature: 27°C; UV detection 254 nm. (From Ref. 15.)

investigations of the corresponding brush-type CSPs where methyl substituents in the naphthyl group increase stereoselectivity. Chromatographic data for the separation of the enantiomers of several profens and some test solutes are given in Table 1. As was stated for the brush-type CSPs, column efficiency is improved and analysis times are shorter in subFC than in normal phase HPLC.

Table 3 Separation of the Enantiomers of 3-(1-Naphthyl)-4-methylene-1,3-oxazolidin-2-one at Different Temperatures on PolyWhelk-O (250 × 4.6 mm)

Temp (°C)	k'_1	k'_2	α	Rs
60	0.54	0.94	1.74	4.86
40	0.49	0.98	2.00	6.73
28	0.49	1.08	2.20	8.51
0	0.56	1.56	2.79	11.9
-20	0.66	2.24	3.39	13.8

Operating conditions: Mobile phase: carbon dioxide containing 10% 3:1 methanol/acetonitrile + 0.25% diethylamine; flow rate: 2 mL/min; back pressure: 200 bar; UV detection 254 nm. (From Ref. 15.)

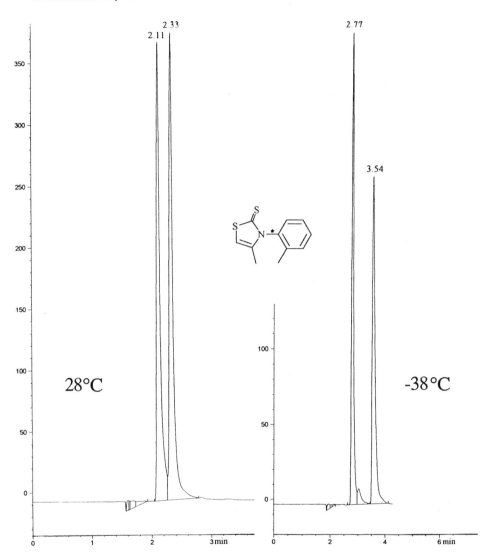

Figure 8 Resolution of 3-(2-methylphenyl)-4-methylthiazoliden-2-thione on Poly-Whelk-O (250 × 4.6 mm) at −38 and 28°C. Operating conditions: Mobile phase: carbon dioxide containing 2% methanol; flow rate: 2 mL/min; back pressure: 200 bar; UV detection at 254 nm.

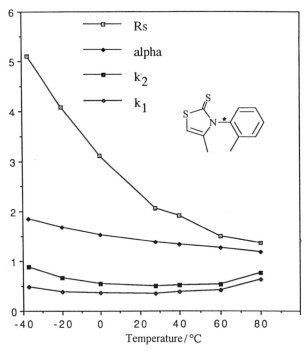

Figure 9 Variable-temperature separation of the enantiomers of 3-(2-methylphenyl)-4-methylthiazoliden-2-thione on PolyWhelk-O (250 × 4.6 mm). Operating conditions: Mobile phase: carbon dioxide containing 2% methanol; flow rate: 2 mL/min; back pressure: 200 bar; UV detection at 254 nm.

As is shown in Fig. 7, the time needed for the resolution is often in the 2–3 min range.

Several phosphonates were used to compare the performance of the Whelk-O 1 and the PolyWhelk-O columns under the same conditions in HPLC. Systematically, the PolyWhelk-O exhibited higher enantioselectivity, reduced capacity factors, and shorter retention times due to reduced surface area and to more complete coverage of residual silanol groups that retain but do not distinguish between the enantiomers [15]. Because of a change of physical and chemical properties of the mobile phase, a direct comparison of HPLC and subFC is somewhat difficult. Nevertheless, resolution and enantioselectivity of enantiomers on the Whelk-O 1 or the PolyWhelk-O was found to be superior under subFC conditions in most cases, as is exemplarily shown for some α-naphthyl-1-ethylamine carbamates in Table 2.

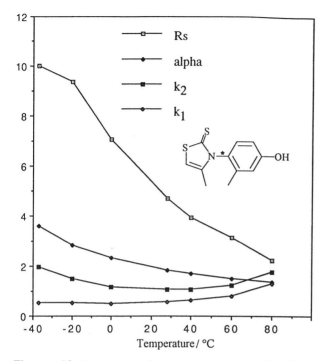

Figure 10 Separation of the enantiomers of 3-(2-methyl-4-hydroxyphenyl)-4-methylthiazoliden-2-thione at different temperatures on PolyWhelk-O (250 × 4.6 mm). For operating conditions, see Fig. 9.

Like its brush-type analog, the PolyWhelk-O proves to be very useful for separating the enantiomers of a variety of compounds of different classes in subFC and can be used over a wide temperature range. As is shown for several heterocyclic atropisomers, the use of cryogenic temperatures may be advantageous for the enantioseparation of configurationally labile compounds or to improve enantioselectivity, as is shown in Table 3 and Figs. 8–9 and 10. Resolution of the atropisomers increases as a consequence of enhanced selectivity at lower temperatures.

In contrast to polysaccharides or protein-derived phases, CSPs based on small selectors provide fast adsorption-desorption kinetics. Accordingly, high efficiency may be retained even at cryogenic temperatures, resulting in both increased enantioselectivity and increased resolution. It also should be noticed that under subFC conditions the retention of both enantiomers decreases with an increase in temperature owing to better solvability in the mobile phase. How-

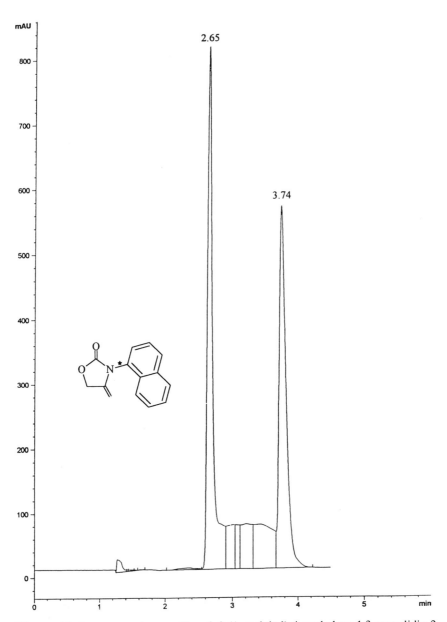

Figure 11 Interconversion profile of 3-(1-naphthyl)-4-methylene-1,3-oxazolidin-2-one on PolyWhelk-O (250 × 4.6 mm) at 80°C. Operating conditions: Mobile phase: carbon dioxide containing 10% methanol; flow rate: 2 mL/min; back pressure: 200 bar; UV detection at 254 nm.

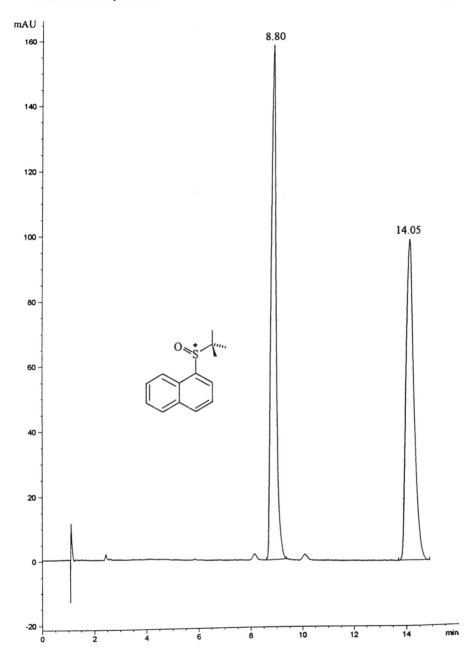

Figure 12 Chiral separation of tert-butyl-1-naphthylsulfoxide on PolyWhelk-O (250 × 4.6 mm). Operating conditions: Mobile phase: carbon dioxide containing 5% methanol; flow rate: 2 mL/min; back pressure: 200 bar; temperature: 90°C; UV detection at 254 nm.

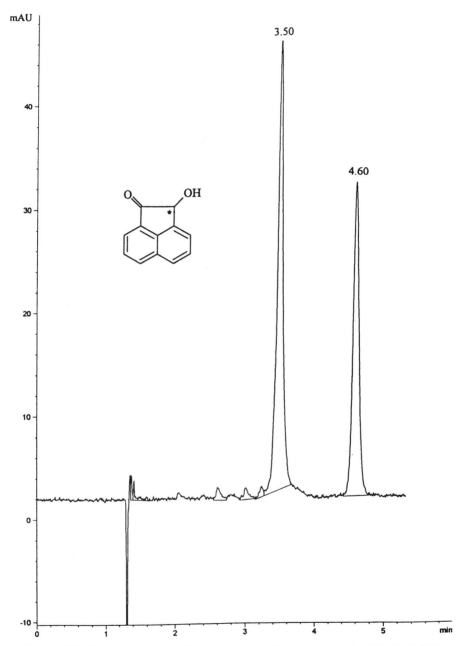

Figure 13 Enantioseparation of 2-hydroxyacenaphthen-1-one on PolyWhelk-O (250 × 4.6 mm). Operating conditions: Mobile phase: carbon dioxide containing 20% methanol; flow rate: 2 mL/min; back pressure: 200 bar; temperature: 26°C; UV detection at 254 nm.

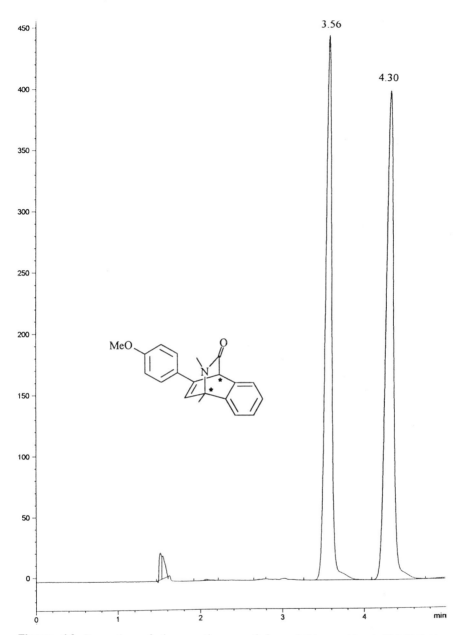

Figure 14 Separation of the enantiomers of 2-aza-5,6-benzo-bicyclo[2.2.2]-8-(4-methoxyphenyl)-1,2-dimethyl-7-octene-3-one on PolyWhelk-O (250 × 4.6 mm). Operating conditions: Mobile phase: carbon dioxide containing 10% methanol; flow rate: 2 mL/min; back pressure: 200 bar; temperature: 28°C; UV detection at 254 nm.

Table 4 Separation of the Enantiomers of Some Hydantoins on PolyWhelk-O (250 × 4.6 mm)

R₁	R₂	Condition	k'₁	α	Rs
Cl	H	A	3.04	1.72	9.52
OH	H	B	0.36	3.03	8.46
OMe	H	B	0.43	3.21	12.4
H	Me	A	1.50	1.13	1.38
OMe	i-Pr	B	0.28	2.79	5.63

Operating conditions: Mobile phase: carbon dioxide containing 5% methanol (A) and 20% methanol (B); flow rate: 2 mL/min; back pressure: 200 bar; temperature 28°C; UV detection 254 nm. (From Ref. 15.)

ever, in the SFC mode, retention can increase at higher temperatures as a result of the lower density and reduced solvation power of the mobile phase. The influence of temperature on the separation of chiral compounds using other CSPs is discussed in Chapter 8.

In contrast to subFC under cryogenic conditions, high temperatures may be necessary either to investigate the rate of interconversion of atropisomers or to work above the isoenantioselective temperature. As is shown in Fig. 11, competitive interconversion and resolution of 3-(1-naphthyl)-4-methylene-1,3-

Table 5 Separation of the Enantiomers of Chlorohydrins on PolyWhelk-O (250 × 4.6 mm)

Ar	k'₁	α	Rs
4-methoxy	0.74	1.36	2.07
2-(5-methoxynaphthyl)	1.98	1.62	8.08
2-(4-methoxynaphthyl)	1.01	1.29	3.66
phenyl	0.49	1.16	0.78

Operating conditions: Mobile phase: carbon dioxide containing 1% methanol; flow rate: 2 mL/min; back pressure: 200 bar; temperature 26°C; UV detection 254 nm. (From Ref. 15.)

Table 6 Enantioseparation of Several Arylsulfoxides on PolyWhelk-O (250 × 4.6 mm)

R	Ar	k'_1	α	Rs
dodecyl	phenyl	1.98	1.35	4.90
isopropyl	4-methylphenyl	1.97	1.24	3.36
isopropyl	4-methylphenyl	0.76	1.08	0.78
tert-butyl	phenyl	1.56	1.28	3.67
methyl	4-chlorophenyl	0.81	1.10	1.03
methyl	perchlorophenyl	3.34	1.60	9.67
methyl	1-naphthyl	4.29	1.24	4.10
methyl	2-naphthyl	4.41	1.22	3.99

Operating conditions: Mobile phase: carbon dioxide containing 5% methanol (A) and 10% methanol (B); flow rate: 2 mL/min; back pressure: 200 bar; temperature: 26°C; UV detection 254 nm. (From Ref. 15.)

oxazolidin-2-one on the column yields in temperature-dependent plateaus. The corresponding ratational energy barrier can be obtained by computer simulation of the elution profiles [16,17].

To prove the broad applicability of the PolyWhelk-O, a variety of different compounds, e.g., carbamates, amides, hydantoins, sulfoxides, ketones, alcohols, heterocycles, and chlorohydrines, have been separated by pSFC, chromatograms, and retention data are exemplarily shown in Figs. 12–14 and Tables 4–8. The PolyWhelk-O is also suitable for the resolution of some underivatized

Table 7 Chiral Separation of a Series of 1-Arylalkanols on PolyWhelk-O (250 × 4.6 mm)

R	Ar	k'_1	α	Rs
n-butyl	1-naphthyl	4.87	1.14	2.78
methyl	1-naphthyl	5.21	1.09	2.12
ethyl	1-naphthyl	4.85	1.09	1.90
isopropyl	3,4-dimethoxyphenyl	5.25	1.07	1.14
methyl	1-(2,3-dimethylnaphthyl)	9.64	1.18	3.52
methyl	2,4,6-trimethylphenyl	1.57	1.13	1.38

Operating conditions: Mobile phase: carbon dioxide containing 1% methanol; flow rate: 2 mL/min; back pressure: 200 bar; temperature: 0°C; UV detection 254 nm. (From Ref. 15.)

Table 8 Separation of the Enantiomers of some 3,5-Dinitrobenzamides on PolyWhelk-O (250×4.6 mm)

Ar	Condition	k'_1	α	Rs
4-methylphenyl	A	0.91	2.47	13.1
4-nitrophenyl	A	0.66	2.08	8.09
2-methylphenyl	A	0.76	2.07	8.83
3,4-dichlorophenyl	A	0.90	3.38	20.9
3,4-dimethylphenyl	A	1.04	3.14	18.9
4-chlorophenyl	A	0.83	2.92	15.6
3,5-bis(trifluoro-methyl)phenyl	B	0.78	2.13	8.79
3,4-dichlorophenyl	B	0.60	10.1	30.5

Operating conditions: Mobile phase: carbon dioxide containing 20% methanol (A) and 5% methanol (B); flow rate: 2 mL/min; back pressure: 200 bar; temperature 26°C; UV detection 254 nm. (From Ref. 15.)

amines and carboxylic acids. Additives such as diisopropylamine or citric acid improved band shapes and column efficiency but were not always necessary, as shown in Tables 9 and 10.

III. CONCLUSION

In a systematic approach, rationally designed CSPs have been developed that are widely useful for the separation of enantiomers. A profound understanding of the chiral recognition mechanism allows a prediction of chiral separations and thus troublesome trial-and-error procedures are limited. For the same reason, the elution order of the enantiomers can be anticipated and related to their absolute configuration. Incorporation of the Whelk-O selector into a polysiloxane backbone improves the chromatographic performance. Both the brush-type Whelk-O 1 and the PolyWhelk-O are stable over a wide temperature range and under subFC and SFC conditions without any loss of efficiency or selectivity. In general, comparison of the chromatographic performance of these CSPs, i.e., enantioselectivity and column efficiency, showed that subFC and SFC on packed columns, respectively, are superior to HPLC.

Table 9 Enantioseparation of Various *trans*-Heterocyclic Primary
Amines on PolyWhelk-O (250 × 4.6 mm)

Structure	Mobile Phase	k'$_1$	α	Rs
	90 % CO_2, 9.8 % MeOH, 0.2 % DIPA	0.56	1.18	0.90
	90 % CO_2, 9.8 % MeOH, 0.2 % DIPA	1.41	1.26	1.48
	80 % CO_2, 20 % MeOH	0.90	1.31	2.44
	90 % CO_2, 9.8 % MeOH, 0.2 % DIPA	0.90	1.17	1.09
	80 % CO_2, 20 % MeOH	0.52	1.21	1.21
	97 % CO_2, 2.94 % MeOH, 0.06 % DIPA	1.86	1.28	1.10
	80 % CO_2, 20 % MeOH	0.55	1.18	0.91
	90 % CO_2, 9.8 % MeOH, 0.2 % DIPA	0.71	1.15	0.90
	90 % CO_2, 9.8 % MeOH, 0.2 % DIPA	0.63	1.29	1.09

Operating conditions: Mobile phase: carbon dioxide-methanol containing
various amounts of diisopropylamine (DIPA); flow rate: 2 mL/min; back
pressure: 200 bar; temperature: 28°C; UV detection 254 nm. (From Ref. 15.)

Table 10 Separation of the Enantiomers of Carboxylic Acids on PolyWhelk-O (250×4.6 mm)

Structure	Mobile Phase	k'_1	α	Rs
Me CO₂H	92 % CO_2, 8 % MeOH, 4.6 g / l citric acid hydrate	4.10	1.19	3.45
Me CO₂H	92 % CO_2, 8 % MeOH, 4.6 g / l citric acid hydrate	1.67	2.16	13.1
Me CO₂H	92 % CO_2, 8 % MeOH, 4.6 g / l citric acid hydrate	0.90	1.22	2.02
Me CO₂H MeO	92 % CO_2, 8 % MeOH, 4.6 g / l citric acid hydrate	0.79	2.09	8.17
Me CO₂H Cl	95 % CO_2, 5 % MeOH, 2.3 g / l citric acid hydrate	1.36	1.12	1.53
Me CO₂H	92 % CO_2, 8 % MeOH, 4.6 g / l citric acid hydrate	1.04	1.87	7.28
CH₃ CO₂H	92 % CO_2, 8 % MeOH, 4.6 g / l citric acid hydrate	4.10	1.19	3.45

Operating conditions: Mobile phase: carbon dioxide containing various amounts of methanol and citric acid hydrate; flow rate: 2 mL/min; back pressure: 200 bar; temperature: 28°C; UV detection at 254 nm. (From Ref. 15.)

REFERENCES

1. P. Mourier, E. Eliot, M. Caude, R. Rosset, and A. Tambuté, *Anal. Chem.*, *57*:2819 (1985).
2. W. H. Pirkle, D. W. House, and J. M. Finn, *J. Chromatogr.*, *192*:143 (1980).
3. C. J. Welch, *J. Chromatogr. A*, *666*:3 (1994).
4. W. H. Pirkle, C. J. Welch, and B. Lamm, *J. Org. Chem.*, *57*:3854 (1992).
5. W. H. Pirkle and C. J. Welch, *J. Chromatogr. A*, *683*:347 (1994).
6. W. H. Pirkle and S. R. Selness, *J. Org. Chem.*, *60*:3252 (1995).
7. W. H. Pirkle and C. J. Welch, *J. Liq. Chromatogr.*, *15*:1947 (1992).
8. W. H. Pirkle, C. J. Welch, and A. J. Zych, *J. Chromatogr.*, *648*:101 (1993).

8. W. H. Pirkle, C. J. Welch, and A. J. Zych, *J. Chromatogr.*, *648*:101 (1993).

9. C. Villani, W. H. Pirkle, *Tetrahedron Asym.*, *6*:27 (1995).

10. C. Villani, W. H. Pirkle, *J. Chromatogr. A*, *693*:63 (1995).

11. W. H. Pirkle and C. J. Welch, *Tetrahedron Asym.*, *5*:777 (1994).

12. R. W. Stringham and J. A. Blackwell, *Anal. Chem.*, *68*:2179 (1996).

13. A. M. Blum, K. G. Lynam, and E. C. Nicolas, *Chirality*, *6*:302 (1994).

14. G. J. Terfloth, W. H. Pirkle, K. G. Lynam, and E. C. Nicolas, *J. Chromatogr. A* *705*:185 (1995).

15. W. H. Pirkle, L. J. Brice, and G. J. Terfloth, *J. Chromatogr. A* *753*:109 (1996).

16. M. Jung and V. Schurig, *J. Am. Chem. Soc.*, *114*:529 (1992).

17. C. Wolf, D. H. Hochmuth, W. A. König, and C. Roussel, *Liebigs Ann.*, 357 (1996).

10

Applications of Packed Column Supercritical Fluid Chromatography in the Development of Drugs

Olle Gyllenhaal, Anders Karlsson, and Jörgen Vessman

Astra Hässle AB, Mölndal, Sweden

I. INTRODUCTION

A. Application of pSFC in the Development of Drugs

The process of developing drugs is time consuming and involves many scientific disciplines. This chapter will discuss applications of packed column supercritical fluid chromatography (pSFC) in the pharmaceutical part of this process (see also Chap. 3). The major role of pharmaceutical research and development is to make a medicine out of a molecule. Here the emphasis will be on analytical problems related to the quality of drug substances and products. In this area, pSFC has now really started to gain momentum. Bioanalytical applications of pSFC will not be discussed here because there are not many examples presented in the literature.

The time between the first quality specification of a substance and the release of a product on the market can very often comprise more than 10 years. This means that data produced in the early part of the development process should withstand scrutiny several years later, and thus such data should be of a high quality from an analytical point of view [1].

Two areas have undergone considerable development in the last two decades: the analysis of impurity patterns of substances and the determination of degradates in formulations. Both areas have matured very much due to the availability of reliable and flexible separation methods. With such methods, it is possible to separate and quantitate impurities and degradates at concentrations lower than 0.1%, which is an important regulatory requirement for the registration process. High-performance liquid chromatography (HPLC, LC) has, in fact, revolutionized the monitoring of the quality of organic substances. With the separation methods now available, we can determine the impurity patterns with a high degree of selectivity and we can also follow the stability of a drug product much easier than ever before. In line with guidelines from authorities, impurities below 0.1% usually need not be identified.

A particular problem related to purity and stability testing is that the major component may be present in a very large excess, up to 1000:1. This means that minor components might be hidden by the major one. Peak purity tests can be of some help (diode array detection or mass spectrometry on-line). However, unexpected impurities coeluting with the major compound have preferably to be revealed by complementary separation systems. These systems should in principle be based on other orthogonal separation mechanisms. This is one of the reasons why there is such an interest in developing alternative separation methods to reversed phase LC, such as pSFC and capillary electrophoresis (CE). One has to bear in mind that any purity test method is, in fact, related to the experience from the method of manufacture. No test procedure can be claimed to cover new production conditions without validation. The impurity patterns have to be well characterized and this requires reproducible and robust analytical methods. The early implications of the profiles are primarily related to safety aspects. However, later on when generic competition starts and when pharmacopoeial standards are developed quality becomes the major matter of concern.

B. The Present Role of SFC in the Pharmaceutical Industry

It is in the above-mentioned perspective that one should place SFC. What about the role of pSFC in the pharmaceutical industry? The acceptance is not yet widespread, probably because the early attempts were with capillary columns. However, the use of packed columns is now definitely on the move [2], and reliable instruments are available on the market. The application areas are mainly covering studies of substance purity and stability, assay of drug products, and monitoring of their stability. A recent review that covers both pSFC and capillary column SFC (cSFC) for the analysis of drugs contains some 90 references up to 1993 [3]. The excellent possibilities available to separate enantiomers are of particular interest. It is in particular in the normal phase mode where we find the superiority of pSFC for the separation of closely related compounds such as stereo- and geometric isomers. As in chiral chromatography, we can in fact talk about a renaissance for normal phase chromatography! The larger differences in interactions, often seen in such systems, play an important role here.

The advantages of pSFC over cSFC are shorter separation times (with still high efficiencies) and better loadability, so that it is generally possible to analyze impurities at a concentration of 0.1% (or below) of the analyte. It has to be emphasized that qualitative information on separated components is highly desirable. Hyphenated techniques are becoming more and more common, as they open possibilities of eludicating structures of unknown compounds. The success of gas chromatography–mass spectrometry (GC-MS) and now LC-MS will most probably be followed by pSFC-MS. It has been stated that it should be easier to remove commonly used phases such as carbon dioxide than common LC mobile phases [4].

C. Comparisons of pSFC with LC and CE

How, then, does pSFC compare with other separation methods? LC is, no doubt, the work horse in the laboratories and reversed phase systems have taken the lead. Some of the *advantages with pSFC* over LC are as follows:

More rapid equilibration, especially in dynamic coating on silica materials
Low viscosity, which allows high flow rate without large loss of efficiency
Low pressure drop, which allows the use of columns in series in order to improve the resolution

It can also be mentioned that the consumption of combustible organic solvents is lower, which is good for the environment and for the economy. It has even been claimed that a single pSFC method can replace several LC methods [5].

When pSFC is compared with CE it is obvious that the competition is much harder here. Maybe quantitation can be more reliable in pSFC as no internal standard is needed and no response dependence on mobility is seen. The possibilities for detecting impurities at the 0.1% level may be more difficult in CE because of loadability problems.

The *disadvantages with pSFC* may at present be that sample volumes are somewhat restricted and that injection techniques sometimes cause problems. Compared with LC one can notice that the signal-to-noise ratio might hamper real trace analysis in pSFC.

The following examples of applications from the development of drugs into medicines will demonstrate the great potential of pSFC for the pharmaceutical industry.

II. APPLICATIONS TO PURITY INVESTIGATIONS

A. Omeprazole and Degradation Products

Omeprazole (Fig. 1) is a fairly polar compound with two protolytic groups, pK_a 5 and 8, respectively. The compound is sensitive to acidic media and thus has to be analyzed under basic conditions. The reduced and oxidized forms of omeprazole are easily chromatographed on various supports in pSFC with aminopropyl-modified silica being the preferred one from the selectivity point of view. The selectivity was less good on diol and cyano-substituted columns and also peak symmetry was inferior to the amino-substituted one [6].

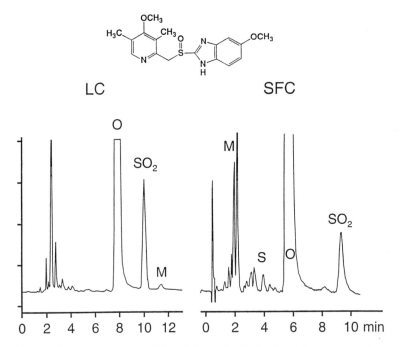

Figure 1 A comparison of LC (left) and pSFC (right) for omeprazole sodium and byproducts when stored under accelerated conditions. LC conditions: Microspher C18 100×4.6 mm id column, 1.0 mL/min of 26% acetonitrile with 0.001 M tetrabutylammonium in phosphate buffer pH 7.4 (ionic strength 0.1). pSFC conditions: Lichrosorb aminopropyl silica 125×4.0 mm id column at 40°C, flow rate 2.0 mL/min of carbon dioxide, 120 µL/min of methanol containing 1% of triethylamine, back pressure 175 bar, UV detection at 300 nm [6]. Identification of peaks: M, methylated omeprazole; S, the sulfide; O, omeprazole; and SO_2, the sulfone, [1,6]. (From *Encyclopedia of Analytical Sciences*, Academic Press, with permission.)

Six different aminopropyl supports were evaluated and the most attractive ones were Lichrosorb and Kromasil. Methanol was chosen as modifier with the addition of a small amount of triethylamine to give basic conditions. Figure 1 shows a comparison with LC. It is remarkable that in pSFC all of the compounds are eluted within 10 min, whereas in LC about 25 min was needed to elute the sulfide (S), which is not seen in the chromatogram to the left in Fig. 1. In addition to these redox-related compounds, a close analog was also tested, the desmethoxylated one, which could be separated in pSFC but with difficulty if present in amounts lower than 0.3% [6].

The behavior of N-methylated omeprazole (compound M in Fig. 1) is also interesting. In LC it elutes late and shows only one peak, whereas in pSFC the two position isomers actually present are well separated within a short retention (Fig. 1).

B. Impurities in a Dihydropyridine

A method using pSFC for a new dihydropyridine drug (clevidipine) has been developed simultaneously with a more regular reversed phase LC one. The compound is used as a hypertensive agent (the structure is given in Fig. 2). A method for the determination of content uniformity based on pSFC has been described for another dihydropyridine, felodipine, isolated from tablets [7].

Using methanol as mobile phase modifier, the selectivity on four different kinds of supports was investigated in order to separate the major compound and possible byproducts such as the corresponding pyridine and carboxylic acid. The data are presented in Table 1. The methanol modifier also contained acetic acid because one of the analytes contained a carboxylic function. The selectivity with diol-modified silica and bare silica was superior to the selectivity obtained with octadecyl and cyano substituents present. The bare silica support was selected for further work.

Under the initial conditions the symmetry of the carboxylic acid was controlled by the presence of acetic acid in the methanol modifier. However, because a modifier gradients is required to elute the wide range of different compounds encountered, this resulted in a severe rise of the baseline at the low wavelength used (240 nm). Furthermore, even with acetic acid present the peak from the dihydropyridine-substituted acid was broad. Data from a closer study reported in Table 2 revealed that polyvalent carboxylic acids, which also contain hydroxy groups, are efficient in improving the peak performance.

Interestingly enough, the symmetry is only slightly affected [8]. The positive effect of 0.004 M of, e.g., citric acid as additive in the methanol modifier is

Table 1 Selectivity Factor (α) Between the New Dihydropyridine Clevidipine and Some Analogs on Different Silica Supports

Support	Pyridine	Nitrile	Acid
Peak no	1	4	5
Si-60[a]	0.16	1.65	2.70
Nitrile[b]	0.23	1.64	2.11
RP-18[b]	0.40	1.21	1.60
Diol[c]	0.16	1.59	4.45

Conditions: flow rate 2.0 mL/min of carbon dioxide with 8% of methanol containing 0.35 M of acetic acid as modifier, column temperature 40°C, back pressure 150 bar, and UV detection at 240 nm. 5 μL injected of a mixture of the solutes in dichloromethane. The structures are shown in Fig. 2b.
[a]Superspher, [b]Lichrosorb, and [c]Lichrospher (all columns 125 × 4 mm id).

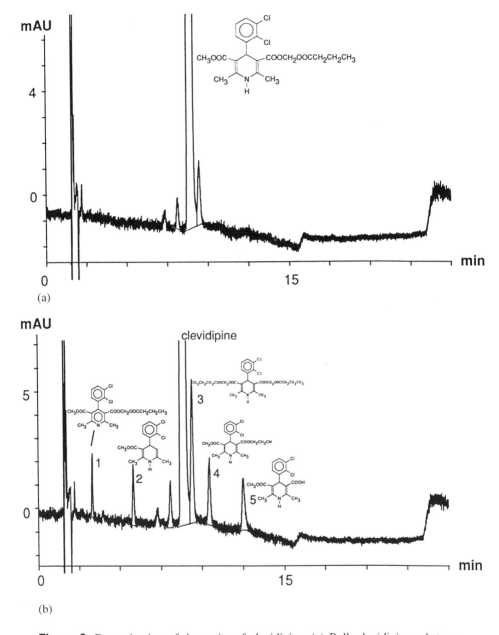

Figure 2 Determination of the purity of clevidipine. (a) Bulk clevidipine substance and (b) with possible impurities added at the 0.1% w/w level. Conditions: Hypersil bare silica column 200 × 4.6 mm id at 50°C, 1.5 mL/min of carbon dioxide containing 5% of methanol, back pressure 150 bar, and UV detection at 240 nm. Modifier gradient 0.6%/min from 5% to 10% then 2%/min to 20%. 5 μL injected of a 4 mg/mL solution in dichloromethane. Aid to peak identification: 1, a pyridine; 2, a decarboxylated acid; 3, a symmetrical ester; 4, a nitrile; 5, an acid.

Table 2 Comparison of Acidic Additives in the Methanol Modifier

Acid	N[a]	k'[b]	α[c]	Asf[d]	pK_a[e]
No acid	2600	22.4	3.14	2.0	—
Cyclohexanoic	2700	22.1	3.06	2.1	4.8
Lactic	6900	18.2	2.48	2.1	3.8
Succinic	3600	19.3	2.69	2.2	4.21
Tartaric	8200	17.8	2.44	2.0	2.93
Citric	8100	17.8	2.46	2.0	3.14

Conditions: Si-60 Superspher 125 × 4 mm id column column at 40°C, flow rate 2.0 mL/min of carbon dioxide with 10% of methanol containing 0.004 M of acid, back pressure 150 bar, and UV detection at 240 nm.

    The selectivity factor (α) calculated from the capacity factor of the dihydropyridine acid vs. an aprotic dihydropyridine (nitrile). Ten minutes equilibration time between each acid, no washout, run order as in the table [8]. 5 μL injected of a 0.2 mg/mL solution in dichloromethane. For structures, see Fig. 2b.

[a]N = number of theoretical plates
[b]k' = capacity factor of the acid
[c]α = separation factor ($k'_{acid}/k'_{nitrile}$)
[d]Asf = asymmetry factor, and
[e]pK_a = acid dissociation constant

persistent for a long time. Citric acid has been used as efficient acidic modifier by others [9–11] and the long-term positive effect observed [11]. If the peak becomes broad, 1 h was sufficient to restore the column efficiency for this weak carboxylic acid (pK_a = 7). Studies of van Deemter curves show that the C term is affected by the presence of citric acid in the mobile phase. The decreased slope at higher flow rates indicates that the kinetics of the adsorption/desorption process is faster after citric acid treatment.

    A closer study of different bare silica supports was necessary because the separation of another possible by product, the symmetrical ester, was imperative. As shown in Table 3, there is a marked difference not only in the plate number (N) on the three supports studied but also in selectivity. The best selectivity was obtained with the Hypersil column.

    Because the resolution of this minor component was insufficient when the parent drug was present in a 1000-fold excess, a statistical experimental design was used for further optimization with flow, percentage of methanol, outlet pressure, and column temperature as variables. Analysis of the reduced full factorial design showed that pressure and flow were of minor importance,

Table 3 The Resolution Between a Dihydropyridine and Its Symmetrical Ester on Different Bare Silica Supports

Silica	α[d]	k'[e]	N[f]	Rs[g]
Hypersil[a]	1.146	6.46	9500	5
Kromasil[b]	1.099	8.90	9500	3
Si-60[c]	1.083	25.9	4500	n.m.[h]

Conditions: columns at 40°C, flow rate of carbon dioxide 1.5 mL/min with 5% of methanol as modifier, back pressure 150 bar, and UV detection at 240 nm. 5 µL injected of a 1.8 mg/mL solution in dichloromethane of the dihydropyridine containing 2% by area of the symmetrical ester. Structures of the dihydropyridines in Fig. 2.

Column dimensions: [a]200 × 4.6 mm id, [b]150 × 4.6 mm id, and [c]125 × 4.0 mm id.

[d]$\alpha = k'_{\text{symmetrical ester}}/k'_{\text{dihydropyridine}}$
[e]k' = capacity factor of the dihydropyridine
[f]N = number of theoretical plates
[g]Rs = resolution ($2 \times (t_{R2} - t_{R1})/(p_{w1} + p_{w2}$, and
[h]n.m. = not possible to measure

whereas modifier (as expected) and column temperature were important [12]. The column temperature was hereafter increased from 40°C to 50°C. The final method includes a gradient of 0.6%/min of methanol from 5% to 10% and then a steeper increase of 2%/min to 20% in order to elute other possible byproducts. A representative chromatogram is shown in Fig. 2a. Addition of possible byproducts at the 0.1% w/w level (Fig. 2b) shows that it is possible to detect impurities and the degradates at this regulatory required level [13]. The method compares favorably with a corresponding reversed phase LC method [14]. Assay of the dihydropyridine can also be performed using 2-propanol as modifier [15].

C. Impurities in Metoprolol

Metoprolol is a medium polar amino alcohol used as an antihypertensive and antiangina drug. It was introduced in the 1970s. Metoprolol and several analogs can be analyzed by cSFC [16] on both apolar and polar stationary phases. However, selectivity or column quality was a problem. The poor symmetry of the peaks was more ascribed to column wall activity than any reaction with supercritical carbon dioxide because nitrous oxide gave similar chromatograms. Because sample loading capacity is a problem with capillaries, a more promising way is to use packed columns. Throughout the years, some papers dealing with the separation of β blockers on packed columns have been published [17,18]. Actually, a wide number of publications are dedicated to the separation of enantiomers of amino alcohols of pharmaceutical interest. There is a great

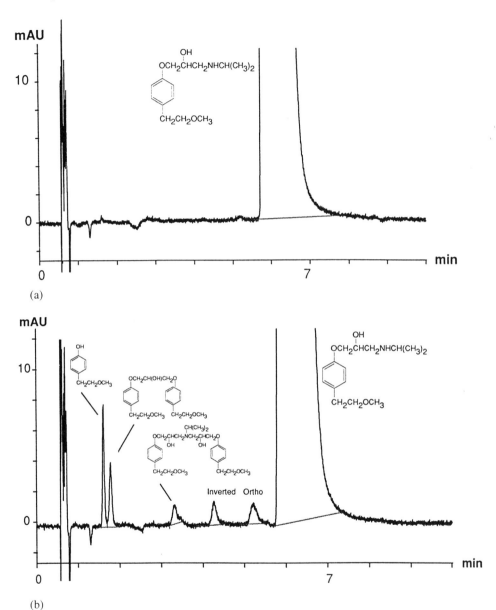

Figure 3 Purity analysis of metoprolol tartrate. (a) Bulk substance and (b) with five possible impurities added at 0.1% w/w. Conditions as in Table 4, 5 μL injected of a 25 mg/mL solution in dichloromethane. The structures of the peaks Ortho and Inverted are given in Fig. 11.

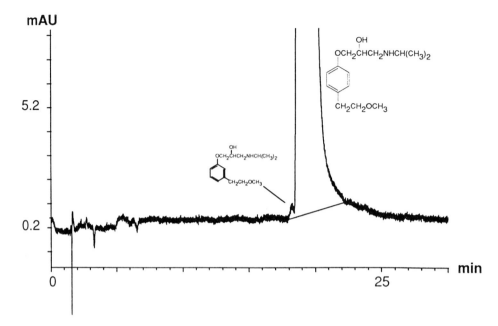

Figure 4 Separation of metoprolol from its meta isomer added at the 0.1% w/w level using three diol columns in series. Conditions as in Table 4 except a flow rate of 2.5 mL. 5 μL injected of a 43 mg/mL solution of metoprolol tartrate in dichloromethane.

Table 4 Influence of the Solid Phase on the Separation Factor (α) Relative to Metoprolol for Some Amino Alcohols

	Chromatographic columns		
Characteristics	Aminopropyl silica, TEA[a]	Aminopropyl silica, TEA[a]/HAc[b]	Diol silica, TEA[a]/HAc[b]
Inverted	0.49	0.74	0.68
Tertbutyl	0.78	0.81	0.75
Ortho	0.79	0.84	0.82
Meta	0.94	0.95	0.95
Unsubst	0.97	1.00	0.89
Biphenyl[c]	1.71	1.57	1.57
Aldehyde[c]	n.m.	n.m.	1.69
Prolonged (diol)	n.m.	2.8[c]	2.24

[a] Triethylamine; 0.07 M in the methanol,

[b] Acetic acid; 0.35 M in the methanol.

[c] Substituent in para position of the aromatic ring in lieu of the methoxyethyl group of metoprolol. n.m. = not measured. Other conditions: flow rate 2.0 mL/min carbon dioxide with 10% of methanol as modifier, back pressure 150 bar, UV detection at 273 nm. 5 μL of sample solution in dichloromethane injected. The structures are shown in Figs. 4, 5, and 11.

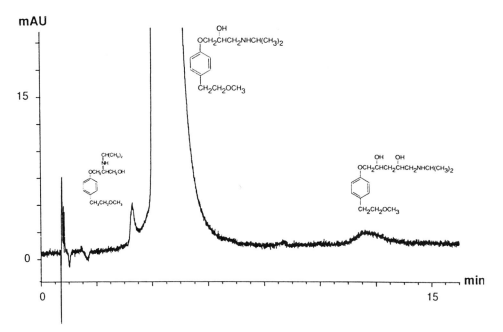

Figure 5 Separation of metoprolol and two impurities. Conditions: Lichrosorb NH$_2$ 125 × 4 mm id column at 40°C, flow rate 2.0 mL/min of carbon dioxide with 10% of methanol containing 0.35 M of acetic acid and 0.07 M of triethylamine, back pressure 150 bar, and UV detection at 273 nm. About 0.3% and 0.6% by area of the minor peaks. 5 μL injected of a 25 mg/mL solution of metoprolol tartrate in dichloromethane.

number of impurities, which are more or less likely to be found in metoprolol bulk substance. From our earlier experience with polar drugs, and from the literature, diol- and aminopropyl-modified silica were investigated as possible column supports in combination with an amine added to a polar modifier. The selectivity factors of the related substances relative to metoprolol, using 10% of methanol containing 0.07 M of triethylamine as modifier at 40°C, are shown in Table 4.

Because we have observed that alkaline methanol gradually degrades the silica support, the use of a fivefold molar excess of acetic acid over amine was also used (Table 4) with only a slight change in selectivity.

The linearity of the system was investigated in the range 80 μg/mL to 25 mg/mL (0.4–125 μg injected) and found to be adequate although the shape of the metoprolol peak broadened for 10 μg injected. This affects the detection of the meta analog of metoprolol, which is obscured by the parent peak. Other possible impurities could be detected at the 0.1% w/w level in metoprolol

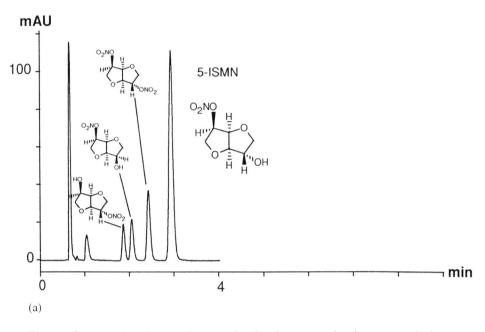

Figure 6 Isosorbide 5-mononitrate and related compounds chromatographed on various supports. (a) Hypercarb (2.0 mL/min, 40°C), (b) Hypersil bare silica (1.0 mL/min, 45°C) and (c) Lichrosorb RP-8 (1.0 mL/min, 50°C). Back pressure 150 bar, UV detection at 214 nm, and 6% of methanol as modifier in the carbon dioxide.

tartrate [19] or succinate as exemplified in Fig. 3b. By coupling three diol columns in series the meta isomer could also be separated from metoprolol and detected at this low level (Fig. 4). The pressure drop over these three columns was only about 90 bar. The advantage of coupling several columns in series has been reported by others [20,21, Chapter 6].

Three compounds are particularly difficult to separate in LC systems using silica materials: the meta isomer, the inverted isomer, and the recently identified diol impurity, where the β side chain is prolonged with one 2-hydroxypropyloxy group [22]. Therefore, the separation of these three structurally related compounds was studied with pSFC and CE. As shown in Fig. 5, the complementary separation obtained with pSFC is noteworthy. With CE, the diol was also separated from metoprolol (500 times excess) and also the inverted form [23]. A comparison of these three separation techniques has recently been reported for the assay of losartan in tablets [24].

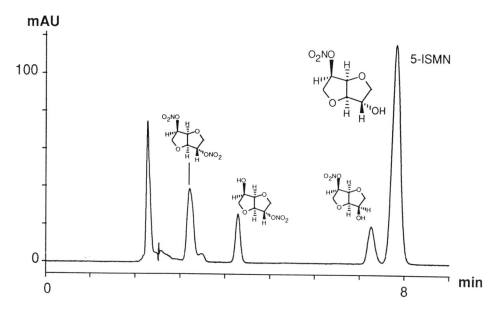

(b)

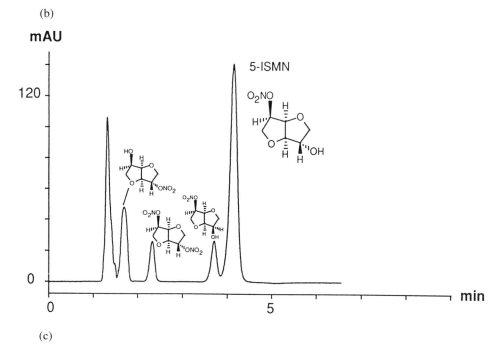

(c)

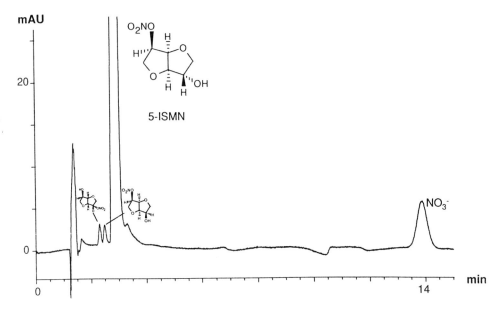

Figure 7 Analysis of isosorbide 5-mononitrate (5-ISMN) bulk substance with related compounds added. Conditions: Hypercarb 100 × 4.6 mm id 40°C, flow rate 1.0 mL/min of carbon dioxide containing 20% of methanol and 5 mM of tetrabutylammonium hydrogen sulfate, back pressure 150 bar, and UV detection at 214 nm. 5 μL injected of a 10 mg/mL solution of 5-ISMN in methanol.

D. Isosorbide-5-mononitrate

This compound is the main metabolite of isosorbide dinitrate and is used as a drug to prevent angina. Among byproducts that can be formed during synthesis or upon storage, the most important are the dinitrate, the 2-isomer, and the isomannide ester. Furthermore, nitrate and nitrite are also possible contaminants or degradation products that preferably should be controlled by chromatography. In reversed phase LC the corresponding analysis has been hampered by the fact that the presence of tetrabutylammonium in the mobile phase tends to degrade the silica column packing.

Three chromatographic supports were examined (Fig. 6). On octyl and unsubstituted silica the separation of isosorbide 5-mononitrate and the isomannide 2-nitrate is crucial but can easily be improved by increasing the column temperature above 40°C. On the other hand, a higher column temperature is detrimental for the nitrite peak. Hypercarb was chosen because this support is useful to separate structurally related compounds and also because of stability when using basic additives in the organic modifier.

Owing to its polar nature as the anion of a strong acid, nitrate was eluted using a counterion such as tetrabutylammonium as hydrogen sulfate. A concentration of 5 mM in 20% of methanol as modifier is adequate to obtain a reasonably short retention time. Thus the ion pair principle opens an interesting tool for the chromatographic control of anions in pSFC. An early work using the ion pair approach was reported by Steuer et at. [25]. A practical example of the separation of isosorbide 5-mononitrate containing impurities added at the 0.1% w/w level is given in Fig. 7. Hypercarb is considered as a rugged column support [26].

III. SEPARATION OF POSITIONAL AND GEOMETRIC ISOMERS BY pSFC

It is very often necessary to develop efficient chromatographic systems for the separation of structurally related compounds that usually appear in synthesis mixtures. For that purpose, pSFC is a good complement to other chromato-

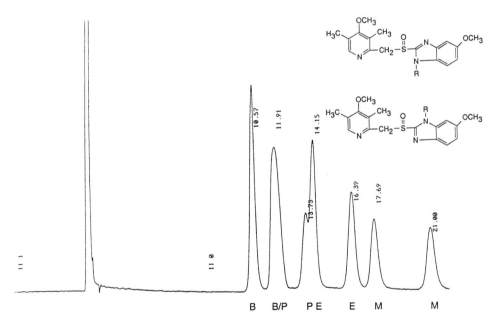

Figure 8 Separation of N-alkylated positional isomers of omeprazole. Conditions: Si-60 Superspher 125×4 mm id column at 40°C, flow rates 1.50 mL/min of carbon dioxide and 60 μL/min of 0.07 M triethylamine in methanol, respectively. Back pressure 150 bar and UV detection at 272 nm. Identification of peaks: R in structural formula = B, butyl; P = propyl; E = ethyl; and M = methyl [27,28].

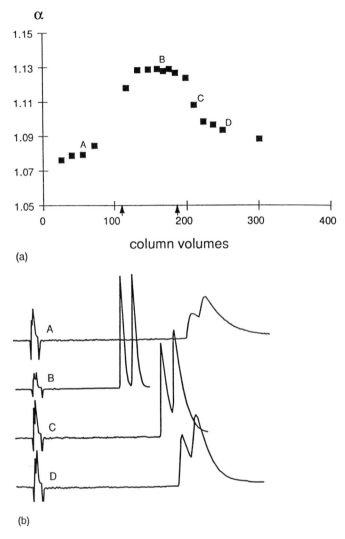

(a)

(b)

Figure 9 Equilibration in pSFC. (a) The selectivity factor (α) as a function of column volumes and (b) chromatograms taken as indicated by the letters A, B, C, and D in (a). Conditions: Superspher Si 60 125 × 4 mm id column at 40°C, flow rates 1.5 mL/min carbon dioxide and 60 μL/min methanol containing no or 0.07 M of triethylamine, respectively, back pressure 150 bar, and UV detection at 272 nm. The arrows indicate change of modifier from pure methanol to methanol with triethylamine and vice versa. 20 μL injected of a solution of 1 mg/mL in dichloromethane of N-methylated omeprazole, 5- and 6-isomers.

graphic methods [5], as will be shown through the following applications, where pSFC is used for the separation of positional isomers or diastereoisomers.

Such an application is presented in Fig. 8. The separations of two positional isomers of four N-alkylated omeprazole compounds, whose structures are given in Fig. 8, were chromatographed using bare silica as the stationary phase. A pSFC system was developed that gives separations between all the positional isomers (see figure). The retention as well as the separation factors increased with decreasing hydrophobicity of the alkyl substituent, butyl < propyl < ethyl < methyl [27,28]. Triethylamine was used as additive in the methanol modifier in order to improve the shape of the peaks [6,27,28].

The equilibration time, when introducing a modifier including triethylamine, is fairly short, i.e., less than 25 volumes. And vice versa, i.e., the positive effect of the amine additive is quickly lost. This is illustrated in Fig. 9 by both the graph of α vs. column volumes in 9a and the chromatograms in 9b.

As mentioned above, the porous graphitic carbon material (PGC, commercially available as Hypercarb), which has a flat surface [26], has been used in LC to separate positional isomers [29] as well as diastereoisomers [30], e.g., to separate isomers of a product with a prolonged β chain in metoprolol [31]. The

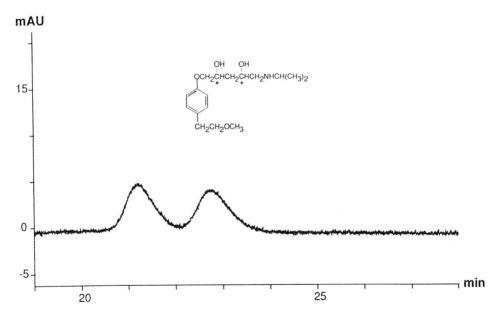

Figure 10 Diastereoisomeric separation of an aminodiol. Conditions: Hypercarb 100 × 4.6 mm id column at 30°C, flow rate 2.0 mL/min of carbon dioxide containing 20% of 0.24 M *N,N*-dimethyloctylamine in methanol, back pressure 150 bar, and UV detection at 273 nm. 5 µL injected of a dichloromethane solution.

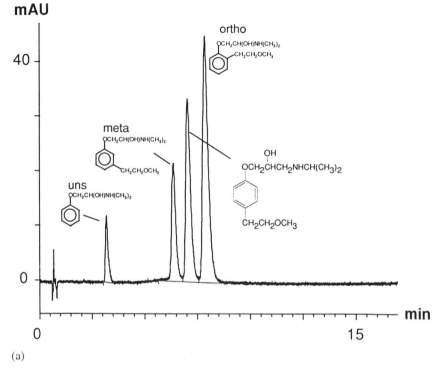

(a)

Figure 11 Separation of metoprolol and analogs on Hypercarb or diol support. Conditions: (a) Hypercarb 100 × 4.6 mm id column at 30°C, modifier 20% methanol with 0.24 M *N,N*-dimethyloctylamine and (b) Lichrospher diol 125 × 4 mm id column at 40°C with 10% methanol containing acetic acid 0.35 M + 0.07 M triethylamine as modifier. Flow rate 2.0 mL/min, back pressure 150 bar, and UV detection at 273 nm. For the structures of the early peaks in (b), see Fig. 3b.

compound has two chiral centers and four possible stereoisomers. The separation of the two diastereoisomers (RR & SS and RS & SR) was successfully performed using Hypercarb as the chromatographic support (Fig. 10). So far, we have failed to separate these diastereoisomers using achiral silica phases. On the other hand, they separate in CE [23]. The possibility of using Hypercarb for the separation of positional isomers of metoprolol was also tested. Four compounds, one without substituent on the aromatic ring, the other with an ethylmethoxy substituent in ortho, meta, or para position, were separated on Hypercarb using pSFC (Fig. 11a).

In order to improve the peak performance, triethylamine was added to the mobile phase. The reason for the improved resolution, due to improved column

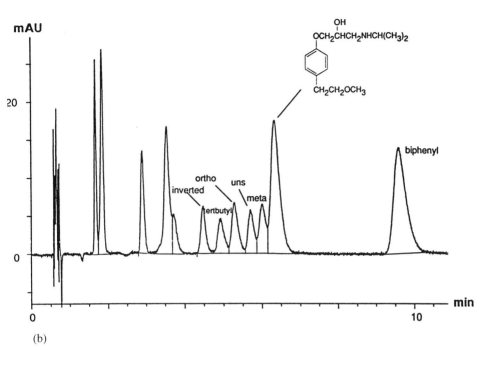

(b)

efficiencies, is probably that the amines must be uncharged when chromatographed on Hypercarb. The separation of the same compounds was also possible with a silica-based support (Fig. 11b). The two chromatograms reveal differences in selectivity between the two chromatographic supports, which are related to different retention mechanisms. While the silica surface separates compounds due to polar attraction, the PGC support retains the analytes selectively due to π-π interaction and dispersion force attractions [26,32]. The three-dimensional structure of the compounds can also affect the selective retention due to differences in the number of possible contact points between the solute and the extremely flat PGC surface [32].

IV. SEPARATION OF ENANTIOMERS AND DIASTEREOISOMERS BY pSFC

Chiral separations have been in the focus for some 15 years and a number of chromatographic supports have been presented for LC. Similarities to normal phase LC regarding the interaction between a chiral selector and a racemic mixture make pSFC an interesting technique for the separation of enantiomers and also for the resolution of diastereoisomers. In fact, the possibilities with pSFC seem in certain cases to be outstanding.

Direct enantioselective separation can be obtained using either a chiral stationary phase (CSP) or a chiral mobile phase additive (CMPA). When using a CSP, differences in adsorption constants of the two enantiomers to the immobilized chiral selector are the reason for the chiral discrimination [33]. When a chiral selector is dissolved in the mobile phase enantioselectivity can be

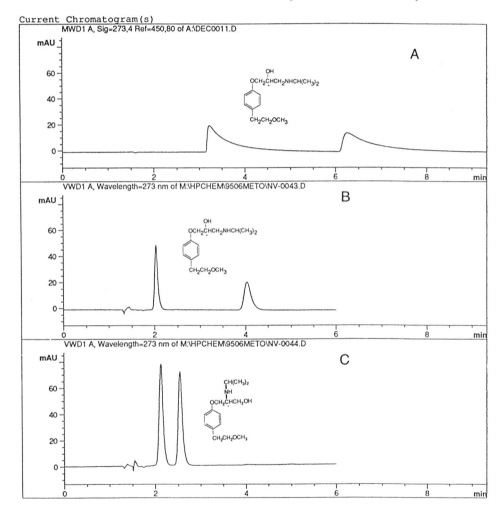

Figure 12 Enantioresolution of aminoalcohols on Chiralcel OD. Conditions: Chiralcel OD 250 × 4.6 mm id column at 25°C, flow rate 2.0 mL of carbon dioxide with methanol containing 0.35 M of acetic acid and 0.07 M of triethylamine, back pressure 150 bar, and UV detection 273 nm. (A) Metoprolol (20% methanol only, no additives), (B) metoprolol (25% methanol with additives), and (C) inverted metoprolol (25% methanol with additives). 5 μL injected of 1–2 mg/mL in methanol or dichloromethane.

obtained both by differences in formation constants of the diastereoisomeric complexes and also by differences in adsorption constants of the formed diastereoisomeric complexes to the achiral stationary phase [34].

A. Chiral Stationary Phases

Several different kinds of CSPs have been used in pSFC to separate stereoisomers (Chapters 8 and 9). According to previously recorded results [18,35–38], we have separated the enantiomers of several aminoalcohols using Chiralcel OD as the stationary phase and added dimethyloctylamine (DMOA) or triethylamine (TEA) to improve the peak shape. The differences in enantioselectivity between metoprolol and its inverted isomer (structure in Fig. 12C) is an interesting observation. Simultaneous separation of the diastereoisomers and the enantiomers of an aminodiol is given in Fig. 13. In our department, an ongoing project is using pSFC to determine the influence of the structure of the aminoalcohols on the enantioselective retention on Chiralcel OD or Chiralpak AD using pSFC and statistical experimental design.

As pSFC really is the separation technique of choice, when dealing with hydrophobic analytes we have chosen pSFC to separate the enantiomers of several dihydropyridines. In order to study and also to optimize the enantioselective resolution an experimental design was used. To acquire a high enantioselective resolution for clevidipine, and the most important related dihydropyridine, two different enantioselective stationary phases were coupled in series, i.e., Chiralcel OD and Chiralpak AD (see also Chapter 6). Enantioselective resolution for the dihydropyridine of pharmacological interest and its major degradation product formed during hydrolysis, as well as some other analogs, is given in Fig. 14.

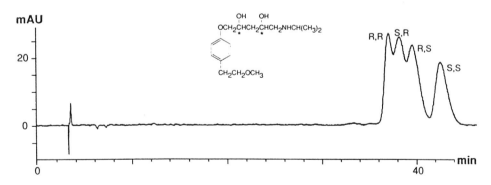

Figure 13 Simultaneous separation of diastereoisomers and enantiomers of an aminodiol. Conditions: Chiralcel OD 250 × 4.6 mm id column at 33°C, flow rate 1.0 mL/min of carbon dioxide with 8% of methanol containing 0.35 M of acetic acid and 0.07 M of triethylamine, back pressure 150 bar, UV detection at 273 nm. 5 μL injected of a 9 mg/mL solution in methanol.

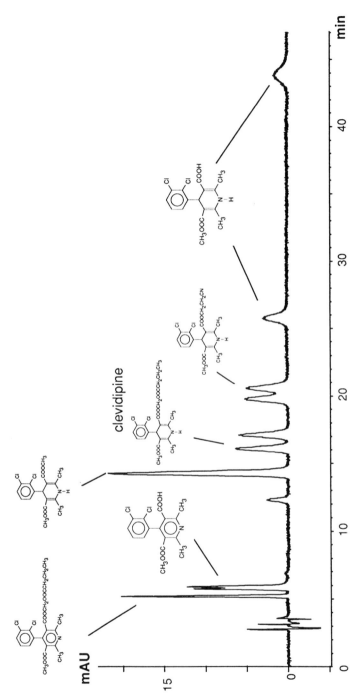

Figure 14 Enantiomeric separation of a new dihydropyridine, clevidipine, and some of its byproducts using coupled columns. Conditions: Chiralcel OD and Chiralpak AD columns at 30°C, each column 250×4.6 mm id, flow rate: 2.0 mL/min of carbon dioxide with 16% 2-propanol containing 10 mM of N,N-dimethyloctylamine, back pressure 100 bar, and UV detection at 240 nm. 5 μL injected of a solution containing about 0.1 mg/mL each of the solutes in methanol.

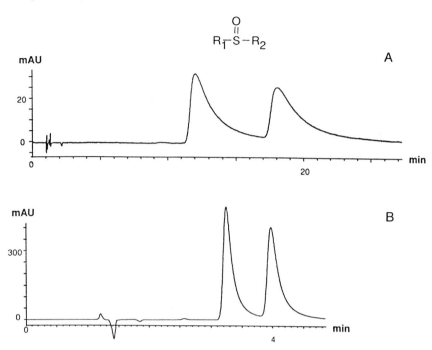

Figure 15 Influence of dimethylsulfoxide on enantioresolution of a racemic sulfoxide. Conditions: Whelk-O1 250 × 4.6 mm id column at 30°C, flow rate 2.0 mL/min of carbon dioxide with 30% of modifier, back pressure 150 bar, and UV detection at 252 nm. (A) 20 mM *N,N*-dimethyloctylamine in methanol as modifier, (B) 20 mM *N,N*-dimethyloctylamine in methanol-dimethylsulfoxide 3:2.5 μL injected of a 1 mg/mL solution in methanol.

Previous investigations have shown that the addition of an organic modifier to the mobile phase can have an effect on the chiral recognition [39]. In a recent application, different organic modifiers were added to the carbon dioxide in order to improve the enantioselective resolution of a racemic sulfoxide using Welk-O 1 as the enantioselective support. When dimethylsulfoxide (DMSO), was added to the mobile phase (Fig. 15), the enantioresolution increased because of an improved peak performance. DMSO, with its sulfoxide function, competed with the enantiomers about the limited amount of the adsorption site(s) which is responsible for the tailing peaks. This resulted in considerably shorter retention times but decreased the separation factor only slightly. Another separation of a sulfoxide is given in Chapter 9.

A similar chiral stationary phase material, i.e., immobilized dinitrobenzylphenylglycine (DNBPG), was used to simultaneously separate the positional

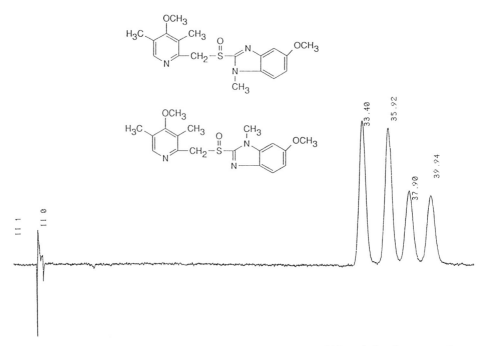

Figure 16 Enantiomeric separation of positional isomers of N-methylated omeprazole [28]. Conditions: Bakerbond DNBPG (dinitrobenzylphenylglycine) 250 × 4.6 mm id column at 40°C, flow rates 1.50 mL/min of carbon dioxide and 60 uL/min of 0.07 M triethylamine in methanol, back pressure 150 bar, and UV detection at 272 nm. 20 μL injected of a 1 mg/mL solution in dichloromethane.

isomers and the enantiomers of another chiral sulfoxide, N-methylated omeprazole (Fig. 16) [28].

B. Chiral Mobile Phase Additive

Direct enantiomeric separations have been obtained in LC using chiral selectors dissolved in the mobile phase. Several racemic analytes have been separated into their enantiomers using a chiral protolyte dissolved in an organic solvent [40], i.e., in normal phase LC. Improvement of the enantiomeric resolution was observed when using porous graphitic carbon as the achiral stationary phase together with chiral N-blocked dipeptides dissolved in the mobile phase [41]. An N-blocked dipeptide, *N*-benzyloxycarbonyl-glycyl-L-proline, was used as the chiral selector for the separation of the enantiomers of propranolol in pSFC on a silica support [25].

mAU

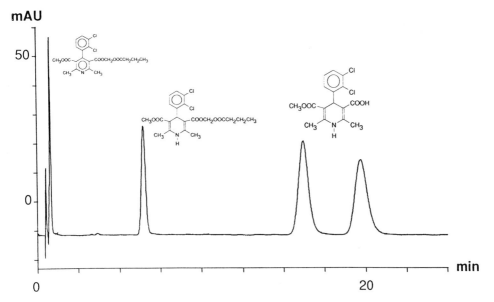

Figure 17 Enantioselective resolution of a dihydropyridine substituted acid using chiral ion pair pSFC. Conditions: Hypercarb 100×3 mm id column at 70°C, flow rate: 1.0 mL/min of carbon dioxide with 20% of methanol containing 1 mM of Z-L-arginine and 100 µg/mL of Tween 60, back pressure 150 bar, and UV detection at 240 nm. 5 µL injected of a 0.1–1 mg solution of each solute/mL of dichloromethane.

We used another additive, *N*-benzyloxycarbonyl-L-arginine (Z-L-arg), to separate the dihydropyridine clevidipine from its major degradation product, an acid formed during hydrolysis. The improved separation of the two compounds is due to ion pair formation between the acidic impurity and Z-L-arg. Furthermore the two enantiomers of the acid were also resolved as diastereoisomeric ion pairs using Hypercarb as the achiral stationary phase (Fig. 17).

For a more comprehensive overview of this topic, see Chapter 8.

V. CONCLUSIONS

pSFC is changing gear. With the advent of new robust and reliable instruments, the advantages of pSFC over LC have become apparent. Among those are the short running times and the efficient separations. Moreover, the possibilities for rapid changes (e.g., as required for method development) from one separation condition to another are remarkable as is the low pressure drop when columns are coupled in series. This latter opportunity gives the analyst a separation system with a very high number of theoretical plates and enables the analysis of very small amounts of closely related compounds in drugs (i.e., isomers at the 0.1% level). These advantages are all dependent on the use of modified carbon dioxide as mobile phase and well-known LC stationary phases under supercritical or subcritical conditions. The future for pSFC is bright and in the coming years this technique can be an established separation technique in the pharmaceutical industry besides LC, GC, and CE. The examples discussed above can really convince the reluctant or hesitating analyst that pSFC really is an alternative to use.

ACKNOWLEDGMENTS

We thank Cecilia Garell, Sophie Hermansson, and Kristina Öhlén for valuable assistance.

REFERENCES

1. J. Vessman, Pharmaceutical analysis overview, *Encyclopedia of Analytical Sciences*, Academic Press, London, p. 3798 (1995).
2. T. A. Berger, *Packed Column SFC*, RSC Chromatography Monographs, Cambridge, 1995.
3. A. Salvador, A. Jaime, G. Becerra, and M. De La Guardia, Supercritical fluid chromatography in drug analysis: a literature survey, *Fresenius J. Anal. Chem.*, *356*: 109 (1996).
4. J. D. Pinkston and T. L. Chester, Putting opposites together: SFC MS. Guidelines for successful SFC/MS, *Anal. Chem.*, *67*: 650A (1995).
5. K. Anton, M. Bach, and A. Geiser, Supercritical fluid chromatography in the routine stability control of antipruritic preparations, *J. Chromatogr.*, 553: 71 (1991).
6. O. Gyllenhaal and J. Vessman, Packed-column supercritical fluid chromatography of omeprazole and related compounds. Selection of column support with triethylamine- and methanol-modified carbon dioxide as the mobile phase, *J. Chromatogr.*, *628*: 275 (1993).
7. J. T. B. Strode III, L. T. Taylor, A. L. Howard, D. Ip, and M. A. Brooks, Analysis of felodipine by packed column supercritical fluid chromatography with electron capture and ultraviolet absorbance detection, *J. Pharm. Biomed. Anal.*, *12*: 1003 (1994).
8. O. Gyllenhaal and A. Karlsson, "The Use of Dynamically Coated Bare Silica for the SFC of a Dihydropyridine Substituted Acid—Improved Peak Performance." Proceedings from the 6th International Symposium on Supercritical Fluid Chromatography and Extraction, Uppsala, Sweden, September 1995, p. 70.
9. A. Giorgetti, N. Pericles, H. M. Widmer, K. Anton, and P. Dätwyler, Mixed mobile phases and pressure programming in packed and capillary column supercritical fluid chromatography: a unified approach, *J. Chromatogr. Sci.*, *27*: 318 (1989).
10. T. A. Berger and J. F. Deye, Separation of benzene polycarboxylic acids by packed column supercritical fluid chromatography using methanol-carbon dioxide mixtures with very polar additives, *J. Chromatogr. Sci.*, *29*: 141 (1991).

11. L. Karlsson, T. Buttler, and L. Mathiasson, Retention and peak shape of acidic substances in capillary supercritical fluid chromatography with binary and ternary mobile phases, *J. Microcol. Sep.*, *4*: 423 (1992).

12. O. Gyllenhaal and A. Karlsson, "Chemometric Optimization of the Separation of Two Dihydropyridine Analogues by Packed Column SFC," Proceedings from the 6th International Symposium on Supercritical Fluid Chromatography and Extraction, Uppsala, Sweden, September 1995, p. 67.

13. O. Gyllenhaal and J. Vessman, "Packed Column SFC for the Purity Control of a Dihydropyridine," Proceedings from the 6th International Symposium on Supercritical Fluid Chromatography and Extraction, Uppsala, Sweden, September 1995, p. 68.

14. O. Gyllenhaal, L. Tekenbergs and J. Vessman, "Comparison of the Purity Control of a Dihydropyridine by LC and SFC," 6th International Symposium on Supercritical Fluid Chromatography and Extraction, Uppsala, Sweden, September 1995, p. 66.

15. L. Karlsson, O. Gyllenhaal, A. Karlsson, and J. Gottfries, Packed column SFC of a new dihydropyridine drug based on direct injection of samples, *J. Chromatogr.*, *749*: 193 (1996).

16. O. Gyllenhaal and J. Vessman, Capillary supercritical fluid chromatography of aliphatic amines. Studies on the selectivity and symmetry with three different columns using carbon dioxide and nitrous oxide as mobile phase, *J. Chromatogr.*, *516*: 415 (1990).

17. R. J. Ruane, G. P. Tomkinson, and I. D. Wilson, The detection of ^{14}C-propranolol following supercritical fluid chromatography using in-line radioactivity detection, *J. Pharm. Biomed. Anal.*, *8*: 1091 (1990).

18. C. J. Bailey, R. J. Ruane, and I. D. Wilson, Packed-column supercritical fluid chromatography of β-blockers, *J. Chromatogr. Sci.*, *32*: 426 (1994).

19. O. Gyllenhaal and J. Vessman, "Packed Column SFC for the Purity Control of Metoprolol Tartrate," Proceedings from the 6th International Symposium on Supercritical Fluid Chromatography and Extraction, Uppsala, Sweden, September 1995, p. 69.

20. T. A. Berger and W. H. Wilson, Packed column supercritical fluid chromatography with 220 000 plates, *Anal. Chem.*, *65*: 1451 (1993).

21. T. A. Berger, Feasibility of screening large aqueous samples from thermally unstable pesticides using high efficiency packed column supercritical fluid chromatography with multiple detectors, *Chromatographia*, *41*: 471 (1995).

22. M. Erickson, K.-E. Karlsson, B. Lamm, S. Larsson, L. A. Svensson, and J. Vessman, Identification of a new by-product detected in metoprolol tartrate, *J. Pharm. Biomed. Anal.*, *13*: 567 (1995).

23. S. Wendsjö, personal communication (1996).

24. R. C. Williams, M. S. Alessandro, V. L. Fasone, R. J. Boucher, and J. F. Edwards, Comparison of liquid chromatography, capillary electrophoresis and super-critical fluid chromatography in the determination of Losartan potassium drug in Cozaar tablets, *J. Pharm. Biomed. Anal.*, *14*: 1539 (1996).

25. W. Steuer, M. Schindler, G. Schill, and F. Erni, Supercritical fluid chromatography with ion-pairing modifiers. Separation of enantiomeric 1,2.-amino alcohols as diastereomeric ion pairs, *J. Chromatogr.*, *447*: 469 (1988).

26. J. H. Knox, B. Kaur, and G. R. Millward, Structure and performance of porous graphitic carbon in liquid chromatography, *J. Chromatogr.*, *352*: 3 (1986).

27. O. Gyllenhaal and J. Vessman, "Packed Column SFC of Substituted Benzimidazole Isomers on Bare Silica. Studies on the Equilibration of the System in the Presence of Polar and Amine Modifiers," Proceedings of the 2nd European Symposium on Analytical SFC and SFE, Riva del Garda, Italy, May 27–28 1993, pp. 231–232.

28. O. Gyllenhaal, A. Karlsson, and J. Vessman, Packed column SFC on N-alkylated omeprazole isomers. Selection of column support and organic modifier followed by equilibration studies with bare silica as support, Manuscript.

29. Q. H. Wan, P. N. Shaw, M. C. Davies, and D. A. Barret, Chromatographic behavior of positional isomers on porous graphitic carbon, *J. Chromatogr.*, *697*: 219 (1995).

30. W. C. Chan, R. Micklewright, and D. A. Barret, Porous graphitic carbon for the chromatographic separation of *O*-tetraacetyl-β-D-glucopyranosyl isothiocyanate-derivatised amino acid enantiomers, *J. Chromatogr.*, *697*: 213 (1995).

31. A. Karlsson and B. Lamm (in preparation).

32. J. H. Knox and B. Kaur, Carbon in liquid chromatography, *High-Performance Liquid Chromatography* (P. R. Brown and R. A. Hartwick, eds.), John Wiley and Sons, New York, pp. 189–222 (1989).

33. S. G. Allenmark, *Chromatographic Enantioseparation: Methods and Applications*, Ellis Horwood, Chichester (1989).

34. C. Pettersson, A. Karlsson, and C. Gioeli, Influence of enantiomeric purity of a chiral selector on stereoselectivity, *J. Chromatogr.*, *407*: 217 (1987).

35. N. Bargmann, A, Tambuté, and M. Caude, Chiralité et chromatographie en phase supercritique, *Analusis*, *20*: 189 (1992).

36. L. Siret, N. Bargmann, A. Tambuté, and M. Caude, Direct enantiomeric separation of β-blockers on ChyRoSine-A by supercritical fluid chromatography. Supercritical carbon dioxide as transient in situ derivatizing agent, *Chirality*, *4*: 252 (1992).

37. P. Biermanns, C. Miller, V. Lyon, and W. Wilson, Chiral resolution of β-blockers by packed column supercritical fluid chromatography, *LC-GC Int.*, *11*: 744 (1993).

38. N. Bargmann-Leyder, A. Tambuté, and M. Caude, A Comparison of LC and SFC for cellulose- and amylose-derived chiral stationary phases, *Chirality*, *7*: 311 (1995).

39. K. Anton, J. Eppinger, L. Fredriksen, E. Francotte, T. A. Berger, and W. H. Wilson, Chiral separations by packed-column super- and subcritical fluid chromatography, *J. Chromatogr.*, *666*: 395 (1994).

40. C. Pettersson and G. Schill, Separation of enantiomeric amines by ion-pair chromatography. *J. Chromatogr.*, *204*: 179 (1981).

41. A. Karlsson and C. Pettersson, Separation of enantiomeric amines and acids using chiral ion-pair chromatography on porous graphitic carbon, *Chirality*, *4*: 323 (1992).

11

Packed Column Supercritical Fluid Chromatography in the Development of Polymer Additives

Claire Berger

Ciba Specialty Chemicals, Inc., Basel, Switzerland

I. INTRODUCTION

The use of sub-/supercritical fluids for chromatography and/or extraction is applicable to a wide variety of compound classes. Of all the application areas that have been investigated to date, polymers and polymer additives have largely benefited from the use of sub-/supercritical fluids [1–4]. This chapter focuses on the use of packed column supercritical fluid chromatography (pSFC) in the analysis of polymer additives, one of the leading areas where supercritical fluid chromatography (SFC) as well as a related technique, supercritical fluid extraction (SFE), are presently finding widespread acceptance and use [4].

Polymer additives are found in plastics, rubbers, surfactants, textiles, and cosmetics. They are usually added at very low amounts (0.01–1.0%) to polyolefins or other polymeric materials to protect them against deterioration during processing at high temperature (200–300°C) and/or to improve their aging characteristics [5]. With increasing use of plastics, there is also an increased need for analytical information about the kinds of polymer additives, their concentration in polymers, and their likelihood of ending up in the food chain or in the environment. Typical problems to be solved in analytical departments of polymer and polymer additive manufacturers include:

Identification of unknown additive systems in commercial and developmental polymers for competitive intelligence

Determination of the composition of each additive from the early research and development stages to quality control in production

Determination of the composition of various process streams for optimization, troubleshooting, and control of production processes

Identification and quantification of related compounds and migrating species for notification to the authorities and application for food contact approval

Determination of additive levels for production control of polymers during their manufacturing

Stability of polymer additives during incorporation into polymers or during the service life of the final product [5–7]

In general, several analytical techniques operating simultaneously are required to perform a full analysis of an additive package in a polymer or to determine the migrating species in food simulants. Capillary supercritical fluid chromatography (cSFC), and pSFC have proven to be suitable for the purity control of most of the additives used in the manufacture of polymers (Table 1), which represents a major advantage over the state of the art comprising a system with dedicated instruments and methods for each polymer additive or group of polymer additives [8].

In this chapter, the performance of pSFC for determination of additives in polymers is reviewed, compared with competitive techniques, and illustrated through examples obtained during a key phase of the development of polymer additives.

II. DETERMINATION OF ADDITIVES IN POLYMERS

A. Why Stabilization of Polymers Through Additives?

It is common knowledge that under exposure to UV radiation or heat in the presence of oxygen, polymers are subject to thermooxidative degradation, which leads, through free radical mechanisms, to scission and crosslinking of the macromolecular chains, and consequently to a deterioration of the physical properties of the polymer. To avoid this phenomenon and to extend their service life, polymers are protected by mixtures of polymer additives, which include light stabilizers, antioxidants, metal deactivators, slip agents, mold release agents, plasticizers, antistatic and antiblock agents, flame retardants, biostabilizers, etc. [5].

B. Analytical Requirements for the Determination of Additives in Polymers

The analysis of additives in polymers is rather difficult due to these having a wide variety of physical (i.e., volatility and molecular weight) and chemical (i.e., amides, esters) characteristics, being present at low concentrations in the solid, often insoluble, matrix of the polymer [9]. Consequently, a number of different analytical methods including ultraviolet (UV-vis) spectroscopy, infrared (IR) spectroscopy, nuclear magnetic resonance (NMR), thin-layer chromatography (TLC), high-performance liquid chromatography (HPLC), gel permeation chromatography (GPC), gas chromatography (GC), and SFC have been used to analyze additives in polymers [10]. The direct spectroscopic

Table 1 Nonexhaustive Overview of SFE and SFC in the Analysis of Polymer Additives

Commercial name and use	Chemical structure	Chemical name	Technique	Ref.
Sodium benzoate Fungicide		Sodium benzoate MW = 145	cSFC	22
BHT, butylated hydroxytoluene, Topanol OC Antioxidant		2,6-Di-$tert$-butyl-4-methylphenol MW = 220	cSFC, pSFC, SFE	4, 6, 8, 9, 10, 19, 21, 22, 28, 31
Tinuvin P UV light absorber		2-(2'-Hydroxy-5'-methylphenyl)-2H-benzotriazole MW = 225	cSFC, pSFC	4, 19, 21, 28
Armostat 400 Antistatic agent	OH(CH$_2$)$_2$NH(CH$_2$)$_n$NH(CH$_2$)$_2$OH	N,N'-Bis-(2-hydroxyethyl)-C12-C16-amine MW = 228-334	pSFC	8, 29
BHEB, butylated hydroxyethyltoluene, Antioxidant		2,6-Di-$tert$-butyl-4-ethylphenol MW = 234	SFE, pSFC	10, 31

Table 1 Continued

Commercial name and use	Chemical structure	Chemical name	Technique	Ref.
Vulkanox NKF Antioxidant		2,2'-Isobutyllidenebis (4,6-dimethylphenol) MW = 266	pSFC	28
Irgacure 907 UV curing agent		2-Methyl-1-(4-methyl-thiopheno)-2-morpholinopropane-1-one MW = 279	cSFC	20, 25, 26
Oleamide, Kemamide O, Crodamide OR Slip, antiblock agent		Oleic acid amide MW = 281	cSFC	4, 8, 14
Tinuvin 326 UV light absorber		2-(3'-*tert*-Butyl-2'-hydroxy-5'-methylphenyl)-2H-5-chlorobenzotriazole MW = 316	cSFC, pSFC, SFE	9, 19, 23, 28
Tinuvin 320 UV light absorber		2-(2'-Hydroxy-3',5'-di-*tert*-butylphenyl)-2H-benzotriazole MW = 323	cSFC, pSFC	19, 21, 28

Chimassorb 81, Cyasorb UV 531 Light stabilizer and UV absorber		2-Hydroxy-4-*n*-(octyloxy)-benzophenone MW = 326	cSFC, pSFC	14, 19, 28
Erucamide, Kemamide E, Crodamide ER Slip, antiblock agent		Erucic acid amide MW = 337	cSFC, pSFC, SFE	4, 8, 9, 19, 25, 28, 29, 31
Cyanox 2246, Vulkanox BKF Antioxidant		2,2'-Methylenebis (4-methyl-6-*tert*-butylphenol) MW = 340	cSFC, pSFC	21, 28
Tinuvin 328 Light stabilizer		2-(2'-Hydroxy-3',5'-di-*tert*-amylphenyl)-2H-benzotriazole MW = 351	cSFC, pSFC	19, 28

Table 1 Continued

Commercial name and use	Chemical structure	Chemical name	Technique	Ref.
Tinuvin 327 Light stabilizer		2-(3,5-Di-*tert*-butyl-2-hydroxyphenyl)-5-chlorobenzotriazole MW = 357	cSFC, pSFC	8, 23, 28
Santonox R Antioxidant		4,4'-Thiobis(6-*tert*-butyl-*o*-cresol) MW = 358	cSFC	4
GMS, Atmos 150, Atmer 129, Radiamuls 142 Viscosity booster	$CH_3(CH_2)_{16}$ —$\overset{\displaystyle O}{C}$— $O-CH_2CH\cdot CH_2OH$ OH	Glycerol monostearate MW = 358	cSFC	6, 8
DIOP Plasticizer		Diisooctylphthalate MW = 390	SFE, pSFC	33

Isonox 129 Antioxidant and heat stabilizer		2,2'-Ethylidenebis(4,6-di- *tert*-butylphenol) MW = 406	SFE, pSFC	4, 10, 31
Irganox 1520 Antioxidant		2,4-Bis(octylthiomethyl) -*o*-cresol MW = 424	cSFC	14
Tinuvin 440 UV stabilizer		8-Acetyl-3-dodecyl-7,7,9,9- tetra-methyl-1,3,8- triazaspiro-(4,5)-decane- 2,4-dione MW = 436	cSFC	19
Tinuvin 120 Initiator		2,4-(Di-*tert*-butylphenyl- 3,5-di-*tert*-butyl-4-hydroxy)- benzoate MW = 438	cSFC, pSFC	8, 28

Table 1 Continued

Commercial name and use	Chemical structure	Chemical name	Technique	Ref.
Tinuvin 770 Light stabilizer		Bis(2,2,6,6-tetramethyl-4-piperidinyl)-sebacate MW = 481	cSFC, pSFC, SFE	8, 9, 19, 14, 23
Tinuvin 292 Light stabilizer		Bis-(1-methyl-2,2,6,6-tetramethylpiperidinyl)-sebacate MW = 508	cSFC	19
DLTDP, Irganox PS 800 Antioxidant		Dilauryldithiopropionate MW = 514	cSFC, pSFC, SFE	4, 9, 19, 21, 28
Irganox 1076 Antioxidant		Octadecyl-3-(3,5-di-tert-butyl-4-hydroxyphenyl)-propionate MW = 530	cSFC, pSFC, SFE	4, 6, 8, 9, 19, 14, 21, 22, 28, 29, 31

	Structure	Name / MW	Method	Ref.
Stearyl stearamide Lubricant, slip, antiblock, and mold release agent	$CH_3\text{-}(CH_2)_{16}\text{—}\overset{\displaystyle O}{\overset{\|}{C}}\text{—HN—}CH_2\text{—}(CH_2)_{16}\text{-}CH_3$	Stearyl stearamide MW = 535	SFE, pSFC	31
Topanol CA Antioxidant		1,1,3-Tris(2-methyl-4-hydroxy-5-*tert*-butylphenyl)-butane MW = 545	SFE, pSFC	33
Irganox MD 1025 Metal deactivator		*N,N*-Bis[1-oxo-3-(3,5-di-*tert*-butyl-4-hydroxyphenyl)-propane] hydrazine MW = 553	cSFC	8, 19
Hostanox SE-10, DOS 2 Antioxidant	$CH_3\text{-}(CH_2)_{17}\text{-S-S-}(CH_2)_{17}\text{-}CH_3$	Dioctadecyldisulfide MW = 571	cSFC, pSFC	8, 28

Table 1 Continued

Commercial name and use	Chemical structure	Chemical name	Technique	Ref.
Irganox 245 Antioxidant		Triethylene-glycol-bis-3-(3-*tert*-butyl-4-hydroxy-5-methylphenyl)-propionate MW = 587	cSFC, pSFC	19, 28
EBS Lubricant, processing aid		*N,N*'-Ethylenebisstearamide MW = 593	SFE, pSFC	8, 31
Calcium/Zinc stearate PVC stabilizer		Calcium/zinc stearate MW = 607/632	cSFC	22
Irganox 259 Antioxidant		*N,N*'-Hexamethylenebis (3,5-di-*tert*-butyl-4-hydroxyphenyl)-propionate MW = 610	pSFC	30

Additive / Function	Chemical name	Technique	Ref.
GDS Viscosity booster	Glycerol distearate MW = 625	cSFC	6, 8
HBCD Flame retardant	Hexabromocyclododecane MW = 641	SFE, pSFC	31
Irganox 1035 Antioxidant	2,2'-Thiodiethylenebis[3-(3,5-)di-*tert*-butyl-4-hydroxyphenyl)]-propionate MW = 643	cSFC, pSFC	19, 28
Irgafos 168 Heat stabilizer	Tris(2,4-di-*tert*-butylphenyl)-phosphite MW = 647	cSFC, pSFC	8, 9, 19, 14, 21, 22, 23, 28, 31

Table 1 Continued

Commercial name and use	Chemical structure	Chemical name	Technique	Ref.
DSTDP, Irganox PS 802 Antioxidant, plasticizer		Distearylthiodipropionate MW = 682	cSFC	4, 6, 8, 19, 14, 21
Tinuvin 144 UV light stabilizer		2-*tert*-Butyl-2-(4-hydroxy-3,5-di-*tert*-butylbenzyl)-[bis-(methyl-2,2,6,6-tetra-methyl-piperidinyl)]dipropionate MW = 685	cSFC	19, 14
TNPP Heat stabilizer		Trisnonylphenylphosphate MW = 688	SFE, pSFC	31

Irganox 1425
Antioxidant

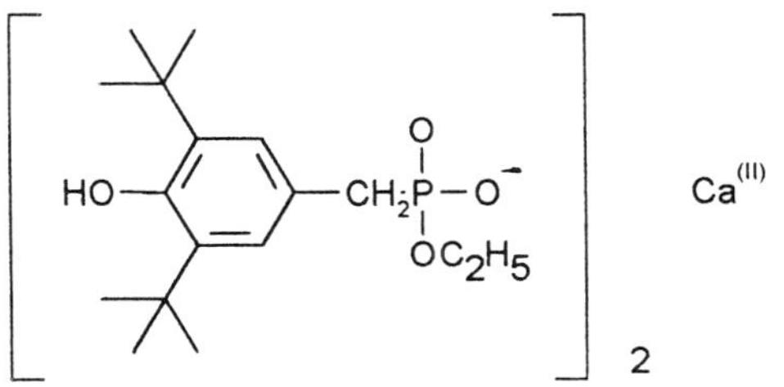

Calcium bis(monethyl-
dibutylhydroxybenzyl)-
phosphonate
MW = 695

pSFC

30

Irganox 1098
Antioxidant

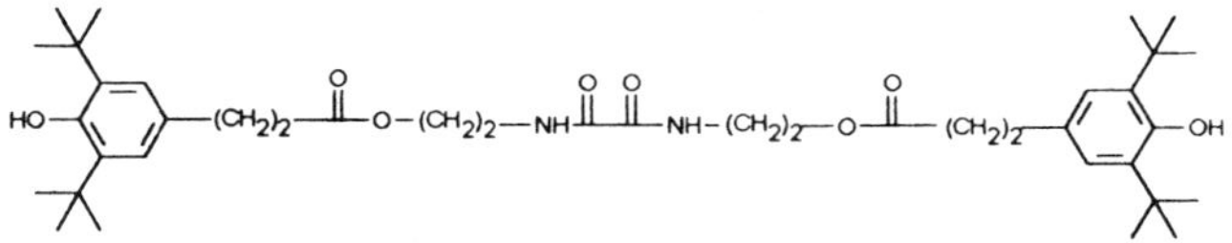

N,N'-Hexamethylenebis(3,5-
di-*tert*-butyl-4 hydroxy-
phenyl)-propionate
MW = 697

pSFC

30

Cyanox 1790
Antioxidant

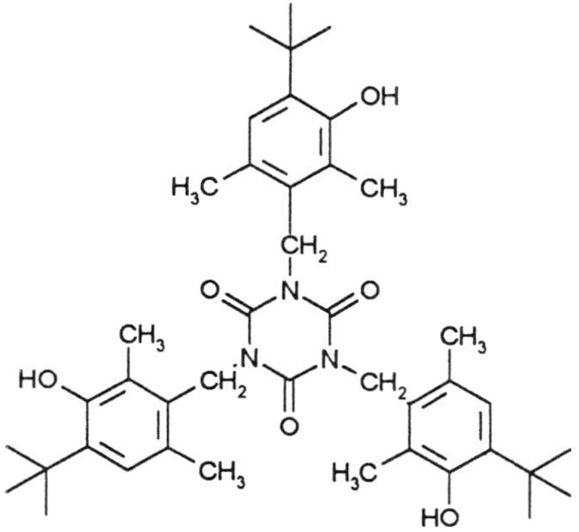

1,3,5-Tris(4-tert-butyl-3-
hydroxy-2,6-dimethylbenzyl)-
isocyanurate
MW = 700

SFE,
pSFC

31

Table 1 Continued

Commercial name and use	Chemical structure	Chemical name	Technique	Ref.
Weston 618 Color and heat stabilizer		Distearylpentaerythritol diphosphite MW = 732	SFE, pSFC	31
Irganox 1330 Antioxidant		1,3,5-Trimethyl-2,4,6-tris(3,5-di-*tert*-butyl-4-hydroxybenzyl)-benzene MW = 774	cSFC, pSFC	19, 21, 22, 28

Irganox 3114
Antioxidant

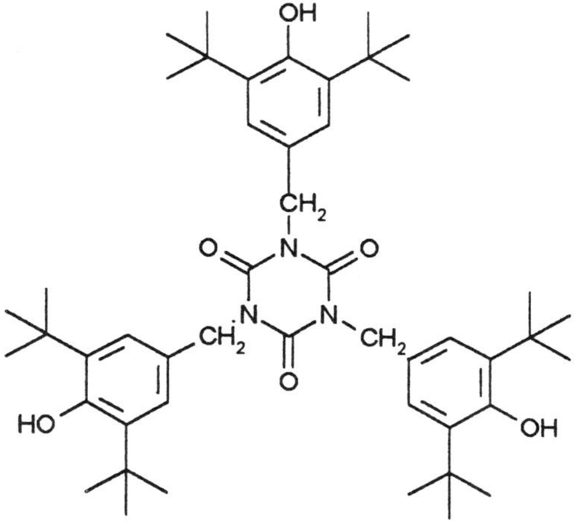

1,3,5-Tris(3,5-di-*tert*-
butyl-4-hydroxybenzyl)
isocyanurate
MW = 784

cSFC,
pSFC

19, 28

Irgafos P-EPQ
Processing
stabilizer

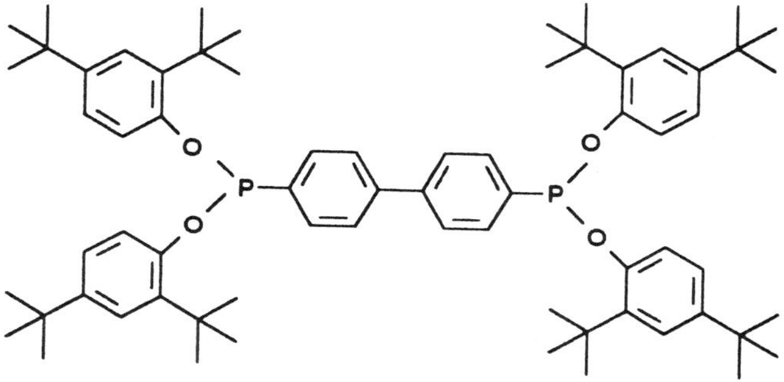

Tetrakis-(2,4-di-*tert*-
butylphenyl)-4,4'-
biphenylene diphosphonite
MW = 1035

cSFC,
pSFC

8

Table 1 Continued

Commercial name and use	Chemical structure	Chemical name	Technique	Ref.
Irganox 1010 Antioxidant		Pentaerythrity-tetrakis-[3-(3,5-di-*tert*-butyl-4-hydroxyphenyl)-propionate] MW = 1178	cSFC, pSFC, SFE	4, 6, 8, 9, 10, 19, 21, 22, 23, 28

methods (UV-vis, IR, fluorescence, phosphorescence, and X-ray fluorescence), though rapid, generally lack specificity and do not always enable quantification, in particular when the absorption frequencies of two or more polymer additives overlap or when the concentrations are too low [5,6].

Therefore, the chromatographic techniques are most commonly applied for quantitative determinations. They all require a cleaning step involving the removal of the polymer additives from the particular matrix prior to the analysis, usually some type of extraction. One of the most common extraction method is liquid or Soxhlet extraction, which is often time consuming and costly. Liquid extraction is usually followed by precipitation of the residual polymeric material and solvent concentration procedures such as evaporation or distillation before analysis [5,6,10]. Due to rising acquisition and disposal costs of organic combustible solvents, limited availability or outright disappearance of some of them, and regulatory restrictions in some locations on generation of laboratory waste, these conventional methods are being replaced by more economical and ecological methods such as the recently introduced accelerated solvent extraction (ASE) or SFE. Chromatographic separations usually enable identification and quantification of all the polymer additives contained in the extract, using either reference retention data or very specific detection methods such as mass spectrometry (MS) or Fourier transform infrared (FTIR) spectroscopy [6,7].

GC is generally limited to the separation of low molecular weight, volatile, thermally stable compounds, although some compounds with a molecular weight greater than 1000 g/mol have been recently analyzed with high-temperature GC [11,12]. Many antioxidants and light stabilizers, which are polar and of high molecular weight, are designed to be reactive and may decompose when exposed to heat in the GC system, even with the now commonly used cold on-column injection technique [5]. In addition, a variety of derivatization methods are available to convert polar, thermally unstable compounds into more stable and/or more volatile derivatives, which can easily be analyzed by GC.

HPLC, though the most convenient technique for separation of thermolabile high molecular weight substances, has some disadvantages over GC including lower separation efficiency, lack of linear mass sensitive detectors, and cost of quality solvent. UV-vis or fluorescence detectors, which are commonly used, require polymer additives to have a chromophoric moiety, while the universal refractive index detector only functions under isocratic conditions. The emerging evaporative light-scattering detector (ELSD), which is described in greater detail in Chapter 4, or HPLC-MS coupling can be used alternatively [13,14]. Both techniques have been commercially available for several years but have not been widely used for the analysis of polymer additives to date. Furthermore, owing to their low solubility in aqueous solutions, polymer additives may need to be chromatographed on nonaqueous reversed phase HPLC systems [15].

A number of reviews and monographs dealing with the analysis of polymer additives and the analytical methods issued by the polymer additive manufacturers constitute the basic literature for determination of additives in polymers [5,16–18]. In this classical literature, a combination of analytical methods is recommended to quantitatively analyze an additive package in a polymer. According to recent publications, the use of SFC for the analysis of additives in polymers could dramatically reduce the analytical effort, in particular when used in combination with SFE. cSFC is by far the most widespread technique for the analysis of polymer additives but applications where pSFC appears to be a attractive alternative are discussed in this chapter.

C. pSFC in Polymer Analysis

The use of SFC in polymer analysis goes back to the pioneer work of Greibrokk and Raynor in the late 1980s [8,19]. For a long time, standard mixtures of commercial polymer additives were used to demonstrate the suitability of SFC for determination of additives in polymers. Real-world problems of the polymer industry have been considered in the literature only recently, mostly boosted through advances of SFE, which allow one to selectively extract additives from the polymer matrix and simultaneously from the low molecular weight oligomers [4,20].

Raynor et al. first evaluated cSFC as a general method for polymer additives characterization with a series of 21 polymer additives commonly used in the stabilization of polyolefins [19]. Greibrokk and Arpino compared cSFC and HPLC for characterization of polymer additives in polyolefin extracts and found cSFC to be superior in terms of both resolution and speed of analysis [8,21]. Furthermore, Knowles et al. enriched the range of polymer additives separated by cSFC with nine commonly used polymer additives extracted from an olefin stream [4]. Some of these could be analyzed by GC, others by HPLC. cSFC was the only technique that could separate all of them in a single run. Since then, many other polymer additives and other matrices have been investigated and successfully analyzed with cSFC, thus demonstrating the suitability of this method for the analysis of polymer additives [6,22,23]. Taking advantage of an easy coupling with a highly sensitive or specific detector such as MS or FTIR, cSFC, which has to cope with limitations such as longer analysis time, difficult elution of polar compounds, lower loadability of the columns, and lower reproducibility and accuracy of quantitative results, is by far the dominant technique in the analysis of polymer additives as illustrated in the abundant cSFC literature [6,14,19–21,24–26].

pSFC, which was explored for polymer analysis early on, is still awaiting a breakthrough for the analysis of polymer additives [8,27–29]. Advantages of pSFC, such as speed of analysis and loadability of the column, which are higher than in cSFC, and the good reproducibility of quantitative results, are determinant factors for the acceptance and validation of a new method in routine quality assurance/quality control (QA/QC). The resolution offered by pSFC is generally sufficient to determine polymer additives in most common polymer samples. In addition, pSFC was in many cases found to be more rapid than the corresponding analysis on the same column in HPLC or reduced the amount of methods to be used [8,24,30]. Some applications have been developed on packed capillary columns that combine the possibility of using equipment designed for cSFC with the benefits of pSFC [8]. This technique is described in greater detail in Chapter 5.

Besides UV-vis absorption, the most commonly used detection techniques in pSFC are ELSD (see Chapter 4) and MS coupling. These techniques are widely used in other application areas and may also be useful for the analysis of polymer additives [24].

Kithinji et al. reported the analysis by pSFC of 19 commercial antioxidants and light stabilizers with a wide range of relative molecular masses and boiling points, present in five commercial polypropylene samples [28]. Using either neat carbon dioxide or methanol-modified carbon dioxide and UV-vis absorbance detection, they found conditions that allowed rapid, well-resolved separations.

Doehl et al. observed that compounds containing amido or amino groups like Oleamide, Erucamide, and Armostat 400 were eluted neither from a ODS-bonded silica microbore nor from a cyanopropyl conventional packed column, with neat carbon dioxide or nitrous oxide as the mobile phase [29]. On the other hand, the disulfide Hostanox SE-10 as well as the aromatic compounds Irganox 1076 and Tinuvin 327 were readily eluted under the same pSFC conditions.

In another study, HPLC and packed capillary SFC (pcSFC) separations were compared for more than 15 polymer additives [8]. Detection in HPLC was by ELSD or refractive index, in pcSFC by ELSD and FID. pcSFC was found to be more universally applicable to these compounds than HPLC. Glyceryl monostearate, an additive without strong chromophores, whereby UV-vis detection is difficult, has been successfully detected with ELSD. The elution pattern was similar in pcSFC and cSFC, according to the molecular weight, in contrast to the pattern obtained in normal phase HPLC, where the triglyceride peak eluted first. N,N'-Ethylbisstearamide, a fatty amide for which no satisfactory HPLC method was available, could be successfully chromatographed with pcSFC. The solubility in neat carbon dioxide was low, but addition of n-propanol, heated injection, and ELSD finally lead to separation of three components in pcSFC.

D. A Performant Analytical Tool: On-Line SFE-pSFC

SFE as a powerful technique to eliminate solvent or matrix effects before chromatographic analysis was combined with pSFC for determination of polymer additives in various types of polymers. The most significant investigations in this area are reviewed in the following section.

1. Polyolefins

Several publications have shown quantitative results for on-line SFE-pSFC from a polyethylene matrix. Low-density polyethylene (LDPE) has been extracted and analyzed with SFE-pSFC with good yields for some of the common polymer additives including BHT, BHEB, Isonox 129, Irganox 1076, Erucamide, and Irganox 1010. The calculated levels of Isonox 129, Irganox 1076, and Irganox 1010 were in good agreement with the assayed values from the polymer manufacturer. Owing to coelution of BHT and BHEB, it was not possible to measure concentrations for these compounds. The relative standard deviation (RSD) was less than 5% for additive amounts ranging from 200 to 250 ppm in the polymer, which is comparable to most conventional techniques [5,10].

In another study, Ashraf-Khorassani et al. used an on-line SFE-pSFC system with a cryofocusing trap to quantitatively analyze Irganox 1076, TNPP, and Weston 618 in polyethylene; Irgafos 168, Cyasorb 3348, and Cyanox 1790 in polyethylene pellets; and BHT, BHEB, Isonox 129, Irganox 1076, and Irganox 1010 in LDPE pellets [31].

Furthermore, Ryan et al. also described a procedure to quantitatively analyze polymer additives in various polyethylene and polypropylene samples from several manufacturers using on-line SFE-pSFC [9]. Successful extractions and analysis were performed on 10 different polymer additives ranging in molecular weight from 220 (BHT) to 1178 (Irganox 1010). Using the area response factor, the measured concentrations of Irgafos 168 and Irganox 1010 were within 5% of the concentration supplied by the manufacturer. Furthermore, either oligomers or polymer additives were selectively extracted depending on the pressure and did not interfere with the chromatographic separation.

2. PVC

Organotin compounds are used as polyvinylchloride (PVC) stabilizers to prevent thermal dehydrochlorination of the polymer during processing and to strengthen the finished product against long-term breakdown by sunlight.

Oudsema et al. demonstrated the applicability of SFE-pSFC for routine analysis of dialkylorganotin stabilizers in rigid PVC materials using formic acid–modified carbon dioxide as extraction solvent and mobile phase for chromatography [32]. Ionic organotin compounds are converted to formate salts during extraction and/or chromatography. The formed salts are relatively inert to further substitution in the formic acid–rich mobile and stationary phase and the ruggedness of the separations obtained by pSFC with formic acid–modified carbon dioxide is very impressive. The results obtained for the quantitative analysis of the dimethyltin additive in PVC by SFE-pSFC are very reproducible:

RSD of 2.9% (n = 6) for a 1.45% additive concentration, calculated as dimethyltin.

In another study, Hunt et al. used off-line SFE-pSFC to determine DIOP and Topanol AC in plasticized PVC [33]. pSFC combined good resolution, rapid analysis time (less than 5 min), and linear calibration over a large range of concentrations (40–960 ppm for Topanol AC and 80–2000 ppm for DIOP). The SFE and the liquid extraction results were in good agreement for DIOP. Methanol was required to extract quantitatively the Topanol AC, which is more polar.

Furthermore, Imahashi et al. directly coupled SFE to semipreparative pSFC for the fractionation of additives in PVC films [34]. The effluent from the pSFC column was separated into five fractions following detection with a multiwavelength photodiode array detector. The individual fractions were then analyzed by HPLC and IR spectroscopy. Fraction collection in general is discussed further in Chapters 3, 12 and 14.

3. Other Polymeric Materials

Ashraf-Khorassani et al. used an on-line SFE-pSFC system with a cryofocusing trap to quantitatively analyze N,N'-ethylbisstearamide in polystyrene; HBCD in styrofoam; stearyl stearamide and Irganox 1010 in Ethofoam; and BHT, Irganox 1076, stearamide, and Erucamide in a polymer film [31].

III. NEW HORIZONS IN THE DEVELOPMENT OF POLYMER ADDITIVES

According to the potential of pSFC for the analysis of additives in polymers as demonstrated in the literature, the goal of our investigations was to use pSFC as an analytical tool to support the process development of polymer additives. Process development is defined as the step whereby a laboratory synthesis is scaled up to an economical production process. Every new polymer additive to be launched on the market or any new production process that may improve the product quality or handling form, production cost, working conditions in the plant, or ecology of the process goes through this step.

A. New Analytical Methods in Process Development

Process development occupies a key position in speeding up the development time of new polymer additives and in reducing production costs of those already commercialized. To cope with this challenge, process development needs an efficient, highly flexible, and highly responsive analytical support, which allows a faster and better understanding of the chemistry and of the process.

In development as in research, process conditions and even raw materials can change from day to day, unlike QA/QC, where the composition of the samples is almost constant and has to be compared with an established specification. The samples to be analyzed in process development always have different compositions. Depending on the stream to be analyzed the concentration of a compound can vary from ppm in some diluted waste streams to more than 99% for the isolated product. Highly reactive raw materials, products, or unstable intermediates may be present in the samples. The sample may contain high molecular weight or polar compounds that do not elute from the column. New related compounds might suddenly appear. Therefore, the requirements of every analytical method used in process development include a high level of reproducibility, accuracy, ruggedness, flexibility, linearity over a wide range of concentration, as well as a good resolution to allow detection of new related compounds. Furthermore, the time invested in method development must be as short as possible.

B. pSFC in the Strategy of Method Development

In the late 1980s Giorgetti et al. [30] published a chromatographic separation developed for routine formulation control of polymer additive mixtures (Fig. 1). The quality control of polymer additives formulations was usually done by HPLC. When Irganox 1425, a calcium salt, was present in the mixture, more than one method had to be applied. Formulation control of these polymer additives mixtures by HPLC was time consuming and required a great deal of solvent. Therefore a pSFC method has been developed to allow analysis of all components of the mixture in one run within 6 min by using a binary modifier with carbon dioxide and pressure programming. According to these promising preliminary results, pSFC was included to our strategy of method development (Fig. 2). This strategy helps optimizing the use of chromatographic resources and improving the responsiveness of the analytical support.

GC is the method of choice for many thermostable polymer additives and related compounds that have enough volatility and a low molecular weight. However, the reproducibility of GC injection is relatively poor and quantitative analysis of high-purity samples (>98%) can be tedious and time consuming. Therefore, the common practice is to use HPLC with an external standard to quantify the main compound for which a reference sample is usually available. The related compounds are analyzed separately in GC with internal standard and the mass-sensitive flame ionization detection. When reference samples are not available, the GC peaks are identified by GC-MS and theoretical factors based on the carbon number can be used to obtain accurate quantitative results after flame ionization detection. Most of the time a full analysis requires two methods.

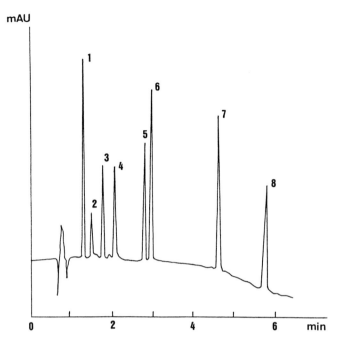

Figure 1 pSFC chromatogram of a synthetic mixture of polymer additives. pSFC conditions: precolumn, 60 × 4 mm, 3 μm, Spherisorb C8; column, 125 × 4 mm, 3 μm, Spherisorb CN; mobile phase, carbon dioxide, 2 mL/min; modifier, 0.15% citric acid in methanol, 0.24 mL/min; temperature, 90°C; pressure gradient, 130–200 bar in 4 min, 200–350 bar in 2 min; detector, UV-vis at 278 nm. 1) Metilox, 2) Irgafos 168, 3) Irganox 1076, 4) Irganox 1425, 5) Irganox 259, 6) Irganox 1330, 7) Irganox 1010, 8) Irganox 1098. Reproduced from the *Journal of Chromatographic Science* by permission of Preston Publications, A Division of Preston Industries, Inc. [30].

For samples that are not amenable with GC, the strategy proposes next to test their solubility in neat carbon dioxide. If they are soluble in neat carbon dioxide, the method of choice will be cSFC.

pSFC can be used to analyze compounds that are not soluble in neat carbon dioxide but can be dissolved in, say, a methanol-carbon dioxide mixture. Good quantitative results are generally obtained in pSFC with UV-vis detection, provided the sample contains compounds having a chromophoric moiety and the peaks are satisfactorily resolved. For compounds that do not absorb in UV-vis detection, ELSD would be a preferable detection alternative. If ELSD is not available, derivatization techniques can be used to add a chromophoric moiety

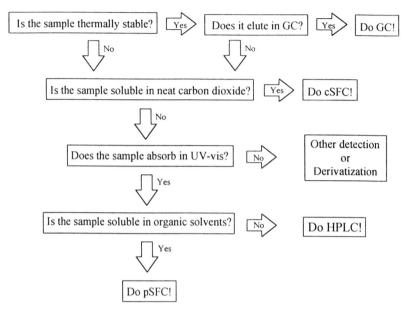

Figure 2 Role of pSFC in the strategy of method development.

to the molecules but they are generally more time consuming and less flexible for quantitative analysis.

HPLC, which usually requires longer analysis time than pSFC for a similar quality of separation and which usually produces more waste solvent, is only recommended for those samples containing compounds having a chromophore, which are too polar or have too high a molecular weight to be eluted in pSFC.

C. Troubleshooting of a Production Process

Troubleshooting of the Tinuvin P production process is used to exemplify the suitability of the strategy of method development described above (Fig. 2) for the analytical support of process development. The method development was carried out with a synthetic mixture of the seven most relevant compounds involved in the process. The mixture contained amino and phenolic compounds. All compounds but one (6) could be analyzed in GC provided a cold on-column injection device was used. All compounds were soluble in neat carbon dioxide and the separation could be alternatively carried out in cSFC. All compounds were soluble in a methanol-carbon dioxide mixture and contained chromophores allowing a separation with pSFC. The major steps of the pSFC method development are discussed in the following sections. Furthermore, several HPLC methods were available but none allowed the separation of all compounds in one run.

1. Choice of the Stationary Phase in pSFC

A few preliminary injections showed that compounds (1) and (2) were very difficult to resolve in pSFC and that the peaks of polar compounds tended to tail. We decided therefore to systematically investigate the influence of the stationary phase on selectivity and peak form for five commercial stationary phases.

A comparison of the retention on a cyanopropyl (CN) and an amino (NH_2) column is shown in Fig. 3. Peaks (1) and (2) which differed only by the number of double bonds present in the molecule, were not totally resolved on the amino column and coeluted on the cyanopropyl column. Peaks (6) and (7), which were well separated on the amino column, coeluted on the cyano column but could probably have been separated through optimization of the pressure and/or modifier gradient.

The elution pattern was greatly modified, reflecting important differences in the selectivity behavior of the two stationary phases involved. On the amino column, which is well suited to separate basic compounds and has a unique selectivity for hydrogen bond acceptors, benzotriazoles (1) and (2) eluted first, followed by the amino compound (3), phenylenediamine (4), p-cresol (5), benzotriazole (6), and, finally, amino-p-cresol (7). On the cyano column, which is well suited for the separation of polar compounds, the amino compound (3) eluted first followed by benzotriazoles (1) and (2) coeluting with p-cresol (5), phenylenediamine (4), and, finally, benzotriazole (6) and amino-p-cresol (7). This elution order reflects an increasing degree of dipole interactions with the stationary phase, which is typical for a cyanopropyl phase.

The peak form was much better on the cyano column, indicating selective interactions with polar compounds. On the amino column, the peaks of phenylenediamine (4) and amino-p-cresol (7) tailed, indicating strong interactions with the stationary phase or insufficient solubility in the mobile phase under the chosen conditions (2% of modifier only).

A comparison of a cyano (CN), a diol, a bear silica, and an ODS-bonded silica column is given in Fig. 4. For this comparison new pressure and modifier gradients were used. Peak (2) and (3), which coeluted on the cyano column, were well separated on the diol column, which is well suited for acidic compounds. Compound (5), which gave a fine symmetrical peak on the cyano column, was totally adsorbed on the diol column, indicating very strong inter-

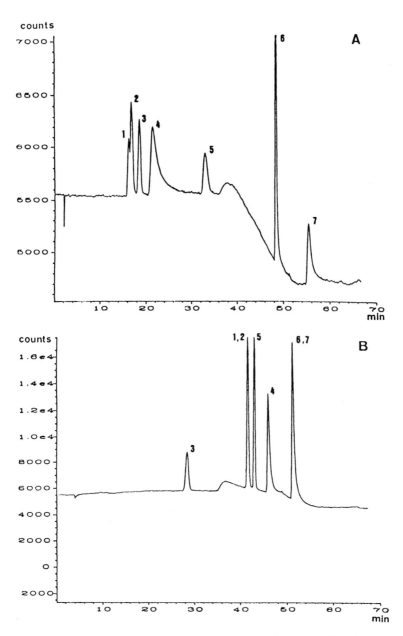

Figure 3 pSFC chromatograms of a synthetic mixture of Tinuvin P and related compounds on different stationary phases. pSFC conditions: column, 2 × 250 × 4 mm, 10 μm, (A) Spherisorb NH$_2$ and (B) CN; mobile phase, carbon dioxide, 2 mL/min; modifier, 2% (35 min) methanol to 5% (15 min) in 10 min; temperature, 100°C; pressure gradient, 130 bar (35 min) to 350 bar (15 min) in 15 min; injection, 5 μL; detector, UV-vis at 286 nm. 1) Tinuvin P, 2) benzotriazole (2), 3) amino compound, 4) phenylenediamine, 5) *p*-cresol, 6) benzotriazole (6), 7) amino-*p*-cresol.

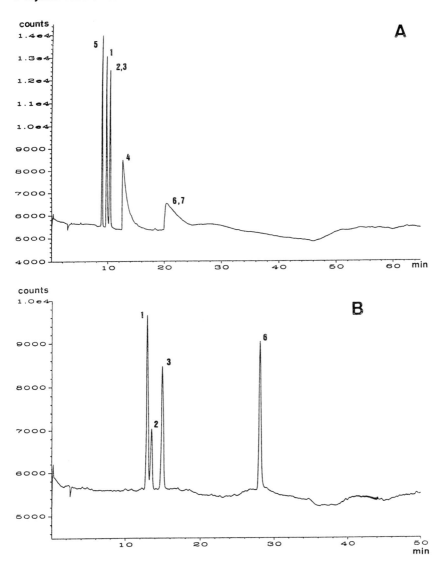

Figure 4 pSFC chromatograms of a synthetic mixture of Tinuvin P and related compounds on different stationary phases. pSFC conditions: column, $2 \times 250 \times 4$ mm, 10 μm, (A) Spherisorb CN, (B) Lichrosorb diol, (C) Spherisorb silica, and (D) Nucleosil ODS; mobile phase, carbon dioxide, 2 mL/min; modifier, 2% (12 min) methanol to 5% (18 min) in 5 min; temperature, 100°C; pressure gradient, 150 bar (25 min) to 350 bar (10 min) in 10 min; injection, 5 μL; detector, UV-vis at 286 nm. Peak identification, see Fig 3.

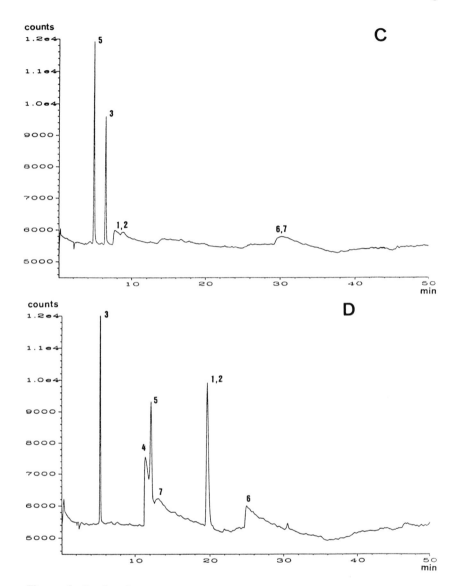

Figure 4 Continued

actions of the stationary phase with hydrogen bond donors. Compound (4) and (7), which gave broad tailing peaks on the cyano column, did not elute at all on the diol column, reflecting again very strong interactions of the stationary phase with these highly functionalized molecules (phenylenediamine and amino-*p*-cresol). The peak form of compound (6) was greatly improved on the diol column, showing selective interactions of this stationary phase with this benzo-triazole. Furthermore, the elution order was similar on the cyano and on the diol column.

On the silica column, only compounds (5) and (3) were properly eluted. These compounds both contained a single active group, capable of building hydrogen bonds with the stationary phase. Broad, tailing peaks were obtained for compounds (1), (2), (6), and (7). Phenylenediamine (4) was totally retained. The elution pattern was not greatly affected, when going from a diol to a silica column. Only compound (3) was moved forward and compounds (1) and (2) backward.

On the ODS column, which is well suited for the separation of nonpolar high molecular weight solutes, the elution order was greatly affected. Compound (3) eluted first; followed by compounds (4), (5), and (7), which were not resolved; then compound (1) and (2), which coeluted; and, finally, compound (6). Symmetrical peaks were obtained for compounds (3), (5), (1), and (2). The compounds containing more than one functional group tailed significantly.

In conclusion, the changes of stationary phase did not affect the analysis time, which was mainly dependent on the modifier and pressure gradients. It also did not greatly affect the resolution of peaks (1) and (2) which could only be improved by increasing the column length as suggested by Berger [35]. However, the nature of the stationary phase had a great influence on the selectivity behavior and on the elution pattern as summarized in Table 2. Attempts to combine a cyano and an amino column did not improve this separation either, but we observed that the order in the series influenced both peak shape and elution pattern. This feature can be used to tune selectivity as illustrated for polymer additives on Fig. 1 and as discussed for other compound classes in Chapter 6.

Symmetrical peaks were obtained on the amino, cyano, diol, and ODS columns for benzotriazoles (1) and (2). In a similar way, the amino compound (3) gave a symmetrical peak on every stationary phase. Phenylenediamine (4)

Table 2 Chromatographic Behavior of Tinuvin P and Related Compounds

Compound	Functional groups	Stationary phase				
		Amino	Cyano	Diol	Silica	ODS
1	-OH	S	S	S	T	S
2	-OH	S	S	S	T	S
3	$-NH_2$	S	S	S	S	S
4	$2 \times -NH_2$	T	T	—	—	T
5	-OH	S	S	—	S	S
6	-OH, N->O	S	T	S	T	T
7	$-OH -NH_2$	T	T	—	T	T

S, symmetrical peak; T, tailing peak; —, no elution.

was best eluted on the cyano column but tailed slightly. *p*-Cresol (5) eluted with a symmetrical peak on the amino, cyano, silica, and ODS columns but was totally retained on the diol column. Symmetrical peaks were obtained on the amino, cyano, and diol columns for compound (6) and none of the stationary phases but one allowed a symmetrical peak for amino-*p*-cresol (7).

In addition, the modifier and pressure gradients were determinant to obtaining symmetrical peaks for the most polar compounds, phenylenediamine (4) and aminophenol (7). For most coeluting pairs apart from compound (1) and (2), conditions could be found to obtain a satisfactory resolution through modification of the modifier or pressure gradient. The best separation was obtained on two 5-μm particle size amino columns in series.

2. Influence of Temperature

In the preliminary study, temperature was found to have a great influence on the separation of benzotriazole (1) and (2), which differ only by the number of double bonds. The best selectivity was achieved at 100°C. Furthermore, we observed that the retention behavior of compound (4) was greatly influenced by the temperature, demonstrating the dominance of volatility over solubility in the retention mechanism of this compound. For this compound only, the retention time decreased for increasing temperatures and peak form was simultaneously improved at higher temperature. The retention behavior of the other compounds was not greatly affected by temperature. However, decomposition may occur at higher temperatures, as seen in Fig. 5 for compound (6) at 120°C.

3. Choice of the Modifier

The polarity of the modifier can also be used to tune the selectivity of a separation. Figure 6 shows a comparison of four different modifiers. First methanol was replaced by ethanol, which has a slightly lower polarity. Neither the elution order nor the peak form was affected. Only a loss of resolution for peaks (1), (2), and (3) which were well resolved with methanol, could be observed, indicating some sensitivity of these compounds to the modifier polarity.

With isopropanol, the resolution was lost for peaks (2) and (3) but could probably have been restored through optimization of the pressure or the modifier content. Compound (7) was no longer eluted but may have been eluted with a higher concentration of modifier.

With acetonitrile, peaks (1), (2), and (3) were badly resolved and compound (4) gave a broad tailing peak. The other compounds did not elute at all. Acetonitrile having a much lower polarity, the concentration of modifier was probably too low to achieve elution of these compounds. With the addition of 0.05% trifluoracetic acid in methanol as modifier, compounds (4) and (7) were totally retained. Fine symmetrical peaks were obtained for the other compounds. The elution order was not modified. No basic additive has been tested.

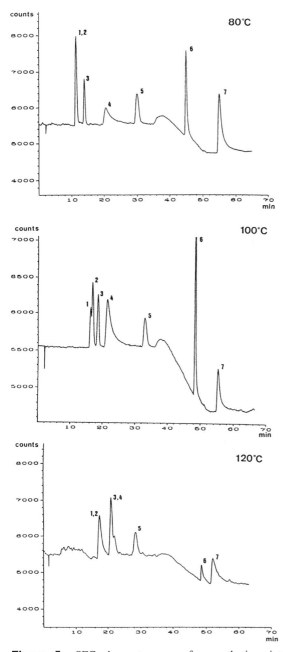

Figure 5 pSFC chromatograms of a synthetic mixture of Tinuvin P and related compounds at different temperatures. Peak identification and experimental conditions, see Fig. 3.

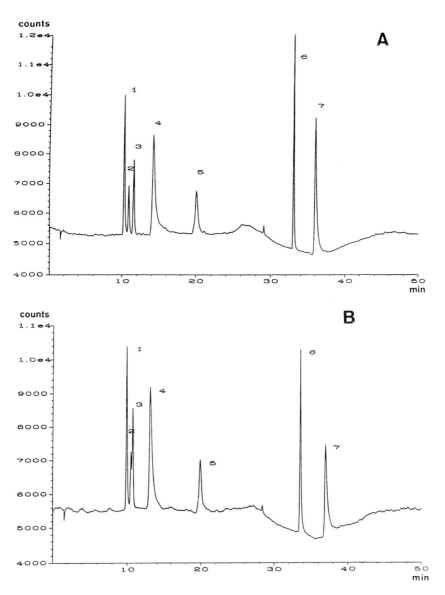

Figure 6 pSFC chromatograms of a synthetic mixture of Tinuvin P and related compounds with different modifiers. pSFC conditions: column, 2 × 250 × 4 mm, 5 μm, Spherisorb NH₂; mobile phase, carbon dioxide, 2.5 mL/min; modifier, (A) methanol, (B) ethanol, (C) isopropanol, (D) acetonitrile, 2% (29 min) to 5% (10 min) in 9 min; temperature, 100°C; pressure gradient, 110 bar (35 min) to 350 bar in 10 min; injection, 5 μL; detector, UV-vis at 286 nm. Peak identification, see Fig. 3.

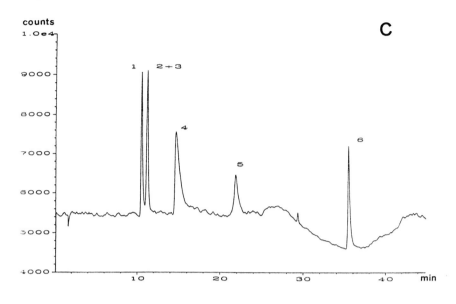

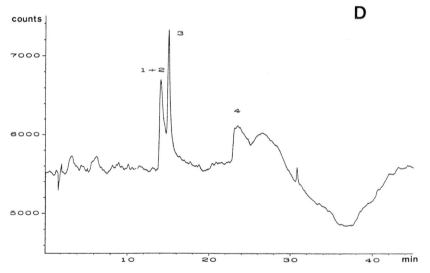

4. Reproducibility and Comparison with Other Techniques

The reproducibilities of peak areas and retention time were calculated for pSFC and cSFC after optimization of the chromatographic conditions (Table 3). Retention times were reproducible in pSFC even with only 2% modifier during the first 15 min of the gradient. The reproducibility of retention times is in good agreement with the data published by Anton et al. on the same type of equipment and improves for higher contents of modifier [35]. In cSFC a very good

Table 3 Reproducibility of the pSFC and cSFC Separation of Tinuvin P and Related Compounds

Compound	Injected amount (% in THF)	pSFC Ret. time (min)	pSFC Ret. time (% RSD)	pSFC Area % (% RSD)	pSFC Limit of detection (ppm)	Injected amount (% in toluene)	cSFC Ret. time (min)	cSFC Ret. time (% RSD)	cSFC Area % (% RSD)	cSFC Limit of detection (ppm)
1	0.02	12.67	±0.7	±1.3	10	0.9	20.33	±0.06	±0.1	50
2	0.02	13.39	±0.7	±2.3	10	0.02	20.65	±0.06	±1.7	N/A
3	0.06	14.44	±0.6	±1.6	10	0.02	12.74	±0.06	±0.9	N/A
4	0.15	19.15	±0.5	±4.7	150	0.02	5.49	±0.06	±2.1	N/A
5	0.14	22.36	±0.2	±1.3	50	0.02	3.99	±0.05	±0.6	N/A
6	0.05	27.61	±0.1	±1.9	10	0.02	26.03	±0.02	±2.9	N/A
7	0.13	34.27	±0.1	±2.0	50	0.02	6.11	±0.07	±1.6	N/A

pSFC conditions: column, $2 \times 250 \times 4$ mm, 5 μm, Spherisorb NH_2; mobile phase, carbon dioxide, 2.5 mL/min; modifier, 2% (15 min) methanol to 5% (15 min) in 5 min; temperature, 100°C; pressure gradient, 150 bar (15 min) to 350 bar (5 min) in 15 min; injection, 5 μL; detector, UV-vis at 286 nm. cSFC conditions, see Fig. 3; peak identification, see Fig. 7; N/A, not available.

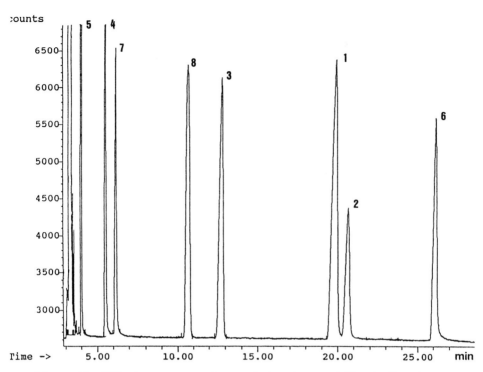

Figure 7 cSFC chromatogram of a synthetic mixture of Tinuvin P and related compounds. cSFC conditions: column, 10 m × 50 μm, 0.25 μm film, SB biphenyl 30; mobile phase, carbon dioxide, 6 μL/min, temperature, 130°C; column inlet pressure gradient, 135 bar (5 min) to 190 bar in 18 min, 190–350 bar in 16 min; injection, 0.5 μL; detector, flame ionization detection, 400°C. Peak identification, see Figure 3.

separation was obtained in less than 30 minutes on a SB biphenyl 30 column (Fig. 7). The retention times were very reproducible with RSD below 0.07%.

The reproducibility of the peak areas was satisfying in pSFC for every peak a part from tailing peak (4), for which the integration could not be performed in a reproducible way. The peak areas in cSFC were slightly better reproducible than in pSFC with RSD at maximal 2.9% (Table 3).

This observation seems to be specific to this separation and cannot be generalized to other separations. According to Anton et al., the relative standard deviation of peak areas in pSFC on this type of equipment can be as low as 0.44% compared to 0.93% in HPLC [36].

The detection limit for Tinuvin P (1) was 50 ppm in cSFC with flame ionization detection, whereas in pSFC it was in the range of 10 ppm for compounds

(1), (2), (3), and (6); 50 ppm for compounds (5) and (7); and 150 ppm for the tailing compound (4).

5. Benefits of the Strategy of Method Development

Applying the strategy of method development illustrated in Fig. 2 allowed optimization of the analytical procedure and improvement of the analytical support as follows: In the particular case of troubleshooting of the Tinuvin P production process, although every compound was eluted in GC, a second method was required as an orthogonal method to verify the accuracy of the quantitative results obtained for compound (6), which might have been reversed to Tinuvin P under heat exposure in the GC injector. Either cSFC or pSFC can be used for this purpose.

According to the strategy of method development, GC proved to be the method of choice for determining Tinuvin P and related compounds in most of the samples. However, for samples having a Tinuvin P content higher than 98%, quantification of the main compound in GC or cSFC with internal standard had a high relative standard deviation for triple replicates. As the presence of non-eluting species was suspected, quantification of Tinuvin P via the related compounds was not possible either. pSFC was favorably used to quantify the main compound in those highly concentrated samples. According to the strategy, there was no need for an HPLC method.

Two chromatographic methods were still needed to complete a full analysis of a sample coming from the Tinuvin P production process. However, a new analytical procedure containing a reworked GC method for the related compounds and a pSFC method for the main compound improved accuracy and analysis time and reduced the solvent consumption 12-fold from 1.15 L/day to less than 0.1 L/day.

D. Characterization of Commercial Polymer Additives

This section shows some examples where pSFC was used as an orientation tool for process development. Qualitative analysis gave information that helped to define the problem and set up an appropriate action plan.

1. Qualitative Analysis of Irganox 1010

A method was required for qualitative comparison of Irganox 1010 samples from different production plants with a minimum development time. Because Irganox 1010 is a well-known molecule in pSFC, we were able to provide the required information within hours whereas implementation of an existing method on our HPLC apparatus and analysis of the samples would have taken at least a day. An example chromatogram is given in Fig. 8.

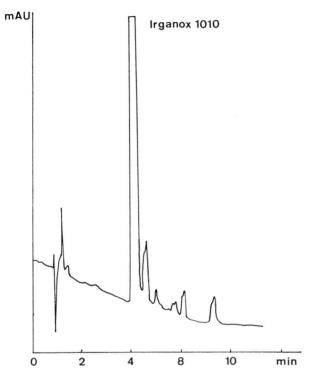

Figure 8 pSFC chromatogram of an Irganox 1010 sample. pSFC conditions: column, 125 × 4 mm, 3 μm, Spherisorb CN; mobile phase, carbon dioxide, 2 mL/min; modifier, methanol; temperature, 60°C; pressure gradient, 200–300 bar in 20 min.

2. Analysis of Chimassorb 944

Chimassorb 944 is a polymeric additive having a very high molecular weight. In addition to GC, which gives information on the volatiles, the most popular method to characterize this polymer additive is GPC. Monomers and low oligomers have not been characterized so far but may play a role in the aging of polymers stabilized with this additive. We have developed a pSFC and a cSFC method allowing to qualitatively characterize monomers and low oligomers in this polymeric additive (Fig. 9). These compounds could not be quantified before whether by GC or GPC. pSFC allowed the elution of compounds with a higher molecular weight compared to cSFC (Fig. 9).

3. Analysis of Wingstay L

A pSFC method was developed to analyze a sample of Wingstay L. Again conditions could be found so that pSFC could favorably replace HPLC (Fig. 10).

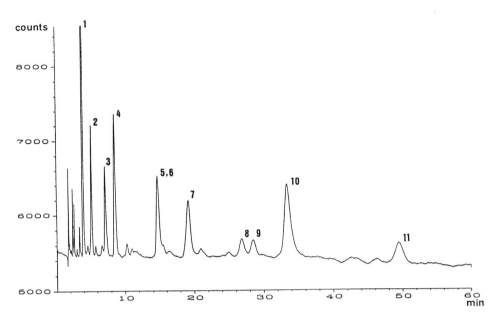

Figure 9A Chromatograms of a Chimassorb 944 sample. pSFC conditions: column, 250 × 4 mm, 10 μm, Nucleosil NH₂; mobile phase, carbon dioxide; modifier 20% methanol; temperature, 100°C; pressure 350 bar; injection, 5 μL; detector, UV-vis at 230 nm.

The better resolution obtained in pSFC allowed to optimize the HPLC method and separate coeluting peaks. The elution pattern was similar in pSFC and in HPLC.

E. Development of New Polymer Additives

This section is dedicated to examples where quantitative analysis was required to investigate the mass balance of the process. Mass balances are necessary to optimize the cost performance of a process.

1. Analysis of Tinuvin 360

A method was needed to support the development of Tinuvin 360. Because Tinuvin 360 has too high a molecular weight to elute properly in GC, a pSFC method has been developed to quantify Tinuvin 360 and the main compounds appearing during synthesis. Due to the tailing of compound (3) the method was not as reproducible as the corresponding HPLC but could probably have been optimized by increasing the temperature. However, pSFC allowed us to discover the existence of isomers that were not known at this stage of the project and to improve the existing HPLC method (Fig. 11).

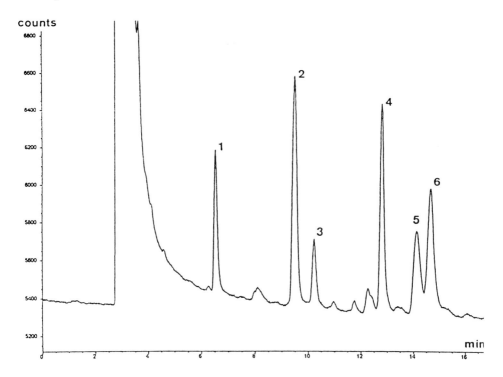

Figure 9B cSFC conditions: column, 10 m × 50 µm, 0.25 µm film, SB methyl 100; mobile phase, carbon dioxide, 6 µL/min; temperature gradient, 130–200°C in 4.6 min; column inlet pressure gradient, 220–380 bar (8 min) in 9 min; injection, 0.5 µL; detector, flame ionization detection, 400°C.

2. Analysis of Tinuvin 1577

In the synthesis of Tinuvin 1577, GC was used to support the process development work (Fig. 12). A new compound (8), which did not elute in GC, even at 310°C, appeared as a new synthesis route was explored to manufacture this polymer additive. Following our strategy of method development (Fig. 2) cSFC was also considered but gave a tailing peak for compound (8), which prohibits quantitative analysis of this compound (Fig. 12).

With pSFC, only one method was required to separate the triazine polymer additive from the intermediates and raw materials involved [37]. All compounds eluted in less than 20 min. vs. 30 min in GC and 40 min in cSFC, as shown in Fig. 12. Very reproducible results were obtained with RSD ranging from 0.7% to 1.5% quite similar to those obtained in GC.

HPLC was used as orthogonal method to verify the accuracy of the pSFC quantitative results, which were found to be in good agreement with those

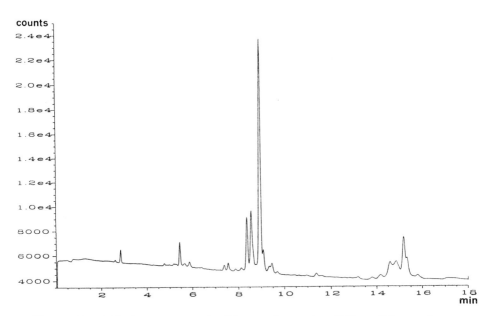

Figure 10 pSFC chromatogram of a Wingstay L sample. pSFC conditions: column, 125 × 4.0 mm, 3 μm, Nucleosil ODS; mobile phase, carbon dioxide, 2 mL/min; modifier 2–5% (5 min) methanol in 10 min; temperature, 90°C; pressure gradient, 90–350 bar (3 min) in 15 min; injection, 5 μL; detector, UV-vis at 230 nm.

obtained in HPLC. But the analysis time in HPLC was twice as long as the analysis time in pSFC as shown on Fig. 12. The detection limit was higher in pSFC than in HPLC and lower than in GC [37]. pSFC really outperformed every other method for this separation with well-resolved and symmetrical peaks for every compound. Consequently, time and capacity were saved in the analytical processes and finally in process development.

3. Analysis of Tinuvin 400

A pSFC method was developed to analyze Tinuvin 400, which is currently analyzed by HPLC, whereas GC is not suitable for this high molecular weight additive and high-temperature GC showed less ruggedness after method transfer.

A pSFC and a cSFC method were alternatively used. In pSFC (Fig. 13) analysis time was approximately the same as HPLC, but the column equilibration time was faster, and the detection limit and reproducibility were as good as in HPLC. The resolution was somewhat better in pSFC than in HPLC. The resolution could have been probably improved further by coupling more columns in series [35], which is not possible in HPLC due to the high pressure drop over a column. In this case, pSFC could favorably replace HPLC by

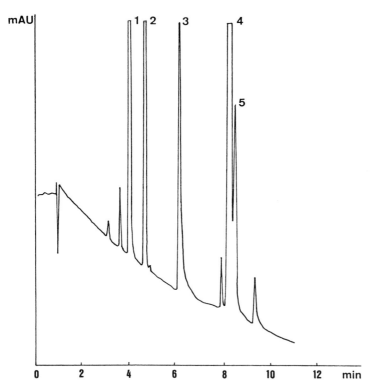

Figure 11 pSFC chromatogram of a synthetic mixture of Tinuvin 360 and related compounds. pSFC conditions: column, 125×4 mm, 3 μm, Spherisorb NH_2; mobile phase, carbon dioxide, 2 mL/min; modifier, methanol + triethanolamine; temperature, 40°C; pressure gradient, 100–300 bar in 13 min. Peak identification: 1) related compound (1), 2) internal standard, 3) related compound (3), 4) Tinuvin 360, 5) related compound (5).

reducing the solvent consumption, increasing the number of analysis per day and improving the quality of the analysis. cSFC was not retained because the resolution was not as good as with a series of packed columns. Tinuvin 400 was totally retained at low temperature and/or at low pressure, and no separation occurred when temperature and pressure were too high. A change of the stationary phase from SB methyl 100 to SB biphenyl 30 did not improve the separation either.

F. Conclusion

According to the presented examples, the proposed strategy allowed optimization of the selection and development of analysis methods for polymer additives. We found no examples where it would have make sense to replace an

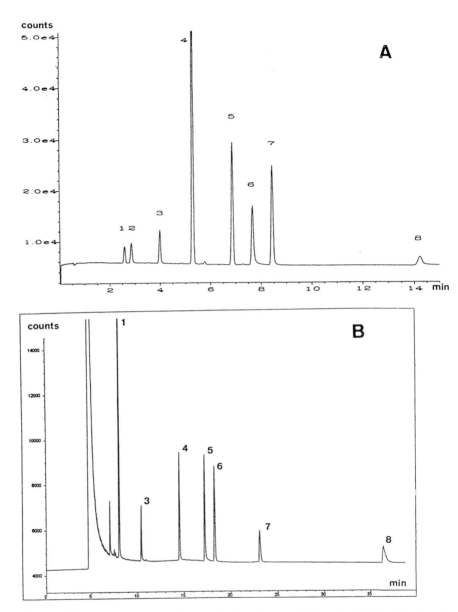

Figure 12 Chromatograms of a synthetic mixture of Tinuvin 1577 and related compounds. (A) pSFC conditions: column, 125 × 4 mm, 3 μm, Nucleosil ODS; mobile phase, carbon dioxide, 1.5 mL/min; modifier, 5% methanol; temperature, 100°C; pressure gradient, 90–350 bar (5 min) in 10 min; injection, 5 μL; detector, UV-vis at 270 nm. Reproduced from the *Journal of Chromatographic Science* by permission of Preston Publications, A Division of Preston Industries, Inc. [37]. (B) cSFC conditions: column, 10 m × 50 μm, 0.25 μm film, SB cyanopropyl 50; mobile phase, carbon dioxide,

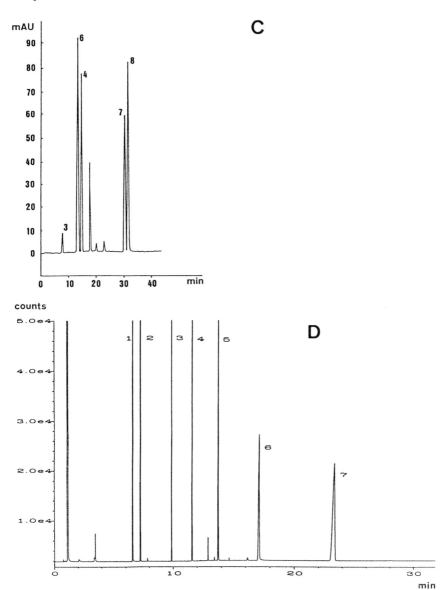

6 µL/min; temperature gradient, 100–200°C (27 min) in 12.5 min; column inlet pressure gradient, 100–380 bar (10 min) in 35 min; injection, 0.5 µL; detector, flame ionization detection, 400°C. (C) HPLC conditions: column, 125 × 4.6 mm, 5 µm, Nucleosil ODS; mobile phase, methanol/water/ethyl acetate; detection UV-vis at 292 nm. (D) GC conditions: column, DB 5; temperature gradient, 40°C (1 min) to 80°C in 5.3 min, 80–310°C (19 min) in 9.2 min; detector, flame ionization detection. Peak identification: 1–6) related compounds (1) to (6), 7) Tinuvin 1577, 8) related compound (8).

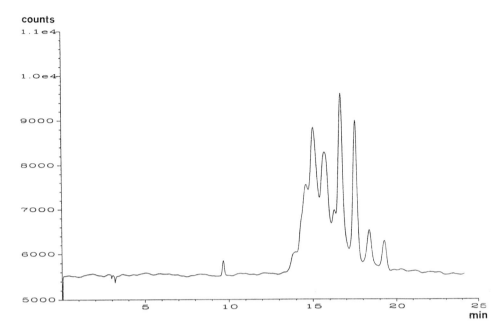

Figure 13 pSFC chromatogram of a Tinuvin 400 sample. pSFC conditions: column, 3 × 250 × 4 mm, 10 μm, Nucleosil ODS; mobile phase, carbon dioxide; modifier 10% methanol; temperature, 120°C; pressure, 270 bar; injection, 5 μL; detector, UV-vis at 290 nm.

existing GC method by a pSFC method. When GC did not work, as, for example, in the separation of Tinuvin P, cSFC is probably the best choice. When cSFC has some limitation, as, for example, in the separation of Tinuvin 1577, pSFC can bring a real improvement. In many cases a combination of two methods was required. For those separations, the HPLC method could be systematically replaced by a pSFC one.

In one case the analytical effort was reduced by pSFC and in another case totally new information was obtained that could not have been obtained either by GC or by HPLC. In most cases pSFC was used as an orthogonal separation technique to validate or improve existing methods (HPLC, GC), to debottleneck the capacity of the laboratory in taking advantage of much shorter equilibration, to reduce method development and/or analysis time, and to reduce the consumption of combustible organic solvent.

IV. OUTLOOK

According to our experience with cSFC and pSFC in process development, we are confident that SFC can be used to solve most of the typical problems encountered in the polymer and polymer additives industry. According to our strategy, we address only problems that cannot be solved efficiently with GC, which remains the first-choice technique in our opinion. The relative potentials of cSFC and pSFC are different depending on the problem addressed, and both techniques are needed in the analysis of polymer additives.

SFE/cSFC/MS is the technique of choice for any kind of analysis of polymer additives in the substrate. It is well suited for

Identification of unknown additives in commercial and developmental polymers for competitive intelligence
Identification and quantification of related compounds and migrating species for notification to the authorities and application for food contact approval
Determination of the additive level in quality control of polymers
Investigations of the stability and aging characteristics of polymer additives during manufacture and service life of the final product

In this area cSFC is the dominant technique because it is easily coupled to MS. However, pSFC on microbore columns and pcSFC are good alternatives to improve the limit of detection and analyze additives present in very low amounts. This is particularly useful for migration studies where extremely low concentrations have to be reported.

pSFC has a great potential in areas where quantitative analysis is needed, such as

Migration studies
Process development
In-process control
Quality control

As demonstrated in this chapter, pSFC is a good substitution method for HPLC, which is currently used in addition to GC to support the process development of polymer additives. The main benefits of pSFC are related to the faster equilibration time, which allows a higher flexibility (responsiveness) and a higher sample throughput (productivity). This is of particular interest in research and development, where isolated samples or very small series have to be analyzed to determine the next step and where the composition of the sample varies greatly from day to day. In addition, pSFC can be used as an orthogonal separation technique to verify the accuracy of GC methods. However, for the time being the incentive is not high enough to justify the cost of commercial equipments and service thereof (e.g., maintenance, education) and additional improvements of the reliability of commercial equipments are needed to attract newcomers.

V. SUMMARY

Plastic products have found a widespread use in modern society. The decisive versatility of these materials is related to the use of polymer additives, which confers to them specific properties depending on the desired future application. Polymer additives fall into a number of general classes including antioxidants, metal deactivators, light stabilizers, curing agents, plasticizers, slip agents, antiblock agents, mold release agents, antistatic agents, flame retardants, and biostabilizers. Some of them, such as plasticizers and mold release agents, have no chromophore. Others, such as antioxidants and UV stabilizers, are designed to be reactive. Some decompose if exposed to heat, and many of them have a high molecular weight. Many commercial polymer additives are not single components but may contain homolog or related compounds in significant amounts. In addition, even if many polymer additives have multiple functions, the number of polymer additives routinely used in the manufacture of, for instance, polyolefins exceeds 20 and covers a large polarity range. The purity and the amount of additive affect the properties of the polymer. Therefore there is a need for analytical methods to characterize additives and to determine their concentration in polymers, typically ppm. For plastic products used in food packaging or water piping, there is also a regulatory need to identify and to determine the amount of polymer additives and related degradation products that can be released to the foodstuff or to the water.

All of this makes the analysis of polymer additives challenging. In most cases, a combination of different analytical techniques operating simultaneously is required. The unique performance of supercritical fluid extraction and super-critical fluid chromatography for determination of polymer additives has been widely demonstrated in the literature since the first polymer additives separation by Raynor et al. in 1987. Although capillary supercritical fluid chromatography is largely dominating the field, packed column supercritical fluid chromatography appears to be similarly attractive for the analysis of polymer additives and to be an ideal replacement for high-performance liquid chromatography. The purpose of this chapter has been to review the current literature and give examples of how pSFC can be used to develop new polymer additives or production processes with greater efficiency.

ACKNOWLEDGMENTS

I thank S. Wachtel, F. Walch, and R. Weber for the pSFC and cSFC work presented in this chapter; G. Langer, G. Ragusa, and A. Higel for the HPLC and GC data on Tinuvin 1577.

REFERENCES

1. F. P. Schmitz and E. Klesper, *J. Supercrit. Fluids*, *3*:29 (1990).
2. M. W. Raynor and K. D. Bartle, *J. Supercrit. Fluids*, *6*:39 (1993).

3. K. Ute and K. Hatada, *Applications of Preparative SFC to Oligomer Analysis and Characterization* (M. Saito, Y. Yamauchi, and T. Okuyama, eds), VCH, New York, p. 231 (1994).

4. D. E. Knowles, and T. K. Hoge, *Applications of Supercritical Fluids in Industrial Analysis* (J. R. Dean, ed.), Blackie, Glasgow, p. 104 (1993).

5. R. Gächter and H. Müller, *Plastics Additives Handbook,* 4th ed., Hanser/Gardner, Cincinnati (1993).

6. D. Dilettato, P. Arpino, K. Nguyen, and A. Bruchet, *J. High Resolut. Chromatogr.,* *14*:335 (1991).

7. T. Buecherl, A. Gruner, and N. Palibroda, *Packag. Technol. Sci.,* 7:139 (1994).

8. T. Greibrokk, B. E. Berg, S. Hoffman, H. R. Norli, and Q. Ying, *J. Chromatogr.,* *505*:283 (1990).

9. T. W. Ryan, S. G. Yocklovich, J. C. Watkins, and E. J. Levy, *J. Chromatogr.,* *505*:273 (1990).

10. M. Ashraf-Khorassani and J. M. Levy, *J. High Resolut. Chromatogr.,* *13*:742 (1990).

11. W. Blum and L. Damasceno, *J. High Resolut. Chromatogr.,* *10*:472 (1987).

12. F. David and P. Sandra, *LC-GC, 11*:282 (1993).

13. D. W. Allen, M. R. Clench, A. Crownson, D. A. Leathard, and R. Saklatvala, *J. Chromatogr. A., 679*:285 (1994).

14. J. D. Vargo and K. L. Olson, *J. Chromatogr., 353*:215 (1986).

15. B. Marcato, C. Fantazzini, and F. Seveni, *J. Chromatogr., 553*:415 (1991).

16. D. A. Wheeler, *Talanta, 15*:1315 (1968).

17. T. R. Crompton, *Chemical Analysis of Additives in Plastics,* 2nd ed., Pergamon Press, Oxford (1977).

18. D. O. Hummel and F. Scholl, *Atlas der Polymer- und Kunstoffanalyse,* Vol. 3, 2nd ed., Carl Hanser Verlag und Verlag Chemie, Munich (1981).

19. M. W. Raynor, K. D. Bartle, I. L. Davies, A. Williams, A. A. Clifford, J. M. Chalmers, and B. W. Cook, *Anal. Chem.,* 60:427 (1988).

20. K. D. Bartle, M. W. Raynor, A. A. Clifford, I. L. Davies, J. P. Kithinji, G. F. Shilstone, J. M. Chalmers, and B. W. Cook, *J. Chromatogr. Sci.,* 27:283 (1989).

21. P. J. Arpino, D. Dilettato, K. Nguyen, and A. Bruchet, *J. High Resolut. Chromatogr., 13*:5 (1990).

22. R. Moulder, J. P. Kithinji, M. W. Raynor, K. D. Bartle, and A. A. Clifford, *J. High Resolut. Chromatogr., 12*:688 (1989).

23. N. J. Cotton, K. D. Bartle, A. A Clifford, S. Ashraf, R. Moulder, and C. J. Dowle, *J. High Resolut. Chromatogr., 14*:164 (1991).

24. K. D. Bartle, A. A. Clifford, N. J. Cotton, J. P. Kithinji, R. Moulder, and M. W. Raynor, *Anal. Proc.,* 27:239 (1990).

25. K. D. Bartle, A. A. Clifford, and M. W. Raynor, *Hyphenated Techniques in Supercritical Fluid Chromatography and Extraction* (K. Jinno, ed.), Journal of Chromatography Library Series, Vol. 53, Elsevier, Amsterdam, p. 103 (1992).

26. M. W. Raynor, A. A. Clifford, K. D. Bartle, C. Reyner, A. Williams, and B. W. Cook, *J. Microcolumn Sep., 1*:101 (1989).

27. K. Anton, N. Periclès, S. M. Fields, and H. M. Widmer, *Chromatographia, 26*:224 (1988)

28. J. P. Kithinji, K. D. Bartle, M. W. Raynor, and A. A Clifford, *Analyst, 115*:125 (1990).

29. J. Doehl, A. Farbrot, T. Greibrokk, and B. Iversen, *J. Chromatogr.*, *392*:175 (1987).
30. A. Giorgetti, N. Periclès, H. M. Widmer, K. Anton, and P. Dätwyler, *J. Chromatogr. Sci.*, *27*:318 (1989).
31. M. Ashraf-Khorassani, D. S. Boyer, and J. M. Levy, *J. Chromatogr. Sci.*, *29*:517 (1991).
32. J. W. Oudsema and C. F. Poole, *J. High Resolut. Chromatogr.*, *16*:198 (1993).
33. T. P. Hunt, C. J. Dowle, and G. Greenway, *Analyst (London)*, *116*:1299 (1991).
34. T. Imahashi, Y. Yamauchi, and M. Saito, *Bunseki Kagaku*, *39*:79 (1990).
35. T.A. Berger and W. H. Wilson, *Anal. Chem.*, *65*:1451 (1993).
36. K. Anton, M. Bach, and A. Geiser, *J. Chromatogr.*, *553*:71 (1991).
37. K. Anton, M. Bach, C. Berger, F. Walch, G. Jaccard, and Y. Carlier, *J. Chromatogr. Sci.*, *32*:430 (1994).

12

Fractionation of Polymer Homologs with Packed Column Supercritical Fluid Chromatography

Koichi Ute

Osaka University, Osaka, Japan

I. INTRODUCTION

With the exception of some compounds of biological origin, macromolecular compounds are inhomogeneous with respect to molecular dimension and structure. Consequently, any discussion concerning macromolecular substances implies a statistical character. This peculiarity sometimes causes ambiguity or complexity to macromolecular science. If polymers with uniform molecular weights ("uniform polymers" or "monodisperse polymers" according to the definition by IUPAC Commission on Macromolecular Nomenclature [1]) were readily available, they could be used as model compounds in the study of the exact molecular structure and physical properties of ordinary, nonuniform polymers [2]. Uniform polymers with definite molecular weights would also serve as reference materials for polymer analysis and characterization, *e.g.*, molecular weight standards for size exclusion chromatography (SEC).

   Fractionation of polymers into the individual homologs by packed column supercritical fluid chromatography (pSFC) is an effective way of obtaining a series of uniform polymers. The applicability of this method to preparation of uniform polymers has been enhanced by the recent progress in polymerization chemistry that provides a variety of polymers suitable for the fractionation, namely, source polymers with controlled molecular weight and well-defined chain structure. In the present chapter, pSFC fractionation of (1) atactic and syndiotactic polystyrenes, (2) isotactic and syndiotactic poly(methyl methacrylate)s, (3) polymers of D- and L-lactides, (4) poly(alkyl isocyanate)s, and (5) poly(oxymethylene) will be described together with the structure and properties of the uniform polymers thus obtained.

Examples of polymer separation by pSFC are also shown in Chapter 3. Large scale pSFC fractionation of other compound classes are discussed in Chapters 14 and 15.

II. PSFC FRACTIONATION OF POLYMER HOMOLOGS AND APPLICATION OF THE UNIFORM POLYMERS TO POLYMER SCIENCE

A. Polystyrene

Many papers, including a comprehensive review by Schmitz and Klesper [3], have been published on pSFC separation of polymer homologs. Among the polymers subjected to pSFC, the most extensively studied is polystyrene. One reason for this may be the commercial availability of polystyrene standards with narrow molecular weight distributions.

In 1969, Jentoft and Gouw reported the first pSFC separation of a standard polystyrene with a viscosity average molecular weight of 578. Fractions were also taken and rechromatographed under the same conditions [4]. Preparative separation of standard polystyrenes has been demonstrated by Klesper and Hartmann [5,6]. The 15 fractions from 1mer to 15mer were isolated from a polystyrene of nominal molecular weight of 2200 by the use of pentane/methanol (9:1) as the supercritical mobile phase and a 5mm $id \times 6$ m column packed with porous silica gel (particle size 37–75 μm) as the stationary phase; more than 11 h was required for each chromatographic run.

Recent developments in pSFC apparatus and packing materials have made it possible to carry out the fractionation in a much shorter time with improved separation [7, 8 and Chapters 3 and 14].

Figure 1A illustrates a pSFC trace of a standard polystyrene with $\bar{M}_n = 2800$. The homologs from about 10mer to 40mer were clearly resolved within a total experimental period of 25 min [9]. This separation was performed on a JASCO Super-200 chromatograph equipped with a Hewlett-Packard 5890 column oven at 200°C and 235 bar. A 10 mm $id \times 250$ mm column packed with nonbonded silica gel (particle size 5 μm, pore size 100 Å; Develosil 100-5, Nomura Chemical Co., Ltd.) was employed. The mobile phase consisted of liquified carbon dioxide and dichloromethane, and the compositional gradient method was applied.

Standard polystyrenes are prepared usually by living anionic polymerization of styrene with alkyllithium in a polar solvent. Such polymerization conditions lead to the formation of polystyrene with low stereoregularity (atactic polystyrene). As a consequence, chromatographic separation of those polystyrenes can be based not only on molecular weight but also on the stereoisomerism in each homolog. When a highly syndiotactic (st-) polystyrene with $\bar{M}_n = 2670$ was subjected to the pSFC under the same conditions as described above, better

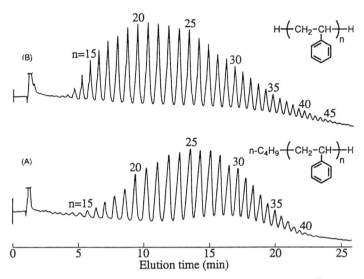

Figure 1 PSFC traces of (A) atactic "standard" polystyrene ($\overline{M}_n$ = 2.8 × 10^3, $\overline{M}_w/\overline{M}_n$ = 1.05) and (B) syndiotactic polystyrene ($\overline{M}_n$ = 2.7 × 10^3, $\overline{M}_w/\overline{M}_n$ = 1.06) [9]. Column, Develosil 100-5 (10 mm *id* × 250 mm); mobile phase, carbon dioxide/ dichloromethane = 60:40 (initial) → 40:60 (30 min); flow rate, 10 mL/min; temperature, 200°C; pressure, 235 bar.

separation than that for standard polystyrene was observed (Fig. 1B) [9]. The bandwidth of each peak in Fig. 1B is narrower than that in Fig. 1A. This sample of *st*-polystyrene was obtained by the polymerization of styrene with a (η^5-C$_5$H$_5$)TiCl$_3$/(i-C$_4$H$_9$)$_3$Al/methylaluminoxane catalyst in toluene [10] followed by catalytic hydrogenation of unsaturated chain ends [9]. Each individual homolog in the *st*-polystyrene was isolated by pSFC. The molecular mass and purity of each fraction was confirmed by field desorption mass spectrometry (FD-MS) as shown in Fig. 2 [9].

B. Poly(methyl methacrylate)

Stereospecific polymerization of methyl methacrylate (MMA) with t-C$_4$H$_9$MgBr in toluene at −78°C proceeds in a living manner to yield highly isotactic poly(methyl methacrylate) (*it*-PMMA) with a narrow molecular weight distribution [11]. On the other hand, stereospecific living polymerization of MMA with a t-C$_4$H$_9$Li/(C$_2$H$_5$)$_3$Al complex in toluene at −93°C gives highly syndiotactic PMMA (*st*-PMMA) [12]. These two kinds of PMMA with quite different stereoregularity have the molecular structure expressed in an identical

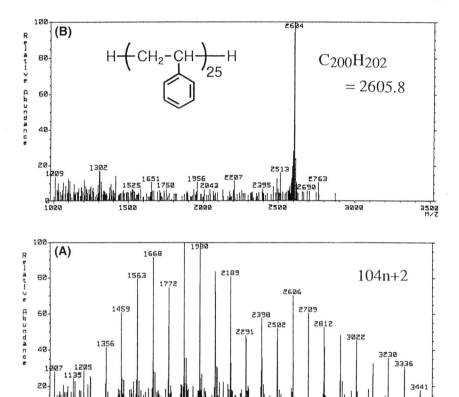

Figure 2 Field desorption mass spectrum of (A) syndiotactic polystyrene ($\overline{M}_n$ = 2.7 × 10^3, $\overline{M}_w/\overline{M}_n$ = 1.06) and (B) the 25mer isolated therefrom by pSFC [9].

formula, t-C_4H_9-$(MMA)_n$-H. PSFC fractionation of the it- and st-PMMAs has been investigated [13–17].

Table 1 summarizes the number average degree of polymerization ($\overline{DP}$), polydispersity index ($\overline{M}_w/\overline{M}_n$), and triad tacticity of the seven PMMA samples used for the experiments (it-A, it-B, it-C, it-D, st-A, st-B, st-C). Both the chromatograph and the column used for the pSFC were the same as those described for the separation of st-polystyrene in the previous section. Supercritical carbon dioxide containing 20–30 vol% ethanol or methanol was used as the mobile phase. The composition of the mobile phase and the fluid pressure (200 bar) were kept constant during each experiment. Negative temperature gradients, which increase the density of the mobile phase and are generally suitable for the PSFC separation of nonvolatile compounds, were applied. Although pSFC sepa-

Table 1 Average Degree of Polymerization ($\overline{DP}$) and Tacticity of the Isotactic and Syndiotactic Poly(methyl methacrylate) Samples Used for pSFC Separation [17]

Sample	$\overline{DP}$	$\overline{M}_w/\overline{M}_n$	Tacticity (%)		
			mm	*mr*	*rr*
it-A	28.6	1.15	96.1	3.9	0.0
it-B	40.8	1.12	95.9	3.5	0.6
it-C	66.5	1.07	97.7	2.3	0.0
it-D	565	1.24	97	3	0
st-A	26.8	1.09	0.3	7.6	92.1
st-B	41.4	1.05	0.0	7.2	92.8
st-C	539	1.19	0	8	92

ration of PMMA was also achieved by the use of a column packed with ODS-treated silica gel [13], it was more effective to use a column packed with nonbonded silica gel for preparative separation because the latter column was capable of separating a larger amount of the sample per chromatographic run [14].

Figure 3A shows a pSFC curve of *it*-B [15]. A small amount of authentic samples of *it*-22mer and *it*-28mer was added to the PMMA as an internal standard of degree of polymerization (DP). The authentic 22 and 28mers were obtained by pSFC fractionation of *it*-A, and their DP values were determined

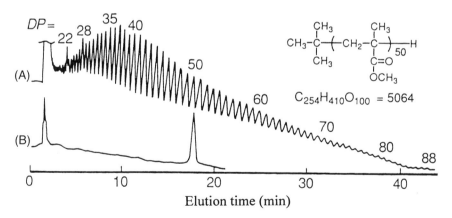

Figure 3 PSFC traces of (A) isotactic poly(methyl methacrylate) (*it*-B in Table 1; $\overline{M}_n$ = 4.1 × 10^3, $\overline{M}_w/\overline{M}_n$ = 1.12) and (B) the 50mer isolated therefrom [15]. Authentic samples of the 22 and 28mers were added to *it*-B as internal standards of DP. Column, Develosil 100-5 (10 mm *id* × 250 mm); mobile phase, carbon dioxide/ethanol = 78/12; flow rate, 11.5 mL/min; initial temperature, 80°C; temperature gradient, −1°C/min; pressure, 198 bar.

clearly by ^{1}H NMR spectroscopy from the relative intensity of the resonances due to the CH_3O groups in the repeating units and the terminal t-C_4H_9 group [14]. Fractionation of the individual homologues from 40mer to 60mer was carried out 45 times. The amount of the sample introduced to the chromatograph was 25 mg (50 µL of 50 w/v% acetone solution), each chromatographic run took about 45 min. Every fraction was purified by repeated fractionation by pSFC because the crude fraction contained a small amount of lower *DP* homologs. The purified *it*-50mer gave a single peak in a chromatogram recorded under the identical conditions (Fig. 3B), indicating the uniformity with respect to *DP*. The total amount of the uniform *it*-50mer obtained was 10.6 mg. In a similar manner, uniform *st*-PMMAs from 14mer to 60mer were isolated from *st*-A and *st*-B.

PSFC is sensitive to the stereoisomerism in polymer homologs. When a mixture of *it*- and *st*-24mers was subjected to pSFC, *it*-24mer eluted faster than *st*-24mer (Fig. 4) [14]. This order of elution was reversed in the pSFC separation on ODS-treated silica gel [13]. These suggest that *it*-PMMA may have a smaller polarity than *st*-PMMA.

Optimization of the operating conditions for pSFC enabled to separate an *it*-PMMA with an $\overline{M}_n$ as large as 6650 (*it*-C) (Fig. 5) [2,16]. Authentic 45mer was added to this PMMA as an internal standard. The individual homologs from 80mer ($C_{404}H_{650}O_{160} = 8068$) to 100mer ($C_{504}H_{810}O_{200} = 10{,}070$) were collected

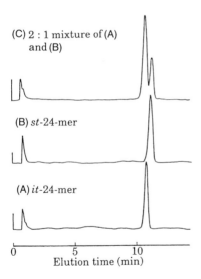

Figure 4 PSFC traces of (A) isotactic 24mer and (B) syndiotactic 24mer of methyl methacrylate and of (C) a 2:1 mixture of the 24mers [14]. Column, Develosil 100-5 (4.6 mm *id* × 150 mm); mobile phase, carbon dioxide/ethanol = 80:20; flow rate, 3 mL/min; initial temperature, 115°C; temperature gradient, −2°C/min; pressure, 198 bar.

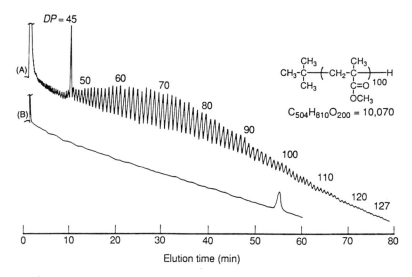

Figure 5 PSFC traces of (A) isotactic poly(methyl methacrylate) (*it*-C in Table 1; $\overline{M}_n$ = 6.7 × 10^3, $\overline{M}_w/\overline{M}_n$ = 1.07) and (B) the 100mer isolated therefrom [2,16]. Authentic 45mer was added to *it*-C as an internal standard of *DP*. Column, Develosil 100-5 (10 mm *id* × 250 mm); mobile phase, carbon dioxide/methanol = 81:19; flow rate, 14.8 mL/min; initial temperature, 80°C; temperature gradient, −0.5°C/min; pressure, 235 bar.

by repeated fractionation with pSFC. The molecular weight of the *it*-100mer was certified by the matrix-assisted laser desorption ionization time-of-flight (MALDI-TOF) mass spectrometry. The MALDI-TOF mass spectrum of the *it*-100mer exhibited a single peak at *m/Z* = 10,096, which is larger than the theoretical value by 26. This discrepancy should be due to the attachment of a sodium or potassium ion to 100mer molecules. The peaks arising from [M + Na]+ (*m/Z* = 10,093) and [M + K]+ (*m/Z* = 10,109) were indistinguishable in this mass range.

The series of uniform PMMAs offers an opportunity to investigate the effects of molecular weight and its distribution on the physical properties of *it*- and *st*-PMMAs. Thermal properties of the uniform PMMAs have been investigated by differential scanning calorimetry (DSC) [14,17]. Glass transition temperature (T_g) of uniform *it*-41mer was determined to be 38.1°C which was higher by 6.6°C than the T_g of a nonuniform *it*-PMMA (*it*-B) with an average *DP* nearly equal to 41. The lower T_g of *it*-B is simply due to the nonuniformity in molecular weight. This phenomenon may be caused by the plasticizing effect of lower *DP* components included in the nonuniform PMMA. A similar effect of molecular weight distribution on lowering T_g was observed between *st*-41mer (T_g = 100.9°C) and *st*-B (T_g = 94.4°C).

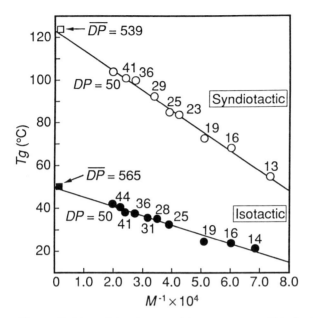

Figure 6 Plots of the glass transition temperature (T_g) of uniform *it*-PMMA (●) and *st*-PMMA (○) samples as a function of the reciprocal molecular weight (M^{-1}) [17]. The T_g data of nonuniform *it*-PMMA (*it*-D in Table 1) (■) and *st*-PMMA (*st*-C in Table 1) (□) are also shown.

When the T_g values of the uniform PMMAs are plotted against the reciprocal of the molecular weight (M^{-1}) as in Fig. 6, the data are well fitted by the relationship presented by Fox and Flory [18]. From a least-squares analysis one obtains the following equations for the *it* and *st* homologs, respectively [17].

$$T_g\ (it) = (49.6 \pm 1.3) - (4.34 \pm 0.26) \times 10^4/M$$
$$T_g\ (st) = (123.3 \pm 1.7) - (9.38 \pm 0.33) \times 10^4/M$$

The T_g data of the nonuniform *it*- and *st*-PMMAs having $\overline{M}_n$ larger than 5×10^4 (*it*-D and *st*-C) also fitted the linear relationships defined by the above equations (cf. Fig. 6). These results indicate that the simple relationships hold for an extensive range of molecular weight ($M > 1 \times 10^3$).

Because of the uniformity in molecular weight along with the high stereoregularity, the uniform *it*-PMMA samples were expected to be highly crystalline. Indeed, the melting temperature of the *it*-41mer crystallized by annealing was found to be 122.3°C, which was higher than that of *it*-B by 2.9°C. A marked difference between the uniform and nonuniform *it*-PMMAs was observed in the heat of fusion per repeating unit (ΔH^*); the ΔH^* of *it*-41mer was 2.2 times as large as that of *it*-B. The degrees of crystallinity for the *it*-41mer and *it*-B were estimated to be 0.79 and 0.35, respectively, on the basis of the ΔH^* values [17]. The reciprocal equilibrium melting temperature (T_m^{-1}) in the Kelvin scale of the uniform *it*-PMMA increased linearly with increasing DP^{-1}, and the relationship can well be represented by the following equation [17]:

$$1/T_m = (2.25 \pm 0.012) \times 10^{-3} + (1.24 \pm 0.11) \times 10^{-2}/DP$$

The results indicate that the crystalline melting behavior of the uniform *it*-PMMA can be explained on the basis of the statistical thermodynamic theory [19] for extended chain crystals.

Although the pSFC and mass spectrometric analysis of the uniform PMMAs proved the purity of the samples with respect to *DP*, there still remains a purity problem at the stereochemical level. The isotacticity of the parent *it*-PMMA (*it*-A and *it*-B) is 96–97% in triads, and thus the *it*-50mer isolated therefrom, should contain approximately one *racemo* diad (*r*) somewhere in its configurational sequence. Recently, a perfectly isotactic PMMA (*mm* ≈ 100%) with a symmetrical chain structure, H-(MMA)$_n$-CH$_3$, has been prepared and fractionated into the uniform *it*-PMMAs from 28mer to 44mer. These uniform *it*-PMMAs showed crystallinity even higher than the uniform *it*-PMMAs obtained from *it*-A and *it*-B [20].

The uniform PMMAs with definite molecular weights are useful as molecular weight standards for SEC. The mixtures of *it*-25mer and *it*-50mer and of *st*-25mer and *st*-50mer, respectively, were subjected to SEC using tetrahydrofuran (THF) as the eluent (Fig. 7A) [15]. It should be noted from Fig. 7A that the elution volume of *it*-50mer is significantly smaller than that of *st*-50mer. Such a small but distinct difference between *it*- and *st*-PMMAs in hydrodynamic volume can be evidenced clearly by the use of uniform *it*- and *st*-PMMAs with exactly the same molecular weight [15]. Calibration curves of high precision and accuracy for *it*- and *st*-PMMAs were made from the elution curves in Fig. 7A. The $\overline{M}_n$ value of *it*-B determined by the SEC (Fig. 7B) using *it*-25mer and *it*-50mer as standards was 4111, which agreed well with the result of end-group analysis by ^{1}H NMR (4138). However, the use of *st*-25mer and *st*-50mer as standards made the $\overline{M}_n$ about 9% larger. A similar situation was observed for the molecular weight determination of *st*-B. In spite of the uniformity in molecular weight, SEC measurements of the uniform PMMAs gave the apparent $\overline{M}_w/\overline{M}_n$ ranging from 1.009 to 1.013. This apparent $\overline{M}_w/\overline{M}_n$ comes from the instrumental spreading of the elution band, and the values show the lower limit of $\overline{M}_w/\overline{M}_n$ detectable in this SEC apparatus [15,16].

It is well known that *it*- and *st*-PMMA chains associate to form a stereocomplex [21]. Despite many investigations reported so far [22], the nature of the complex formation is not well understood. The stereocomplex between uniform *it*- and *st*-PMMAs avoids experimental difficulties arising from molecular weight distribution and offers a straightforward approach to the complexation

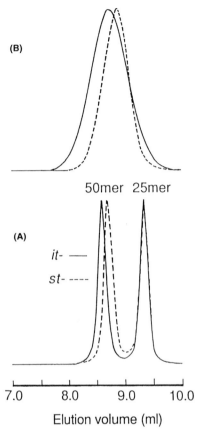

Figure 7 Size exclusion chromatograms of (A) the mixtures of uniform 25 and 50mers (*it*-25mer + *it*-50mer, solid line; *st*-25mer + *st*-50mer, broken line) and of (B) nonuniform *it*- and *st*-PMMAs (*it*-B in Table 1, solid line; *st*-B, broken line) [15]. Column, Shodex KF-803 (8 mm *id* × 300 mm); eluent, THF; temperature, 25°C.

studies. Figure 8 shows an SEC trace for a 1:1 mixture of *it*- and *st*-50mers in THF at 3°C. Besides the elution peak with a small splitting due to stereoisomerism in te 50mers, an extra peak with very low polydispersity appeared at the elution time corresponding approximately to 100mer. This extra peak should be attributed to the stereocomplex between the *it*- and *st*-50mers in a 1:1 stoichiometry [15]. The peak area for stereocomplex decreased as the *DP* of the uniform *it*- and *st*-PMMAs decreased, and no such peak was observed for a mixture of *it*- and *st*-40mers. It is suggested from these results that the minimum length of *it* and *st* sequences required for the stereocomplex formation should

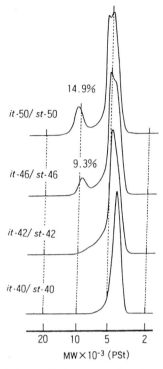

Figure 8 Size exclusion chromatograms of 1:1 mixtures of uniform *it*- and *st*-PMMAs [15]. Column, Shodex KF-803 (8 mm *id* × 300 mm); eluent, THF; temperature, 3°C.

exist at 40–42 [2,16]. Such clear-cut information as described above can hardly be obtained without uniform polymer samples.

C. Poly(lactic acid)

Ring opening polymerization of D- and L-lactides can give rise to the optical antipodes of isotactic poly(lactic acid), namely, poly(D-lactic acid) and poly(L-lactic acid), respectively. Both D- and L-lactides with high optical purity (>>99%) are available. Living polymerization of lactides has recently been reported by several authors [23–25].

Low molecular weight poly(D-lactic acid) prepared by the living polymerization of D-lactide with Al[OCH(CH$_3$)$_2$]$_3$ in toluene [24] was fractionated by pSFC (Fig. 9A) [26]. Supercritical carbon dioxide containing ethanol was used as the mobile phase, and the composition gradient was applied. Nonbonded silica gel (the same column as described in the previous sections) was used as the stationary phase. Temperature and fluid pressure were kept constant

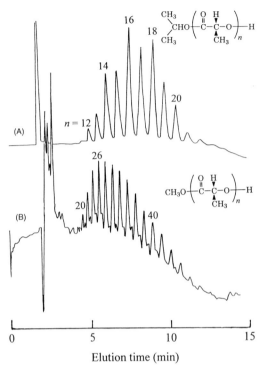

Figure 9 PSFC traces of poly(D-lactic acid)s prepared by polymerization with (A) Al[OCH(CH$_3$)]$_3$ [26] and (B) (salen)AlOCH$_3$. Column, Develosil 100-5 (10 mm *id* × 250 mm); mobile phase, carbon dioxide/ethanol = (A) 97:3 (initial) → 90:10 (15 min), (B) 90:10 (initial) → 85:15 (20 min); flow rate, 10 mL/min; temperature, (A) 65°C, (B) 55°C; pressure, 198 bar.

throughout the separation (65°C, 200 bar). In the pSFC experiments above 65°C, degradation of this polyester sample was found to occur to some extent. The *DP* of each fraction was determined from the end-group analysis by ^{1}H NMR.

In spite of the fact that the poly(D-lactic acid) was prepared from the cyclic dimer of D-lactic acid, the polymer consisted of the homologs with both even and odd numbers of repeating units (cf. Fig. 9A). The formation of the odd *DP* homologs clearly indicates that the living polymerization of lactide with Al[OCH(CH$_3$)$_2$]$_3$ accompanies intermolecular and/or intramolecular transesterification of the polymer chains. Recently, *N,N'*-bis(salicylidene)ethylenediiminoaluminum methoxide [(salen)AlOCH$_3$] has been found to be an efficient initiator for the living polymerization of lactide [25]. PSFC analysis of a poly(D-lactic acid) prepared with (salen)AlOCH$_3$ has revealed that the polymer contained only a small amount of odd *DP* homologs (Fig. 9B) [26].

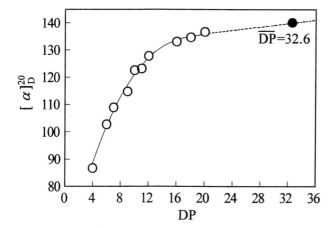

Figure 10 Specific rotation of uniform poly(D-lactic acid)s in chloroform (O) [26]. The value for a nonuniform poly(D-lactic acid) with $\overline{DP}$ of 32.6 is also shown (●).

The uniform poly(D-lactic acid)s obtained by the pSFC fractionation as shown in Fig. 9A were highly crystalline. The D-9mer ($C_{30}H_{44}O_{19}$ = 708.7) gave thin-needle crystals from a solution in ethanol, and X-ray analysis showed the crystal to belong to the *orthorhombic* system of space group $P2_12_12_1$ (a = 12.60 Å, b = 62.21 Å, c = 9.69 Å, Z = 8, D_c = 1.239 g/cm^3).

Specific rotation, $[\alpha]_D^{25}$, of the uniform poly(D-lactic acid)s in chloroform increased with increasing DP toward an asymptotic limit which is practically reached at $DP \sim 20$ (Fig. 10) [26]. The $[\alpha]_D^{25}$ of the 20mer was almost comparable to the value for high molecular weight poly(D-lactic acid) (+151°). The increase of $[\alpha]_D^{25}$ in the low DP region is attributable to an increasing contribution of the conformational chirality (one-handed helix) in the polymer chain.

Blends of two isotactic poly(lactide)s of opposite configurations are known to form a stereocomplex (a polyracemate complex) in bulk. The stereocomplex between high molecular weight ($\overline{M}_v$ = 7.0 × 10^4) poly(D-lactic acid) and poly(L-lactic acid) has a melting point of 230°C, whereas the individual polymers have a melting point of 180°C [27]. When equal quantities of D-16mer and L-16mer were dissolved together in chloroform and then crystallized by slow evaporation of the solution, the crystalline blend showed a sharp melting point at 156.2°C. The melting point was much higher than that of the individual 16mers (77.1°C), indicating the formation of a stereocomplex similar to that between high molecular weight poly(lactic acids)s [26].

D. Polyisocyanates

Polyisocyanates have received a great deal of attention due to their chiroptical properties arising from stiff helical conformation of the polymer chain [28].

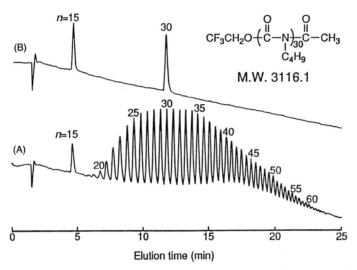

Figure 11 PSFC curve of (A) poly(butyl isocyanate) ($\overline{M}_n = 4.0 \times 10^3$, $\overline{M}_w/\overline{M}_n = 1.20$) and (B) the uniform 30mer isolated therefrom [31]. An authentic sample of the 15mer was added to the polymers as an internal standard of *DP*. Column, Develosil 100-5 (10 mm *id* × 250 mm); mobile phase, carbon dioxide/ethanol = 85:15 (initial) → 70:30 (25 min); flow rate, 10 mL/min; temperature, 90°C; pressure, 198 bar.

However, the lack of efficient synthetic routes for polyisocyanates with controlled molecular weight has hindered the full progress in the research of this polymer. The discovery of the living polymerization of isocyanates with organotitanium compounds [29,30] has made it possible to prepare the low molecular polyisocyanates suitable for pSFC fractionation.

The living polymerization of butyl isocyanate was initiated with $TiCl_3$ OCH_2CF_3 [29] and terminated with acetic anhydride. The poly(butyl isocyanate) with the chain structure $CF_3CH_2O(CONC_4H_9)_nCOCH_3$ was obtained. The acetyl end-capping is necessary because low molecular weight polyisocyanates with the -NH terminus in the polymer chain readily decompose to cyclic trimers (isocyanurates) by the "back-biting" mechanism under these pSFC conditions. A sample of the endcapped poly(butyl isocyanate) ($\overline{M}_n = 3.97 \times 10^3$, $\overline{M}_w/\overline{M}_n = 1.20$) was separated into the individual homologs from 15mer to 55mer by means of pSFC using carbon dioxide/ethanol as the mobile phase at 90°C and 200 bar (Fig. 11) [31]. The *DP* of each homolog was determined from the pSFC curve on the basis of the peak due to authentic 15mer.

Single crystals (*mp* = 98.6~99.9°C) were grown from an ethanol solution of the 12mer ($C_{64}H_{113}O_{14}N_{12}F_3$ = 1131.7). X-ray crystallographic analysis showed the crystal to belong to the *monoclinic* system of space group $P2_1/c$. The crystal structure is shown in Fig. 12. The backbone of the polyisocyanate chain adopts

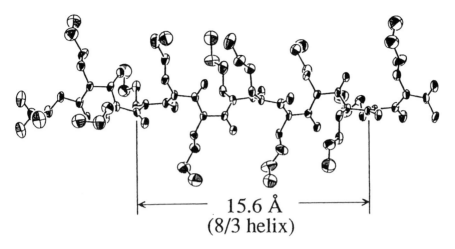

15.6 Å
(8/3 helix)

Figure 12 Crystal structure of butyl isocyanate 12mer [$C_{64}H_{113}O_{14}N_{12}F_{13}$ = 1131.7, *monoclinic $P2_1/c$, a* = 25.247(1) Å, *b* = 22.108(6) Å, *c* = 14.150(2) Å, β = 93.131(6)°, *d* = 1.122 g/cm³, 5304 reflections, *R* = 0.120] [32].

an 8/3 helix with a repeat distance of 15.6 Å [32]. This chain structure coincides with the results of X-ray diffraction studies on powder samples and oriented films of high molecular weight poly(butyl isocyanate) [33].

Ultraviolet spectra of the uniform poly(hexyl isocyanate)s from 6mer to 30mer, which were prepared in a similar manner as described above, were measured in cyclohexane (Fig. 13) [32]. The $n{\rightarrow}\pi^*$ transition due to the carbonyl groups in the polyisocyanate backbone showed the absorption at the wavelength around 248 nm. The molar extinction coefficient per repeating unit (ε) for this absorption band increased with *DP*, and the absorption maximum shifted slightly to the longer wavelength. This indicates that the conjugation between the amide linkages in polyisocyanate backbone increases with increasing *DP*.

Green and his coworkers have demonstrated that poly(hexyl isocyanate) shows induced circular dichroism (CD) when dissolved in optically active chlorinated hydrocarbons. This induced CD was ascribed to an excess of one-handedness in the helical conformation of poly(hexyl isocyanate) [34]. In order to investigate the effects of *DP* on the one-handed helix, solvent-subtracted CD spectra of a series of the uniform poly(hexyl isocyanate)s were measured in (−)-β-pinene (Fig. 13) [32]. Solutions of uniform poly(hexyl isocyanate)s with *DP* ≥ 10 displayed the induced CD at the wavelength range from 245 to 280 nm. The molar ellipticity per repeating unit increased as the *DP* increased. These together with the results of UV spectra suggest that the poly(hexyl isocyanate)

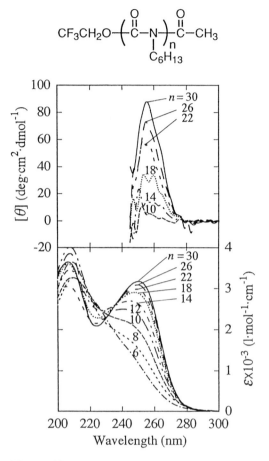

Figure 13 UV and solvent-subtracted CD spectra of uniform poly(hexyl isocyanate)s at 23°C [UV: in cyclohexane, $c = 1.3 \sim 2.8 \times 10^{-2}$ mg/mL; CD: in $(-)$-β-pinene, $c = 2.1 \sim 6.9$ mg/mL] [32].

chains with less than 10 repeating units may hardly form stable one-handed helix in solution.

E. Poly(oxymethylene)

In the middle of the 1920s, Staudinger and his coworkers fractionated low molecular weight poly(oxymethylene) diacetate, $CH_3CO_2(CH_2O)_nCOCH_3$, into the individual homologs from 1mer to 26mer by means of rectification and fractional crystallization. Each fraction was subjected to chemical analysis and molecular weight measurements, and the results agreed well with theoretical

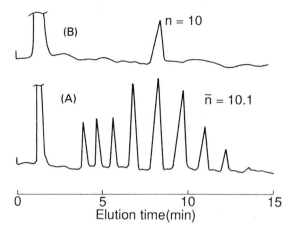

Figure 14 PSFC traces of (A) poly(oxymethylene) diacetate and (B) the 10mer isolated therefrom [37]. Column, Superpak Crest SIL (7.2 mm *id* × 250 mm); mobile phase, carbon dioxide/methanol = 90:10; flow rate, 5.0 mL/min; initial temperature, 130°C; temperature gradient, −2°C/min; pressure, 198 bar.

values. *DP* dependance of the melting point and density of the poly-(oxymethylene) homologs was also investigated [35,36]. These studies provided the basis for the formulation of many concepts essential in macromolecular science.

Recently, purely uniform poly(oxymethylene) diacetates that can be isolated as single crystals have been obtained by pSFC fractionation [37]. Figure 14 shows a pSFC trace of poly(oxymethylene) diacetate prepared from paraformaldehyde and acetic anhydride. During the separation, thermal degradation of the substrate occurred slightly (~3%). Each poly(oxymethylene) homolog obtained by the fractionation showed a sharp melting peak in DSC measurements. The melting points for the 10mer, 11mer, and 12mer were 63.1–66.2°C, 71.2–73.2°C, and 79.9–82.5°C, respectively. The values were significantly higher than those in the literature (10mer, 52–53.5°C; 11mer, 64–65°C; 12mer, 73–75°C) [35,36], which demonstrates improved purity of the poly(oxy-methylene) homologs obtained by the pSFC method.

Single crystals of the uniform poly(oxymethylene) diacetates from 9mer to 13mer were grown from ethanol and subjected to X-ray crystallographic analysis. The crystallographic data are summarized in Table 2. The calculated density (D_c) of the poly(oxymethylene) homologs increased with *DP*. The D_c for the 12mer was almost comparable with the experimentally found density (1.40–1.45 g/cm^3) of high molecular weight poly(oxymethylene). The acetal backbone of the 10mer molecules has been found to assume 2/1 helix in the crystal [37]; the sublattice structure is very similar to that of the metastable *orthorhombic* modi-

Table 2 Crystallographic Data for Oligo(oxymethylene) Diacetates from 9mer to 13mer [37]

Data	9mer	10mer	11mer	12mer	13mer
Formula	$C_{13}H_{24}O_{12}$	$C_{14}H_{26}O_{13}$	$C_{15}H_{28}O_{14}$	$C_{16}H_{30}O_{15}$	$C_{17}H_{32}O_{16}$
Formula wt.	372.33	402.35	432.38	462.40	492.43
Space group	C_c	$P2_12_12_1$		$P2_12_12_1$	
$a(\text{Å})$	46.360(3)	7.610(1)	114.56(1)	7.626(6)	7.7945(5)
$b(\text{Å})$	5.0538(5)	42.244(7)	9.080(2)	47.952(12)	64.686(6)
$c(\text{Å})$	7.7366(8)	5.959(1)	7.838(2)	5.9532(10)	4.5876(3)
$\alpha(\deg)$	90	90	90	90	90
$\beta(\deg)$	91.061(1)	90	90.19(1)	90	90.252(5)
$\gamma(\deg)$	90	90	90	90	90
$V(\text{Å}^3)$	1812.3(3)	1915.7(6)	8153(2)	2177.2(36)	2313.1(3)
$D_c(\text{g/cm}^3)$	1.364	1.395	1.409	1.411	1.414

fication of poly(oxymethylene) [38]. These structural features of the 10mer crystal have enabled comprehensive analysis of the vibrational assignments for the micrometer-sized single crystals of *orthorhombic* poly(oxymethylene) on the basis of polarized infrared and Raman spectroscopy [39,40]. On the other hand, single crystals of the 9mer belongs to the *monoclinic* system, and the helix pitch in the 9mer molecules is between 9/5 and 2/1. The X-ray determination indicates that the crystal structure of low molecular poly(oxymethylene) is highly dependent on *DP*.

III. CONCLUSION

1. Semipreparative pSFC using a 10-mm-*id* × 250 mm column packed with nonbonded silica gel is a versatile means of fractionating low molecular weight polymer homologs. Tens to hundreds of milligrams of each individual homologs with uniform molecular weights (uniform polymers) can be obtained by the repeated (50–100 times) pSFC fractionation. The maximum molecular weight for those uniform polymers exceeds 10,000 (100mer of methyl methacrylate).

2. Prerequisite for the efficient pSFC fractionation is to prepare source polymer with well-defined chemical structure.

3. Uniform polymers obtained by the pSFC fractionation are useful to study the exact molecular structure and physical properties of ordinary, nonuniform polymers. Uniform polymers with definite molecular weights also serve as reference materials for polymer analysis and characterization.

REFERENCES

1. IUPAC Commission on Macromolecular Nomenclature, *Compendium of Macromolecular Nomenclature*, Blackwell Scientific, Oxford, p. 52 (1991).
2. K. Hatada, K. Ute, and N. Miyatake, *Prog. Polym. Sci.*, *19*:1067 (1994).
3. F. P. Schmitz and E. Klesper, *J. Supercritical Fluids*, *3*:29 (1990).
4. R. E. Jentoft and T. H. Gouw, *J Polym. Sci., Polym. Lett. Ed.*, *7*:811 (1969).
5. W. Hartmann and K. Klesper, *J. Polym. Sci., Polym. Lett. Ed.*, *15*:713 (1977).
6. K. Klesper and W. Hartmann, *Eur. Polym. J.*, *14*:89 (1978).
7. M. Saito, Y. Yamauchi, and T. Okuyama (eds.), *Fractionation by Packed-Column SFC and SFE: Principles and Applications*, VCH, New York, (1994).
8. C. Berger and M. Perrut, *J. Chromatogr.*, *505*:37 (1990).
9. K. Ute, T. Takahashi, K. Hatada, and M. Kuramoto, *Polym. Prepr. Jpn.*, *43*:168 (1994).
10. N. Ishihara, T. Seimiya, M. Kuramoto, and M. Uoi, *Macromolecules*, *19*:2464 (1986).
11. K. Hatada, K. Ute, K. Tanaka, Y. Okamoto, and T. Kitayama, *Polym. J.*, *18*:1037 (1986).
12. T. Kitayama, T. Shinozaki, T. Sakamoto, Y. Yamamoto, and K. Hatada, *Makromol. Chem., Suppl.*, *15*:167 (1989).
13. K. Hatada, K. Ute, T. Nishimura, M. Kashiyama, T. Saito, and M. Takeuchi, *Polym. Bull.*, *23*:157 (1990).
14. K. Ute, N. Miyatake, T. Asada, and K. Hatada, *Polym. Bull.*, *28*:561 (1992).
15. K. Ute, N. Miyatake, Y. Osugi, and K. Hatada, *Polym. J.*, *25*:1153 (1993).
16. K. Hatada, K. Ute, T. Kitayama, T. Nishiura, and N. Miyatake, *Macromol. Symp.*, *85*:325 (1994).
17. K. Ute and N. Miyatake, K. Hatada, *Polymer*, *36*:1415 (1995).
18. T. G. Fox and P. J. Flory, *J. Appl. Phys.*, *21*:581 (1950).
19. P. J. Flory, *J. Chem Phys.*, *15*:397 (1949).
20. K. Ute, Y. Yamasaki, M. Naito, N. Miyatake, and K. Hatada, *Polym. J.*, *27*:951 (1995).
21. A. M. Liquori, G. Anzuino, V. M. Corio, M. D'Alagni, P. de Santis, and A. Savino, *Nature*, *206*:358 (1965).
22. J. Spevácek and B. Schneider, *Adv. Colloid Interface Sci.*, *27*:81 (1987).
23. L. Trofimoff, T. Aida, and S. Inoue, *Chem. Lett.*, 991 (1987).
24. Ph. Dubois, C. Jacobs, R. Jérôme, and Ph. Teyssié, *Macromolecules*, *24*:2266 (1991).
25. A. Le Borgne, V. Vincens, M. Jouglard, and N. Spassky, *Makromol. Chem., Macromol. Symp.*, *73*:37 (1993).
26. K. Ute, K. Hatada, *Preprints of 5th SPSJ International Polymer Conference*, p. 269 (1994).
27. Y. Ikada, K. Jamshidi, H. Tsuji, and S.-H. Hyon, *Macromolecules*, *20*:906 (1987).
28. M. M. Green, N. C. Peterson, T. Sato, A. Teramoto, R. Cook, and S. Lifson, *Science*, *268*:1860 (1995).
29. T. E. Patten and B. M. Novak, *J. Am. Chem. Soc.*, *113*:5065 (1991).
30. T. E. Patten and B. M. Novak, *Macromolecules*, *26*:436 (1993)
31. K. Ute, T. Asai, Y. Fukunishi, and K. Hatada, *Polym. J.*, *27*:445 (1995).

32. K. Ute, Y. Fukunishi, R. Niimi, T. Iwakura, and K. Hatada, *Polym. Prepr. Jpn.*, *45*:3284 (1996).
33. U. Shmueli, W. Traub, and K. Resenheck, *J. Polym. Sci., Part A-2*, *7*:515 (1969).
34. M. M. Green, C. Khatri, and N. C. Peterson, *J. Am. Chem. Soc.*, *115*:4941 (1993).
35. H. Staudinger and M. Lüthy, *Helv. Chim. Acta*, *8*:41 (1925).
36. H. Staudinger, R. Signer, H. Johner, M. Lüthy, W. Kern, D. Russidis, and O. Schweitzer, *Ann. Chem.*, *474*:145 (1929).
37. K. Ute, T. Takahashi, K. Matsui, and K. Hatada, *Polym. J.*, *25*:1275 (1993).
38. G. Carazzolo and M. Mammi, *J. Polym. Sci., A*, *1*:965 (1963).
39. M. Kobayashi, T. Adachi, Y. Matsumoto, H. Morishita, T. Takahashi, K. Ute, and K. Hatada, *J. Raman Spectrosc.*, *24*:533 (1993).
40. M. Kobayashi, Y. Matsumoto, A. Ishida, K. Ute, and K. Hatada, *Spectrochim. Acta*, *50A*:1605 (1994).

13

The Use of Supercritical Fluids in Environmental Analysis

Andrei Medvedovici

University of Bucharest, Bucharest, Romania

Agata Kot

Technical University of Gdańsk, Gdańsk, Poland

Frank David

Research Institute for Chromatography, Kortrijk, Belgium

Pat Sandra

University of Ghent, Ghent, Belgium

I. INTRODUCTION

During the last decade, analysts have witnessed remarkable advances in separation sciences, e.g., the introduction of supercritical fluid chromatography (SFC). Moreover, hyphenation of the separation column to spectroscopic techniques has been realized for all forms of chromatography allowing identification of unknowns and sensitive and selective detection of target compounds. On the other hand, developments in sample preparation proceeded somewhat slowly, often hindering full exploitation of the new separation techniques. Methods already in use in the first half of the century, e.g., liquid extraction, Soxhlet extraction, etc., which are time consuming, often an order of magnitude longer than the separation step, and involve the use of large volumes of solvents that are expensive and toxic, are in most cases still applied. Emerging new sample preparation methods are solid phase extraction (SPE), solid phase microextraction (SPME), supercritical fluid extraction (SFE), fractionation by size exclusion chromatography (SEC), accelerated solvent extraction (ASE), to mention a few. In the new developments it is interesting to note the "supercritical fluid" is a key word both in separation sciences and in sample preparation. Considerations on automation, speed, and cost indeed created enormous interest in super-critical fluid technology in the past 10 years, and researchers and companies have spent an increasing amount of effort, time, and money on the development of SFC and SFE. As a supercritical fluid is often referred to as a "supersolvent" with gas-like mass transfer and liquid-like solvating properties, expectations were very high. The gas-like and liquid-like combination indeed should make supercritical fluids very attractive for extraction (SFE) and chromatography (SFC) and, as a logical consequence, for the combined technique SFE-SFC.

In this chapter, the possibilities of using supercritical fluids in environmental analysis will be discussed. However, it is not the aim of this chapter to give a complete overview of the use of supercritical fluids in environmental analysis but rather to present our experiences and results of the last 5 years.

II. PACKED COLUMN SUPERCRITICAL FLUID CHROMATOGRAPHY IN ENVIRONMENTAL ANALYSIS

Gas (GC) and high-performance liquid chromatography (HPLC) are well-estab-lished separation techniques in environmental analysis and both fulfill the requirements put on analytical techniques in general: performance, low operat-ing costs, productivity, and regulatory compliance. In first instance, the success of SFC depends on those criteria. If they are not fulfilled, there is no way for a breakthrough of SFC. Second, environmental analysis means trace analysis. Robust SFC techniques and instrumentation together with columns offering high sample capacity and large volume loadability are needed to meet these requirements. This explains the superior characteristics of packed column SFC (pSFC) compared to capillary SFC (cSFC) for environmental analysis. The future of SFC depends on examples of problems uniquely solved by SFC or "better" solved than with GC or HPLC, and this in the context of requirements for analytical techniques. In the past, the features of SFC have all too often been illustrated with examples that can be better solved by the classical separation methods. Possible advantages of SFC compared to GC can be summarized as follows:

1. More reliable data for polar and thermolabile compounds
2. No need for derivatization
3. Extension of the molecular weight range
4. The possibility to introduce selectivity in the mobile phase

    Advantages of SFC over HPLC are:

1. Faster analysis
2. Higher efficiencies

3. Better performance compared to normal phase HPLC
4. Use of GC type detectors
5. Easier pre- and postcolumn hyphenation, e.g., interfacing with spectroscopic detectors
6. Less toxic mobile phases, less solvent disposal costs

The fundamental aspects of pSFC have recently been summarized in the book of T. Berger [1]. pSFC offers efficiency and selectivity for a number of relevant applications in environmental analysis. By far the best chromatograms for pesticide analysis were presented by Berger et al. [1–3]. By coupling several silica columns organochlorine, organophosphorus, phenylurea, sulfonylurea, triazine, and carbamate pesticides could be separated with high efficiency and selectivity. As an example, the separation of 44 pesticides is illustrated in Fig. 1 [1].

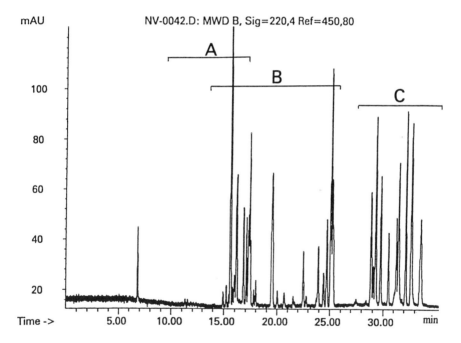

Figure 1 pSFC separation of organochlorine (A), organophosphorus (B), phenylurea and sulfonylurea (C) pesticides. Experimental conditions: column Hypersil silica 5 μm, 140 cm × 4.6 mm id; flow rate 2 mL/min; temperature 60°C; pressure programmed from 80 bar (5 min) to 130 bar at 5 bar/min; modifier methanol from 2% (5 min) to 20% at 1%/min; detection 220 nm; amount of injected sample 20 to 200 ng each analyte [1].

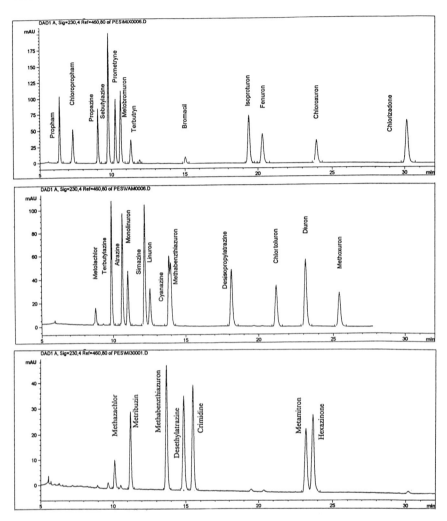

Figure 2 pSFC separations of some representative pesticides on silica. Experimental conditions: column Keystone silica 5 μm, 25 cm × 4.6 mm id; flow rate 2 mL/min; temperature 40°C; pressure programmed from 100 to 250 bar at 5 bar/min; modifier methanol from 1% to 21% at 0.5%/min; detection 230 nm; amount of injected sample ~ 200 ng each compound [37].

The use of very long columns is not required for routine analysis, and 25–50 cm columns in length normally offers enough efficiency to tackle most of the separation problems encountered in environmental analysis. Neat silica and diol phases have proven to be the most useful and versatile phases for the analysis of

Table 1 Reproducibility of Retention Data Obtained for Some Pesticides Analyzed by pSFC[a]

Crt. no.	Pesticide	SFC-DAD analysis	
		k	RSD (%)
1	Propham	1.24	2.6
2	Chlorpropham	1.58	1.9
3	Metolachlor	2.04	3.5
4	Propazine	2.18	1.9
5	Sebutylazine	2.40	1.7
6	Terbutylazine	2.44	2.9
7	Prometryn	2.61	1.5
8	Methazachlor	2.64	1.4
9	Atrazine	2.70	2.6
10	Metobromuron	2.70	1.3
11	Monolinuron	2.83	2.4
12	Terbutryn	2.98	0.9
13	Metribuzin	3.02	1.2
14	Simazine	3.23	2.1
15	Linuron	3.36	2.0
16	Cyanazine	3.82	1.8
17	Methabenzthiazuron	3.87	1.8
18	Desethylatrazine	4.30	0.8
19	Bromacil	4.31	0.7
20	Crimidine	4.52	0.7
21	Desisopropylatrazine	5.34	1.5
22	Isoproturon	5.87	0.5
23	Fenuron	6.18	0.5
24	Chlorotoluron	6.40	1.3
25	Diuron	7.10	1.3
26	Metamitron	7.26	0.5
27	Hexazinone	7.43	0.5
28	Chloroxuron	7.45	0.4
29	Methoxuron	7.89	1.3
30	Chlorizadone	9.63	0.3

[a]Experimental conditions as in Fig. 2.

pesticides [4–7]. This is illustrated in Fig. 2 showing some representative pSFC chromatograms for pesticides on neat silica. Retention reproducibility in pSFC is much better compared to normal phase HPLC. In Table 1, the retention factors (k) and relative standard deviations (RSD%) for a number of pesticides are listed for eight injections over a 3-day period.

Cyanopropyl, short-chain alkanes (C1–C8), and phenyl phases have also been evaluated during our studies for pesticide analysis but the specific selec-

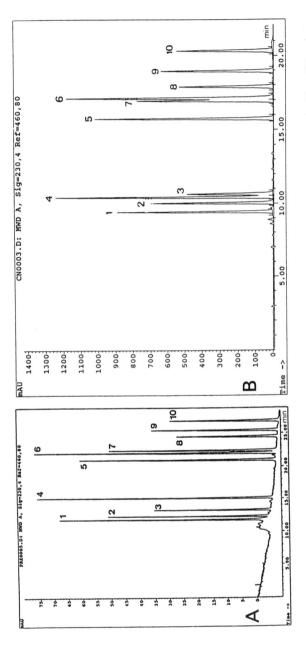

Figure 3 pSFC separation of phenylurea pesticides on 25 cm silica (A) and 40 cm cyanopropyl silica (B) columns. Experimental conditions as in Fig. 2. Peaks: 1, monolinuron; 2, metobromuron; 3, linuron; 4, methabenzthiazuron; 5, isoproturon; 6, fenuron; 7, chlorotoluron; 8, diuron; 9, chloroxuron; 10, metoxuron.

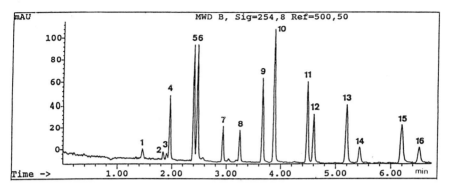

Figure 4 pSFC separation of PAHs on a column tandem: Spherical ODS-HL 3 μm, 15 cm × 4.6 mm id + Chromspher PAH 3 μm, 10 cm × 4.6 mm id. Experimental conditions: flow rate 3 mL/min; temperature programmed from 40°C to 55°C at 2°C/min; pressure programmed from 80 to 200 bar at 50 bar/min; modifier acetonitrile from 1% to 30% at 8%/min; detection 254 nm; amount of injected sample ~ 20 ng each compound [9]. Peaks: 1, naphthalene; 2, acenaphthylene; 3, fluorene; 4, acenaphthene; 5, phenanthrene; 6, anthracene; 7, fluoranthene; 8, pyrene; 9, benzo[a]anthracene; 10, chrysene; 11, benzo[b]fluoranthene; 12, benzo[k]fluoranthene; 13, benzo[a]pyrene; 14, dibenzo-[a,h]anthracene; 15, indeno[1,2,3,c,d]pyrene; 16, benzo[g,h,i]perylene[9].

tivities hardly improved resolution. This is illustrated in Fig. 3 in which the separations of 10 phenylurea on a 25-cm silica (A) and on a 40-cm cyanopropyl silica (B) are compared. Only one important shift is noted, namely, for metha-benzthiazuron (4) representing a compound with a different polarity compared to the other phenylurea.

Surprisingly, octadecyl silica (ODS) also performs very well in pSFC and more especially for PAH analysis. By selectivity tuning between two different ODS phases complete separation of the 16 priority pollutant PAHs was realized in 6 min [9] (see Chapter 6). The separation is shown in Fig. 4 and the column order had no influence on the separation characteristics.

On the other hand, aminopropyl silica has shown unique selectivity for the separation of nitro- and hydroxyl- polynuclear aromatic hydrocarbons (PAH) from normal PAHs. This separation (Fig. 5) has been exploited to fractionate low traces of nitroPAHs from a PAH fraction collected from air particulates [8].

These unique applications illustrate the potential of pSFC for environmental analysis. This together with some of its features, such as the possibility of using nitrogen-phosphorus detection (NPD), electron capture detection (ECD), and spectroscopic detectors, and the ease of pre- and postcolumn hyphenation, provides pSFC with a bright future for environmental analysis.

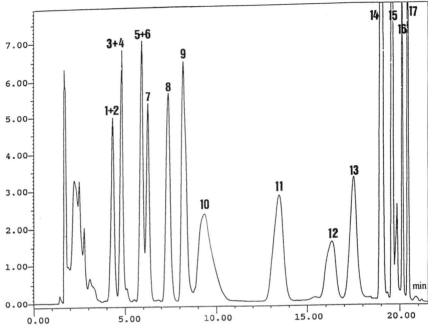

Figure 5 pSFC separation of hydroxyl- and nitro-polynuclear aromatic hydrocarbons. Experimental conditions: column Keystone silica 5 μm, 25 cm × 4.6 mm id; flow rate 2 mL/min; temperature 50°C; pressure 200 bar; modifier acetonitrile-methanol = 1:1 programmed from 1% (5 min) to 2.5% at 0.15%/min, then to 27.5% with 5%/min; detection 220 nm; amount of injected sample ~ 200 ng each compound [8]. Peaks: 1, 2-nitronaphthalene; 2, 3-nitro-1,1'-biphenyl; 3, 1-nitronaphthalene; 4, 2-nitro-1,1'-biphenyl; 5, 1,5-dinitronaphthalene; 6, 2-nitrofluorene; 7, 9-nitroanthracene; 8, 1,3-dinitronaphthalene; 9, 3-nitrofluoranthene; 10, 1-nitropyrene; 11, 2,2'-dinitro-1,1'-biphenyl; 12, 2,7-dinitrofluorene; 13, 2,5-dinitrofluorene; 14, 1-hydroxylnaphthalene; 15, 2-hydroxylnaphthalene; 16, 9-hydroxylphenanthrene; 17, 1,8-dinitronaphthalene [8].

III. SUPERCRITICAL FLUID EXTRACTION IN ENVIRONMENTAL ANALYSIS

SFE is a sample preparation technique the fundamentals of which are embodied within the theory of SFC. In fact, the chromatographic process may be seen as a series of solute extractions between the mobile phase and the stationary phase. SFE was introduced in the middle of the century as a processing technique. Interest in SFE grew rapidly in the late 1960s and early 1970s mainly for industrial scale extractions, e.g., caffeine from coffee, bitter acids from hops, flavors from spices, etc. SFE as a tool in the analytical laboratory was discovered in the early 1980s. Environmental concern finally resulted in a boom of SFE in the late

1980s and early 1990s. Interesting information on the growth of SFE and also of SFC to the present state of the art can be found in Refs. 10–15.

The main task of sample preparation is to separate analytes of interest from a matrix by investing the minimum amount of effort and time (productivity) necessary to provide quantitations with high degrees of accuracy and precision (performance). Traditional approaches to sample preparations are often labor-intensive and require many manual manipulation steps. Analytical scale SFE is slowly gaining its place among the sample preparation methods for solid and semisolid matrices because, compared to classical methods, it offers:

1. Reduced need for organic combustible solvents, which are often toxic, e.g., dichloromethane, and harmful to laboratory personnel and the environment.
2. Reduced extraction times. SFE extraction times are in the order of 30 min, whereas for Soxhlet extraction at least 12 h extraction is required.
3. Automation.
4. Possibility to hyphenate off- or on-line with chromatographic separation techniques.
5. High selectivity allowing direct analysis without any further clean-up or enrichment.

The maturation of SFE and its acceptance by industrial laboratories and government agencies depends, however, on a number of factors. Some important issues are as follows:

1. *Reliable instrumentation.* Modifying a SFC or taking an old piston pump out of the cellar is not the way to construct reliable SFE instrumentation. Nor do statements such as "We recommended for SFE neophytes to purchase a simple, manual system or assembling one from modular components that can be obtained from several suppliers, and experimenting to find out how to extract a certain type of sample" [16] advance acceptance of SFE. Unfortunately, in routine and quality assurance and/or control laboratories, nowadays there is no time or money available to experiment (playground) with SFE instrumentation. Reliable instrumentation, on which a particular application can be worked out or a method validated, is the first prerequisite.
2. *Uniform instrumentation.* Specifications and features of commercially available SFE systems differ too much, i.e., maximum pressure, maximum pump flow rate, restrictor type, restrictor flow rate, collection mode, mode of modifier addition, etc. At present, there is no guarantee that an application optimized and validated on system X can be repeated on system Y. This makes it extremely difficult to develop standardized methods. SFE instrumentation of the future should embrace multiple options for selectivity, collection, etc.

3. *Need for reliable data.* All too often, published data are unrealistic as they are based on spiked samples (ppm level!); on sample sizes that are too small to be representative; no information is given on important sample parameters such as water content, carbon content (both can easily be elucidated with thermogravimetric analysis), etc. An SFE neophyte recently asked us which SFE method to apply, out of the many published, for the extraction of PAHs from soil. At present, there is only one answer: it all depends on. . . . All too often, published data show how good SFE and how bad conventional techniques are. This doesn't help the acceptance of SFE as a valuable technique either. As an example, for the extraction of organotin compounds from a certified reference material CRM 462 sample (Institute of Reference Materials and Measurements, European Commission, Geel, Belgium), we found the same recovery with ultrasonic treatment as with SFE and this in the same time. The latter requires, however, megadollar equipment! Last but not least, SFE should not be applied in those cases where other methods are definitely working better, e.g., most data published on SFE of water matrices.

4. *Need for certified reference samples and collaborative studies.* Spiked and weathered or aged samples only provide information on the solubility in the supercritical medium; real-world samples are required. On the other hand, collaborative studies should be extended to worldwide participation, eventually via instrument manufacturers.

5. *Need for better theoretical understanding of the mechanisms of SFE.*

6. *Exploiting more the selectivity possibilities of SFE.*

A schematic drawing of the instrument used during this work is shown in Fig. 6. The system consists of an extraction cartridge (1), an oven (2), a carbon dioxide pumping module (3), an independent modifier pump (4), a variable restrictor or nozzle (5), a solid phase trap (6), and a pump to deliver the rinse solvents (7). A personal computer provides instrumental control. One of the most interesting features of SFE, namely, its "selectivity," has not been fully exploited until now in the preparation of environmental samples. By optimizing SFE conditions, it should be possible to prepare extracts that are directly amenable to chromatographic analysis. Modern SFE instrumentation contains several parts in which selectivity can be introduced (Fig. 6). In the first instance, the extraction selectivity and efficiency can be controlled by the nature of the supercritical medium (*selectivity 1*). The reasons to add a polar or an apolar modifier to carbon dioxide, normally used as supercritical medium, are threefold: (1) to increase the solubility, (2) to destroy the matrix effects, and (3) to enhance diffusion by swelling of the matrix. In principle, modifiers could also be added to retain unwanted solutes but this counteracts (2) and (3). An illustration of the effect of the addition of a polar modifier on the recovery of polychlorinated biphenyls (PCBs) can be found in the literature [17,18]. The second

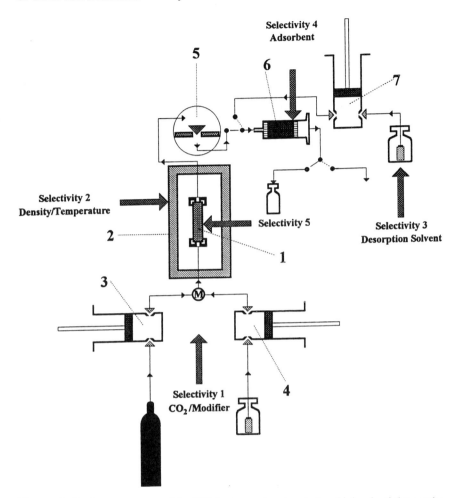

Figure 6 Basic components of the SFE instrument and parts in which selectivity can be introduced.

selectivity concerns the density of the supercritical medium and the temperature (*selectivity 2*) [19]. The importance of the latter parameter is often neglected but, as solubility is also controlled by the vapor pressure, this can be exploited to introduce selectivity as will be illustrated with the determination of PCBs in lipid matrices. After leaching of the sample, the extract is collected on a solid trap filled with an apolar or polar adsorbent that can be selected according to the application (*selectivity 3*). The trap is then rinsed with a solvent, polarity of which can be chosen to desorb the solutes of interest in a selective way (*selectivity 4*). Last but not least, an adsorbent can be added in the extraction thimble (*selectivity* 5). In the first instance, this opens up the possibility of extracting liquids by SFE [20], but it also permits the retention of unwanted polar solutes (fixation) or enhanced recoveries of apolar solutes (exaltation).

The selectivity of SFE will be illustrated in this chapter with the following applications: the determination of hydrocarbons in soil, the analysis of PCBs in cod liver oil, the SF extraction of PAHs from barbecued meat, and the determination of organochloro pesticides (OCPs) in tobacco leaves.

A. Importance of Density and Temperature: The Determination of Hydrocarbons in Soil

In the United States, the procedures for the determination of petroleum hydrocarbons in soil samples are described in EPA methods 418.1, 8015, and 8440. These methods involve liquid extraction of soil samples with Freon 113, followed by GC-flame ionization detection (FID) or direct infrared (IR) analysis. Because of the present controversy on the use of freons, the possibilities of SFE as an alternative method were evaluated (EPA draft method 3560 [21]). In the proposed method, strong extraction conditions (80°C, 0.78 g/mL, 40 min) are used to obtain similar results as with the classical extraction method. The disadvantage of using strong extraction conditions is that other compounds mainly from natural origin, such as plant sterols, are extracted as well and additional clean-up is required. Soil contamination from fuels such as naphtha, gasoline, kerosene, or diesel is reflected by the presence of the hydrocarbons C5–C24 and these soluble compounds can be extracted with high selectivity by

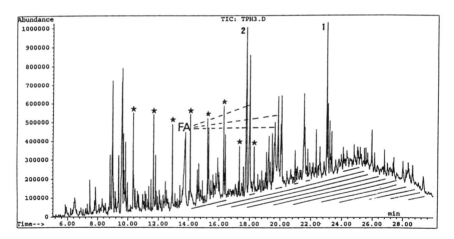

Figure 7 Analysis of an SFE extract of soil at 0.78 g/mL density. Peaks: ★, hydrocarbons; FA, fatty acids; ///, hump of sterols and related products; 1, di(ethylhexyl)-phthalate; 2, diphenylsulfone.

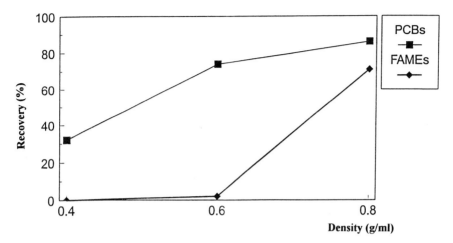

Figure 9 Recovery in function of density (60°C, 15 min, 1 mL/min) [28].

45°C, collection on an ODS trap at 45°C, and rinsing with 1.5 mL *n*-hexane. The first parameter to be varied was density. The results are shown in Fig. 9. At 0.4 g/mL, the recovery for both the PCBs and the fat is low; at 0.6 g/mL, 75% of the PCBs is extracted and only 2% of the fat is coextracted; and at 0.8 g/mL, 88% of the PCBs is extracted but also 72% of the fat. This plot shows the influence of the density on the recovery and the difference in solubility of the two compound classes. This difference can form the basis for further optimization of the selectivity in SFE for that particular case.

From Fig. 9, densities of 0.6 g/mL and 0.5 g/mL were selected as starting points for further optimization because at these densities relatively high recov-

Table 2 SFE of PCBs from Cod Liver Oil: Influence of SFE Conditions on Recovery of PCBs and FAMES [28]

	Density (g/mL[1])	Temp. (°C)	Time (min)	Flow (mL/min[1])	Recovery (%) PCBs	FAMES
1	0.4	60	15	1	32	0
2	0.6	60	15	1	75	2
3	0.8	60	15	1	88	72
4	0.6	60	30	1	83	7
5	0.6	60	15	2	104	10
6	0.5	60	15	2	98	2
7	0.5	80	30	1	(140)	15

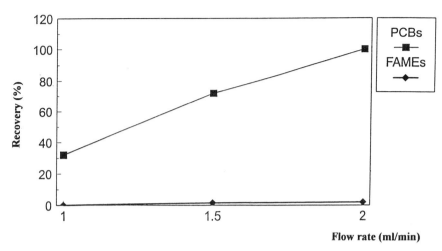

Figure 10   Recovery in function of flow rate (60°C, 15 min, 0.5 g/mL) [28].

eries of PCBs were obtained, whereas the fat recoveries were low. The data obtained are summarized in Table 2 and the following conclusions can be drawn. An increase of the extraction time from 15 to 30 min at constant density (0.6 g/mL), extraction temperature (60°C), and flow (1 mL/min) resulted in a higher recovery for the PCBs (83%), but also in an increased fat recovery (7%). An increase of the flow from 1 to 2 mL/min at 0.6 g/mL, extraction time (15 min), and extraction temperature (60°C), resulted in 104% PCBs and 10% fat. From these data it was clear that flow is an important parameter. Optimized conditions could then be found using a reduced density of 0.5 g/mL at 60°C and an extraction time of 15 min. Under these conditions the influence of flow and temperature on the recovery were evaluated. The results for the flow are shown in Fig. 10.

At a flow of 2 mL/min, the recovery of PCBs was 98% whereas only 2% of the fat is extracted. An increase of the extraction temperature from 60°C to 80°C at 0.5 g/mL, a flow of 1 mL/min, and 30 min extraction time yielded a too high recovery for the PCBs and an intolerable increase of fat recovery (15%). The recovery of the PCBs could not be measured accurately because of contamination of the CGC-ECD system. In this case, increasing the extraction temperature is not the best choice for increasing the recovery because the selectivity of the extraction is lost. With the SFE conditions of 0.5 g/mL density, a temperature of 60°C, extraction for 15 min, and a flow of 2 mL/min, a reliable procedure was developed. Repeatability was good as will be discussed further. Similar results were reported by Bowadt and coworkers [31] for the extraction of low concentrations of PCBs from fish tissues.

C. Selectivity of the Solid Phase Trap and Rinsing Solvents: The Analysis of PAHs in Barbecued Meat

In the previous example, density, temperature, and flow were tuned to fractionate PCBs and lipids. The same SFE conditions could not be applied for the enrichment of PAHs from barbecued meat for two reasons. First, for the quantitative extraction of PAHs higher densities are required to solubilize the four-, five- and six-ring PAHs. Second, matrix effects (fixation) do occur which should be broken. Pork meat was roasted in an open fire in such a way that the meat was partially carbonized (black parts). A layer of 0.5 cm was removed over 3 × 3 cm, ground, mixed, and subjected to Soxhlet extraction and SFE. The PAHs were measured by HPLC analysis on a 25 × 0.46 cm id Vydac PAH column with a water-acetonitrile gradient. Detection was performed by fluorescence with variable-wavelength settings for the different PAHs. The recovery of PAHs by SFE of 250 mg was compared to data obtained by Soxhlet extraction of 1 g material. Soxhlet extraction was performed with dichloromethane during 18 h and the extract was cleaned over silica before HPLC analysis. The PAH concentrations recovered with Soxhlet were considered as 100%. In the SFE optimization procedure, starting conditions were set at a temperature of 60°C, a flow of 2 mL/min, and an extraction time of 5 min static and 30 min dynamic. The trap, kept at 45°C, was filled with silica (Merck 100–200 mesh, conditioned for 16 h at 130°C, deactivated with 3% water) and the nozzle temperature was 45°C. The recoveries for fluoranthene and benzo[a]pyrene, selected as representative for the 16 PAHs enlisted in the priority pollutant list, and of the fat, measured via the FAMEs, were measured at 0.4, 0.6, and 0.8 g/mL carbon

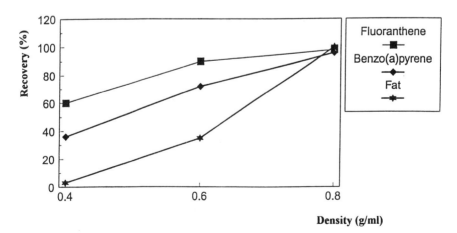

Figure 11 Recovery of PAHs from barbecued meat. Influence of the density on recovery [28].

dioxide density. Figure 11 reveals that fractionation of PAHs from fat is much more difficult than of PCBs from fat [28]. At 0.4 g/mL the recovery of fluoranthene (three-ring) is 60% but only 25% for the five-ring benzo[a]pyrene. The larger the ring, the lower is the solubility in neat carbon dioxide. At 0.6 g/mL recoveries increased to 91% and 73%, respectively, but the fat yield increased to 37%.

For quantitative recovery of the PAHs, a density of 0.8 g/mL is required but under those conditions the fat is coextracted. No conditions could be found to avoid this and therefore optimizing SFE conditions was not successful in this case. We could, however, rely on selectivities 3 and 4, i.e., selection of the trap adsorbent and eluting solvents, for this particular application. Based on the polarity difference between PAHs and triglycerides, column chromatography on silica is a well-established method to enrich PAHs from fat and oil samples. In the case of the barbecued meat, the fat content was relatively low and the fractionation column can be miniaturized to a cartridge as is the case in the SFE instrument. The silica trap was subsequently rinsed with 1.5 mL hexane/dichloromethane (1:1) (solvent A), 1 mL hexane/dichloromethane (1:1), and

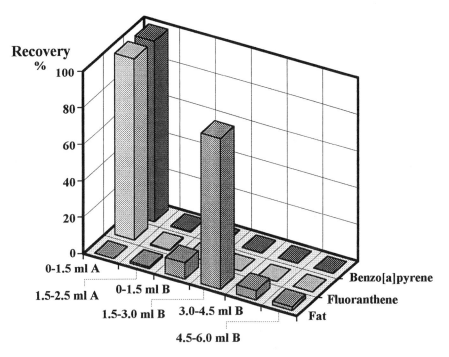

Figure 12 Fractionation of PAHs from barbecued meat in the solid phase trap. Influence of eluting solvents A and B on PAH and fat recovery [28].

4 times 1.5 mL methanol (solvent B). The distribution of fluoranthene, benzo-[a]pyrene, and fat in the different fractions is shown in Fig. 12. The PAHs were completely recovered in fraction 1 without any trace of fat. The quantities of PAHs measured in barbecued meat were in the ppm range, which illustrates that care should be taken in the way of roasting meat in an open fire. Because of the inhomogeneity of the sample, reproducibility studies could not be performed.

D. Fixation Effects in the Extraction Thimble: The Analysis of OCPs in Tobacco Leaves

The fifth selectivity was actually discovered during the development of an SFE procedure for the determination of pesticides in fruit juices. In order to extract this liquid sample, the sample was "solidified" by mixing with diatomeceous earth, i.e., Chromosorb in a ratio of 2:3. The extract obtained was relatively clean and the pesticides could be elucidated by CGC-MS in the full-scan mode [20].

The same principle was applied to the analysis of organochloro pesticides (OCPs) in tobacco leaves [33] but for a different reason. The classical procedure to analyze OCPs in tobacco leaves consists of (1) Soxhlet extraction with dichloromethane, (2) clean-up over Florisil silica, and (3) dual-column (SE-54/OV-1701) CGC-ECD analysis. Not only is this procedure very labor-intensive and time consuming but moreover, and notwithstanding the high selectivity of the ECD detector for OCPs and the dual-column approach, it gives more than 100 peaks. Initial SFE experiments by spiking an OCP reference sample on filter paper indicated that the compounds are completely soluble at densities of 0.5 g/mL and a temperature of 70°C. Other conditions were: extraction 5 min static and 30 min dynamic, flow rate 1 mL/min, nozzle temperatures 45°C, trap material octadecyl silica at 25°C, and rinsing solvent 1 mL hexane. These moderate extraction conditions applied to 0.5 g tobacco leaves from Cuba gave a very complex chromatogram with, as ascertained by CGC-MS, alkaloids, unsaturated fatty acids and ketones, aromatic compounds, and so forth as main components. Lowering the SFE power would result in too low recoveries for the OCPs. The tobacco leaves were therefore mixed with 1 g silica (Merck 100–200 mesh, conditioned for 16 h at 130°C, deactivated with 3% water). With the above-described SFE conditions, the chromatogram indeed was much cleaner but the recovery for OCPs was lowered to 30–70% because of the strong adsorption power of silica. The SFE conditions were therefore adapted and an extraction temperature of 80°C and a density of 0.75 g/mL were tried out. Upon spiking the tobacco sample in the 10 ppb level with OCPs, all pesticides enlisted in EPA method 608 could be recovered for 100% with the exception of the more polar endosulfan sulfate for which the recovery was only 50%. The latter compound could be 100% extracted applying 10% methanol as modifier, but the selectivity enhancement by the use of silica in the extraction cartridge disap-

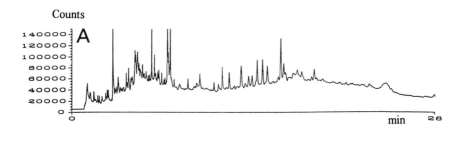

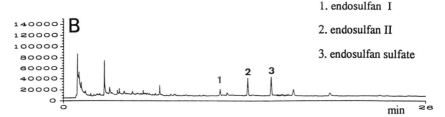

Figure 13 Analysis of tobacco leaves without (A) and with (B) silica in the extraction thimble [33].

peared by the high polarity of the modifier. For that particular compound, its accurate determination required ion monitoring mass spectroscopy. The other OCPs can be quantified by SFE-CGC-ECD without addition of methanol as modifier as illustrated in Fig. 13. Figure 13A shows the analysis of tobacco leaves without addition of silica gel in the cartridge and Fig. 13B shows the analysis with addition of silica gel.

The analyses were performed by CGC-ECD on a 30 m × 0.25 mm × 0.25 μm HP Ultra 2 column, the temperature of which was programmed from 50°C to 300°C at 10°C/min. The carrier gas was hydrogen at 30 cm/s and splitless injection was applied. The concentrations of endosulfan I and II were 190 ppb and 472 ppb, respectively, and for endosulfan sulfate 430 ppb. As the recovery of endosulfan sulfate for a spiked sample is only 50%, an exact concentration could not be advanced. The repeatability of the method was not outstanding, i.e., RSD% ≈ 10.5 for $n = 5$, which is mainly because of the inhomogeneity of the sample and the small sample amounts used for extraction.

E. Considerations on the Ruggedness of SFE

The SFE instrument is completely automated with its entire operation controlled by a personal computer. This provides not only productivity but also an excellent performance in terms of reliability and repeatability. The reproducibility

Table 3 Supercritical Fluid Extraction Conditions for PCBs from Soil

Instrument	HP 7680 SFE
Sample matrix	Soil
solutes	PCBs
sample weight	500 mg
Supercritical fluid (SFC/SFE grade)	Carbon dioxide
Density	0.75 g/mL
Flow	2 mL/min
Extraction cell volume	7 mL
Extraction cell temperature	60°C
Modifier used	No
Static extraction time	5 min
Dynamic extraction time	30 min
Solid phase trap	Octadecyl silica gel
Trap temperature	20°C
Nozzle temperature	45°C
Rinse solvent	Hexane
Rinse volume	2 × 1.5 mL
Rinse flow	1 mL/min

could not be evaluated as we had only one system available in our laboratory and round-robin tests have not been organized until now. The repeatability of SFE was checked by analyzing PCBs in a contaminated soil sample nine times over a 3-day period. The extraction conditions are summarized in Table 3. After extraction, the PCBs were collected in two 1.5 mL portions of hexane. The portions were combined and diluted to 5 mL. Then 250 ng octachloronaphthalene was added as internal standard (IS). The analyses were performed using CGC with ECD. A typical chromatogram is shown in Fig. 14. The presence of PCBs is easily detected, whereas little interference of coextracted compounds is noted. Note that the total analysis time is only 80 min (5 min static extraction, 30 min dynamic extraction, 5 min rinsing, and 40 min chromatographic analysis). The repeatability data are summarized in Table 4. For the most abundant PCB congeners (numbering according to Ballschmiter [34]: PCB nos. 28, 52, 101, 118, 153, 138, and 180) the relative peak areas (area PCB/area IS), which were also corrected for variations in sample weight, are listed. The relative standard deviation is better than 5% in all cases. This is excellent for this type of environmental analysis, whereby ppb concentrations are measured and the total analytical repeatability is taken into account.

The accuracy of SFE was also checked by Sandra et al. [25] for the same compounds by analysis of a certified sediment sample (CRM 392, BCR, Brussels). Results were good as reflected by 100 ppb for PCB 28, 283 ppb for PCB

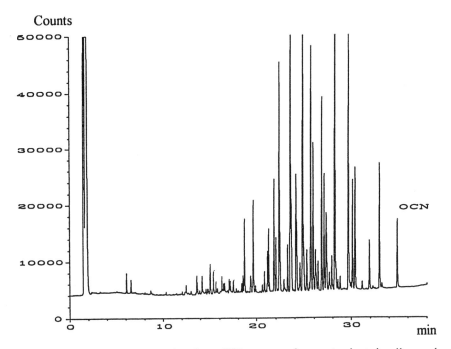

Figure 14 CGC-ECD analysis of an SFE extract of a contaminated soil sample. Experimental conditions are given in the text.

153, and 295 ppb for PCB 180 by SFE and by 100 ± 10 ppb for PCB 28, 288 ± 18 ppb for PCB 153, and 313 ± 24 for PCB 180 by traditional methods.

We have to note, however, that SFE generally requires much smaller sample sizes than conventional methods. Those small sample sizes may not be representative of the overall sample, biasing the analytical result. This is also one of the reasons why off-line SFE often is preferred over on-line SFE for inhomogeneous samples such as plant material [19].

IV. SUPERCRITICAL FLUIDS IN HYPHENATED SYSTEMS

Sample preparation is the top priority for application of reliable and cost-effective automation. Solid phase extraction (SPE) is more and more replacing liquid-liquid extraction for the enrichment of organic micropollutants from water samples, e.g., PAHs, organochloro pesticides, phenols, phenylureas, etc. Besides a better performance compared to traditional methods, SPE has the distinct advantages of easy automation and the possibility of on-line coupling

Table 4 Repeatability of SFE for the Extraction of PCBs from a Contaminated Soil Sample [26]

PCB	SFE1/IS	SFE2/IS	SFE3/IS	SFE4/IS	SFE5/IS	SFE6/IS	SFE7/IS	SFE8/IS	SFE9/IS	MEAN	SD	RSD
PCB 28	0.4554	0.4637	0.4480	0.5076	0.4390	0.4556	0.4619	0.4578	0.4696	0.4621	0.0193	4.17 %
PCB 52	0.2067	0.2133	0.1943	0.2020	0.1856	0.1965	0.1912	0.1996	0.2015	0.1990	0.0083	4.16 %
PCB 101	1.0875	1.0908	1.0309	1.1249	1.0771	1.1819	1.0707	1.0995	1.1627	1.1029	0.0469	4.25 %
PCB 118	3.3869	3.4410	3.2105	3.4979	3.1974	3.5269	3.3487	3.4429	3.4605	3.3903	0.1182	3.49 %
PCB 153	5.3066	5.4613	5.1071	5.5734	5.1068	5.6365	5.3321	5.4469	5.4285	5.3777	0.1849	3.44 %
PCB 138	5.2923	5.4275	5.0707	5.5433	5.0780	5.6278	5.2875	5.4121	5.4523	5.3546	0.1917	3.58 %
PCB 180	7.6645	7.9754	7.4162	8.1001	7.4729	8.3106	7.7359	7.9227	8.0248	7.8470	0.2967	3.78 %
OCN	1.0000	1.0000	1.0000	1.0000	1.0000	1.0000	1.0000	1.0000	1.0000	1.0000	0.0000	0.00 %

with the separation system. The literature is well documented on hyphenation of SPE with different chromatographic techniques [35–38] and a full range of adsorbents have been applied, the most important of which are octadecyl silica and polystyrene-divinylbenzene resins. Until recently, SFC was not regarded as either a quantitative or a trace analysis technique but the introduction of reliable instrumentation, and the revival of pSFC has changed this situation and opened the technique for precolumn hyphenation with the sample preparation step.

The combination of solid phase extraction (SPE) with carbon dioxide desorption (SFE) and supercritical fluid chromatography (SFC) has some advantages over its capillary GC (SPE-CGC) and HPLC (SPE-HPLC) counter-parts. Compared to CGC, there is no need for complex interfacing to volatilize

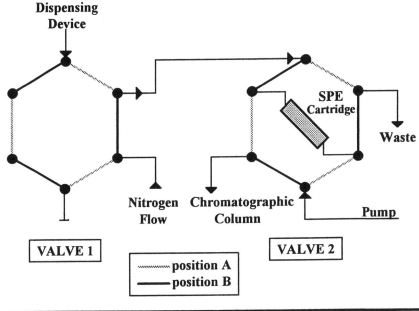

Operation	Position Valve 1	Position Valve 2
Conditioning, Loading, Washing	A	B
Drying	B	B
Injection	A	A

Figure 15 System configuration for a single cartridge system for SPE-SFC-DAD (based on a Gilson 233 XL sample preparation unit) [38].

and remove the desorption fluid, while compared to HPLC the selectivity of desorption can be better controlled.

Two SPE-pSFC-DAD combinations have recently been described [37,38]. The first system, based on the Gilson 233 XL sample preparation station (Fig. 15), is a simple single-cartridge approach that can be applied if the sample load is rather low, whereas the second system, based on the Merck-Hitachi automatic SPE preparator (OSP-2A), is a multicartridge approach allowing one to monitor unattended organic micropollutants in 16 water samples (Fig. 16).

The best adsorbents for SPE-SFE-pSFC in terms of recoveries (adsorption), peak shapes, carryover (desorption kinetics), and total chromatographic profile (selectivity of desorption) are polystyrene-divinylbenzene (PSDVB) resins.

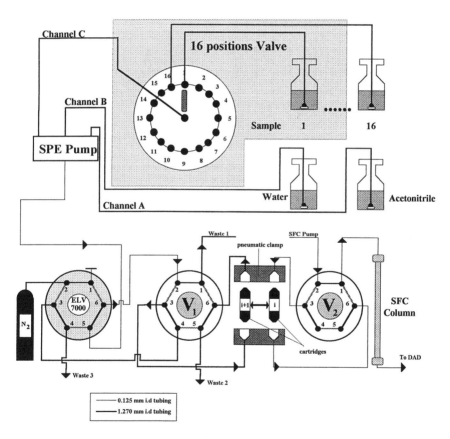

Figure 16 Experimental setup for routine on-line SPE-SFC-DAD analysis on a multiple cartridge system (based on Merck OSP 2A sample preparator) [38].

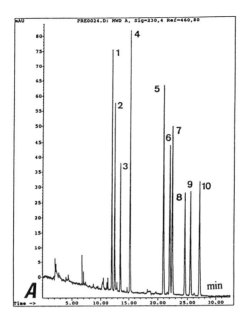

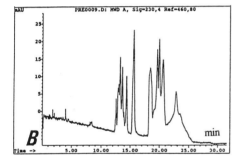
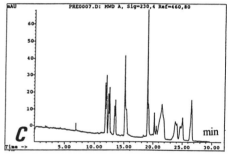

Figure 17 Influence of the drying time using nitrogen gas on the SFC analysis for the phenylurea. (A) SPE-pSFC-DAD of spiked water (10 ppb) with 15 min drying. (B) without drying. (C) 5 min drying. Peaks: 1, monolinuron; 2, metobromuron; 3, linuron; 4, methabenzthiazuron; 5, isoproturon; 6, fenuron; 7, chlorotoluron; 8, diuron; 9, chloroxuron; 10, metoxuron. Experimental conditions: Column Keystone silica 5 μm, 250 mm L × 4.6 mm id; flow rate 2 mL/min; temperature 40°C; pressure programmed from 100 to 250 bar at 5 bar/min; modifier methanol from 1% (2 min) to 21% at 0.5%/min; detection 230 nm [20].

Table 5 Recoveries for Some Pesticides Analyzed by On-line SPE-pSFC-DAD[a]

		RSD (%) for peak area values		Recovery
No.	Pesticide	SPE-pSFC-DAD	pSFC-DAD	(%)
1	Propham	1.5	1.6	94.0
2	Chlorpropham	2.5	1.3	93.7
3	Metolachlor	3.9	2.4	91.6
4	Propazine	0.5	2.3	100.8
5	Sebutylazine	1.3	1.4	90.6
6	Terbutylazine	2.2	0.6	103.2
7	Prometryn	0.9	2.7	103.8
8	Methazachlor	2.0	3.2	95.2
9	Atrazine	0.8	0.7	112.2
10	Metobromuron	2.2	1.5	88.5
11	Monolinuron	0.7	0.9	105.9
12	Terbutryn	2.3	1.4	98.1
13	Metribuzin	1.1	0.6	102.1
14	Simazine	0.5	0.5	106.1
15	Linuron	1.6	0.4	102.6
16	Cyanazine	1.1	0.4	105.4
17	Methabenzthiazuron	0.6	0.7	102.3
18	Desethylatrazine	0.3	0.6	107.5
19	Bromacil	1.4	3.1	104.1
20	Crimidine	0.4	0.8	108.5
21	Desisopropylatrazine	0.4	1.0	105.7
22	Isoproturon	0.2	0.9	106.0
23	Fenuron	0.4	1.3	99.3
24	Chlorotoluron	0.8	0.8	104.3
25	Diuron	0.9	0.6	101.6
26	Metamitron	1.2	0.8	90.6
27	Hexazinone	1.0	1.0	90.9
28	Chloroxuron	4.3	1.9	45.3
29	Methoxuron	0.9	0.9	98.8
30	Chlorizadone	4.2	0.9	51.2

[a] SPE material, Lichrolut EN, 60 µm.

Lichrolut EN with 60 µm particle size (Merck) is a highly hydrophobic material with excellent adsorption characteristics for polar herbicides and pesticides.

Besides the nature of the adsorbent, parameters controlling the adsorption are the sample loading flow rate and the sample volume. This was studied in depth and is described in Ref. 38. The most critical step in the SPE-pSFC procedure is, however, the drying step with nitrogen gas. In fact, without drying the adsorbent between adsorption and desorption, the SPE-pSFC hyphenation is not working at all because of excessive peak distortion. This is illustrated for the enrichment of phenylurea (Fig. 17). Residual water traces result in a desorption process with

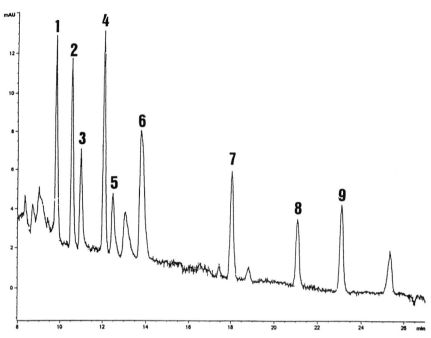

Figure 18 SPE-pSFC-DAD of some pesticides at the 0.24 ppb level in water. Peaks: 1, terbutylazine; 2, atrazine; 3, monolinuron; 4, simazine; 5, linuron; 6, cyanazine; 7, desisopropylatrazine; 8, chlorotoluron; 9, diuron. Experimental conditions as in Fig. 2 [37].

two nonmiscible fluids (carbon dioxide and water), which gives peak broadening and splitting. For 60-μm particles a drying period of 15 min is optimal.

In Table 5, recoveries at the 20 ppb level for 30 pesticides are given together with the % RSDs for a series of 8 experiments. For comparison, the % RSDs are also presented for 5 direct injections via a 5-μL sample loop of the same absolute amounts. With the exception of chloroxuron and chlorizadone, the recoveries are very good, whereas for all pesticides the % RSDs fall in the range of the chromatographic % RSDs, illustrating the excellent repeatability of SPE-pSFC-DAD.

The detectability in SPE-pSFC-DAD is illustrated in Fig. 18, showing the SPE-pSFC-DAD analysis of 240 ppt spiked in tap water of metolachlor (% RSD 2.0), terbutylazine (% RSD 9.1), atrazine (% RSD 8.9), monolinuron (% RSD 1.2), simazine (% RSD 4.1), linuron (% RSD 11.5), metabenzthiazuron (% RSD 2.6), desisopropylatrazine (% RSD 3.3), chlorotoluron (% RSD 9.4), and diuron (% RSD 3.5). The relative standard deviations were calculated for five experiments.

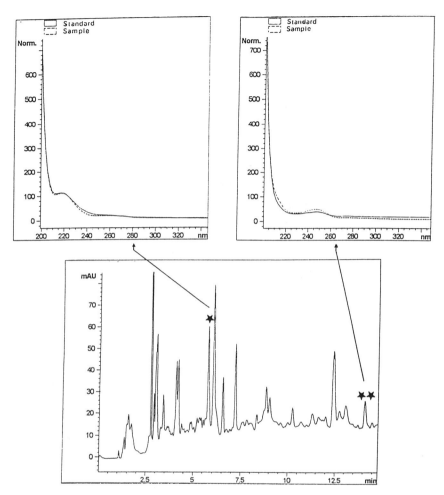

Figure 19 Chromatogram showing the analysis of a river water sample. Chromatographic conditions as in Fig. 2, except column which was Hypersil HP 200 × 4.6 mm id, 5 μm SPE material—ODS 10 μm. Sampled volume 200 mL. Peaks: (★) atrazine; (★★) diuron. Insert: recorded DAD spectra with standard spectra [38].

An interesting feature of carbon dioxide desorption is its selectivity. Only the solutes that are soluble in the fluid will be desorbed and enter the column. This means that polar and ionic solutes that often occur in real-world samples, e.g., humic acids, fulvic acids, lignine, detergents, etc., in surface, ground, and river waters, will not be introduced into the column. Compared to SPE-HPLC-DAD, where the desorption fluid is highly hydrophilic and desorbs some of these solutes as well, clean chromatograms with flat baselines are obtained. This

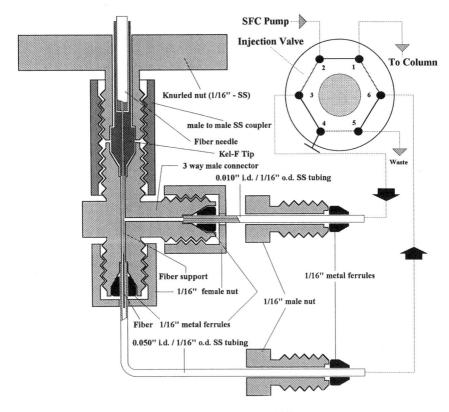

Figure 20 SPME-pSFC injection port modification [44].

renders qualitative and quantitative analysis easier. This is illustrated in Fig. 19 showing the analysis of water collected from the river Lys in July 1995. In this particular case, and this in order to emphasize the selectivity of carbon dioxide desorption and to allow comparison with published SPE-HPLC-DAD chromatograms, we used a 10-μm ODS cartridge. Atrazine and diuron could easily be elucidated through their retention times and DAD spectra (inserts of Fig. 19) and quantified. The concentrations were 1.38 ppb for atrazine and 0.95 ppb for diuron with for triplicate analysis RSD% of 4.8 and 6.1, respectively. Desethylatrazine was present at 0.28 ppb.

Solid phase microextraction (SPME) can also be applied in pSFC with supercritical fluid desorption. SPME is a solventless sample preparation method, integrating sampling, extraction, and injection in a single step. After sorption on a polymer-coated fiber of the analytes of interest from an aqueous (liquid) or head space (gaseous) medium, the analytes are desorbed commonly thermally

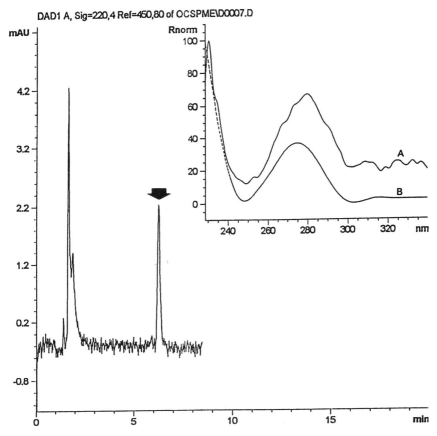

Figure 21 SPME-pSFC-DAD analysis of 50 ppb of 4-octylphenol in water. Insert: recorded UV spectrum (A) compared to reference spectrum (B). Experimental conditions: column HP APSG (aminopropyl silica) 5 μm, 20 cm × 4.6 mm id; flow rate 2 mL/min; temperature 40°C; pressure 200 bar; modifier methanol 5% isocratic; detection 220 nm; initial sample volume 10 mL; SPME: 100 μm polydimethylsiloxane-coated fiber Supelco [44].

by placing the fiber in a hot GC inlet [39–41]. Liquid [42] and supercritical fluid desorption [43,44] have also recently been described.

A simple modification of an external loop injection valve of a pSFC instrument allowing insertion of the fiber and desorption of the sorbed compounds by the supercritical fluid is presented in Fig. 20. The applicability is illustrated with the analysis of 50 ppb 4-octylphenol in an aqueous sample (Fig. 21). 4-Octylphenol originates in water samples from the biodegradation of nonionic detergents, emulsifiers, dispersing agents, etc.

V. CONCLUSIONS

Supercritical fluids possess excellent characteristics for application in environmental analysis. On the one hand, pSFC has proved to be a reliable separation technique and on the other hand solutes of interest can be selectively extracted from solid materials by SFE. Supercritical fluids also have a bright future in environmental analysis in terms of the combination of SFE with SFC, which results in fully automated environmental monitoring systems.

REFERENCES

1. T. A. Berger, *Packed Column SFC* (R. M. Smith, ed.), Publ. The Royal Society of Chemistry, Thomas Graham House, The Science Park, Cambridge, U.K., p. 205 (1995).
2. T. A. Berger, and W. H. Wilson, *Anal. Chem.*, 65:1451 (1993).
3. T. A. Berger, W. H. Wilson, and J. F. Daye, *J. Chromatogr. Sci.*, 32:179 (1994).
4. S. Shah, and L. T. Taylor, *J. Chromatogr.*, 505:293 (1990).
5. T. A. Berger, *J. Chromatogr. Sci.*, 32:25 (1994).
6. M. E. Mc Nally, and J. R. Wheeler, *J. Chromatogr.*, 453:63 (1988).
7. J. R. Wheeler, and M. E. McNally, *J. Chromatogr.*, 410:343 (1987).
8. A. Medvedovici, F. David, G. Desmet, and P. Sandra, *J. Microcolumn Sep.*, (1997).
9. A. Kot, F. David, and P. Sandra, *J. High Resolut. Chromatogr.*, 17:272 (1994).
10. B. A. Charpentier and M. R. Sevenants, *Supercritical Fluid Extraction and Chromatography*, ACS Symposium Series 366, American Chemical Society, Washington, DC (1988).
11. M. Perrut (ed.), *Int. Symposium on Supercritical Fluids, Proceedings*, Société Française de Chimie, Institut National Polytèchnique de Lorraine (1988).
12. K. P. Johnston and J. M. L. Penninger, *Supercritical Fluid Science and Technology*, ACS Symposium Series 406, American Chemical Society, Washington, DC (1989).
13. S. B. Hawthorne, *Anal. Chem.*, 62:633A (1990).
14. Special Issues: Supercritical Fluid Chromatography, *J. Chromatogr. Sci.*, 6–7: (1989).
15. M. Verschuere, *Supercritical Fluid Extraction as Sample Preparation Technique for Chromatography* (in Dutch), Ph. D. dissertation, University of Ghent, Belgium (1994).
16. S. B. Hawthorne, *Anal. Chem.*, 66:369A (1994).
17. V. Janda and P. Sandra, *Hydrochemia '90*, 1:295 (1990).
18. S. B. Hawthorne, D. J. Miller, D. E. Nivens, and D. C. White, *Anal. Chem.*, 64:405 (1992).
19. M. Verschuere, P. Sandra, and F. David, *J. Chromatogr. Sci.*, 30:338 (1992).
20. P. Sandra, A. Kot, A. Medvedovici, and F. David, *J. Chromatogr. A*, 703:467 (1995).
21. V. Lopez-Avila, J. Benedicto, N. S. Dodhiwala, R. Young, and W. Beckere, *J. Chromatogr. Sci.*, 30:335 (1992).
22. S. B. Hawthorne, and D. J. Miller, *J. Chromatogr.*, 403:63 (1987).

23. F. I. Onuska, and K. A. Terry, *J. High Resolut. Chromatogr.*, *12*:527 (1989).
24. S. B. Hawthorne, and D. J. Miller, *J. Chromatogr.*, *403*:63 (1987).
25. F. David, M. Verschuere, and P. Sandra, *Fres. J. Anal. Chem.*, *344*:479 (1992).
26. F. David, R. Soniassy, M. Verschuere, and P. Sandra, *Int. Labmate*, *18*:6 (1993).
27. J. J. Langenfield, S. B. Hawthorne, D. J. Miller, and J. Pawliszyn, *Anal. Chem.*, *65*:338 (1993).
28. F. David, A. Kot, E. Vanluchene, E. Sippola, and P. Sandra, "Optimising Selectivity in SFE and Chromatography for the Analysis of Pollutants in Different Matrices," Proceedings of the Second European Symposium on Analytical Supercritical Fluid Chromatography and Extraction, Riva del Garda, Italy (P. Sandra and K. Markides, eds.), Huethig Verlag, Heidelberg, pp. 10–15 (1993).
29. S. Bøwadt, B. Johansson, F. Pelusio, B. Larsen, and C. Rovida, *J. Chromatogr.*, *662*:428 (1994).
30. S. Bøwadt and B. Johansson, *Anal. Chem.*, *66*:667 (1994).
31. S. Bøwadt, B. Johansson, P. Fruekilde, M. Hansen, D. Zilli, B. Larsen, and J. de Boer, *J. Chromatogr.*, *645*:189 (1994).
32. P. Sandra, A. Kot, E. Sippola, and F. David, *Int. Environm. Techol.*, *4(2)*:4 (1994).
33. P. Sandra, A. Kot, J. Castanho, and F. David, *Int. Environ. Techol.*, *4(5)*:10 (1994).
34. K.-H. Ballschmiter and M. Zell, *Fresenius Z. Anal. Chem.*, *302*:20 (1980).
35. J. J. Vreuls, G. J. de Jong, R. T. Ghijsen, and U. A. Th. Brinkman, *J. Assoc. Anal. Chem.*, *77*:306 (1994).
36. T. H. M. Noij and M. M. E van der Kooi, *J. High Resolut. Chromatogr.*, *18*:535 (1995).
37. P. Sandra, A. Medvedovici, A. Kot, and F. David, *Int. Environm. Techol.*, *5(3)*:10 (1995).
38. P. Sandra, A. Medvedovici, A. Kot, L. Vilas Boas, and F. David, *LC-GC International*, *9*:540 (1996).
39. C. Arthur and J. Pawliszyn, *Anal. Chem.*, *62*:2145 (1990).
40. D. Louch, S. Motlagh, and J. Pawliszyn, *Anal. Chem.*, *64*:1187 (1992).
41. D. Potter and J. Pawliszyn, *Environ. Sci. and Tech.*, *28*:298 (1994).
42. J. Chen and J. Pawliszyn, *Anal. Chem.*, *67*:2530 (1995).
43. Y. Hirata and J. Pawliszyn, *J. Microcol. Sep.*, *6*:443 (1994).
44. A. Medvedovici, F. David, and P. Sandra, *J. High Resolut. Chromatogr.* (in press).

14

Preparative Supercritical Fluid Chromatography: Grams, Kilograms, and Tons!

Pascal Jusforgues

Prochrom R&D, Champigneulles, France

Mohamed Shaimi and Danielle Barth

Ecole Nationale Supérieure des Industries Chimiques, Nancy, France

I. INTRODUCTION

Since the pioneer work of Klesper et al. [1] in 1962, packed column supercritical fluid chromatography (pSFC) has been envisaged not only as an analytical tool but as a production tool within preparative SFC (Prep-SFC). Indeed, Prep-SFC has a high theoretical potential to separate components of a mixture, to separate with ease the purified components from the chromatographic eluent, and, eventually, to recycle the eluent on-line. For example, when carbon dioxide is the chosen eluent, a simple isothermal depressurization will drastically reduce its solvating power and the purified sample will precipitate. In this chapter, we will try to give an overview of the development of Prep-SFC (still quite limited) and to show that its theoretical potential remains intact now that the technological difficulties have been addressed. We will focus our attention not on the chromatographic behavior of supercritical fluids (SF), described in other chapters of this book, but on the surrounding technical key points that will allow the use of Prep-SFC as a reliable and efficient tool.

II. PREPARATIVE FROM LABORATORY TO FACTORY

First, we must establish that the name "Prep-SFC" covers several technologies and addresses different goals depending on the scale of production envisaged. Prep-SFC can be used to purify micrograms or metric tons of sample. In this

403

chapter, we will distinguish four different scales of production, as discussed in the following sections.

A. From Micrograms up to Milligrams

Samples of this size are mostly used for molecular structure elucidation. Their purification requires analytical chromatographs that have been slightly modified to collect (quantitatively or not) the pure products at column outlet. This technique is more relevant for analytical chromatography and is not described here. Chapters 3 and 12 give examples of such separations.

B. From Tens of Milligrams up to Tens of Grams

Samples of this size can be used, for example, for toxicological trials or as synthesis intermediates. The injected crude mixture is often difficult to obtain and the quantitative collect of the purified samples is a key point of the purification process. A purification study at this scale is also a necessary preliminary step to bigger scale purification. The technology used in such equipment is a blend of analytical and industrial technology. For example, one can decide to recycle the eluent or not. This technique is described hereunder as Lab-Prep-SFC.

C. From Hundreds of Grams up to Hundreds of Kilograms

Samples of this size are already an industrial production in some cases. But the equipment to produce such quantities can also be used as pilot to make studies prior to increased production. The technology used is necessarily industrial. This technique is described hereunder as Pilot-Prep-SFC, for which a further example is given in Chapter 15.

D. From Hundreds of Kilograms up to Metric Tons

Productions of that scale are industrial. Often, the equipment used will be dedicated to only one production and, thanks to that specialization, big savings will be made on investment costs. This technique is described herein as Production-Prep-SFC, and a further example is given in Chapter 15.

III. BIBLIOGRAPHY

From 1962 to the mid 1980s, the field of Prep-SFC matured slowly as an extension of analytical pSFC. In 1982, Pilot-Prep-SFC and Production-Prep-SFC started after the patent of Perrut [2]. A pilot equipment was built and the process feasibility was demonstrated [3–6]. A review of Prep-SFC, from 1962 to 1989,

Table 1 Recent Applications Of Lab-Prep-SFC

Sample purified	Sample size	Column dim. id × L (mm)	Trapping yield
Milbemycin α_2 from microbial cells extract [34]	100 and 500 mg	20 × 250	Quantitative
Racemics resolution [20,21]	1–10 mg	10 × 250	70–90%
Steroids purification [22,23]	8.5 mg/injection	8 × 250	94%
Fullerenes fractionation by hyphenated SF extraction/pSFC [24]	2.4 mg	4.6 × 250	Not given
Methylmethacrylate oligomers fractionation [25]	50 mg	10 × 250	Not given
Lemon oil fractionation by adsorption-desorption pSFC [26,27]	0.5 g	7.2 × 50	Not given
Tocopherols fractionation [28,29]	100 mg	2 columns 10 × 250	30–50%
Caffeine extract [28,30]	200 mg	10 × 250	Not given
Fatty acids esters (EPA, DHA) separation [12]	350 mg	10 × 125	50%

is given by Berger et al. [7]. Since 1990, very little has been published on the subject, i.e., about 29 publications on Lab-Prep-SFC [8–36], 5 publications on Pilot-Prep-SFC [5,37–40], and several posters appearing at specialized symposia. The most significant ones are summarized in Tables 1 and 2. From these tables, it seems that Prep-SFC has raised little interest. This is the case in public research laboratories, (the main source of publications) because Prep-

Table 2 Recent Applications of Pilot-Prep-SFC

Sample purified	Sample size	Column dim. id × L (mm)	Purity trapping yield
Synthesis intermediate purification [4]	2.5 g/h	60 × 1000	98.9% Not given
Fatty acids esters (EPA 56%, DHA 31%) separation [5]	40 g/h	60 × 1000	EPA 95%; DHA 85% 90%
Racemics resolution [34]	30 g/day	60 × 1000	94% and 96% enantiomeric excess Not given
Fatty acids esters (EPA) purification [35]	Not given	130 id, length not given	Purity: 90%+ Not given

SFC requires heavy investment in equipment that was not very reliable until recently. However, in private industry, several companies have continued their work on the subject and technology is now available for small Lab-Prep-SFC [41,46,], for large Lab-Prep-SFC [42], and for all sizes from Lab-Prep-SFC to Prod-Prep-SFC [43]. A company [44] proposes an optimized process for Prod-Prep-SFC production of EPA (eicosapentaenoic acid) fatty acid ester from fish oil, which is described in Chapter 15. Several pharmaceutical companies use routinely Lab-Prep-SFC for sample preparation at the gram scale (unpublished).

IV. TECHNOLOGY

Prep-SFC, as described here, is one of the different processes gathered under the generic name of preparative elution chromatography (which does not include such processes as simulated moving bed or frontal chromatography) whose principle is represented in Fig. 1. An eluent (gas, liquid, or supercritical) flows through the chromatographic column. Pulse injections of crude mixture are periodically introduced in the eluent. After nondestructive detection at the column outlet, the purified fractions of sample are periodically directed to the different traps. The sample is separated from the eluent that is recycled. Depending on the process, recycling of the eluent can be done on-line or off-line. The cycle time is the time elapsed between two successive injections. The success of Prep-SFC depends on the control of some key points as illustrated through examples below.

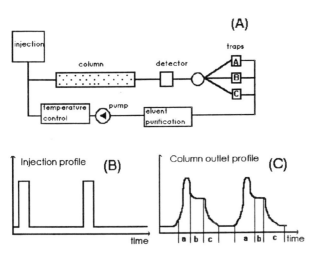

Figure 1 Principle of preparative elution chromatography: (A) scheme, (B) injection profile, (C) column outlet profile.

A. Operating Parameters

The collect of the different fractions cannot be based on the peak shape because high productivities require injection of large quantities and the resulting peaks at the column outlet are often not resolved. Consequently, the collection of fractions is made essentially on a time basis (which can be adapted by some threshold detection techniques). Thus, it is important that the chromatogram shape be perfectly reproducible. This can be obtained only through a precise control of the operating parameters: temperature, pressure, composition, and flow rate. As we will see in Secs. V and VI, this objective has been attained and the reproducibility of peak parameters is better than 1% over hundreds of successive injections.

B. Injection

In analytical pSFC, injection of the sample is often made through a usual injection loop. The sample is dissolved in a liquid phase that is introduced in the flowing supercritical eluent. In Prep-SFC, this technique can be used as long as the injected quantity is not too large. When the sample size is large, a peak broadening is observed that is due to the limited kinetic of dissolution of the liquid phase in the supercritical eluent. A diphasic mixture then reaches the head of the column and results in a broadening of the peaks and, consequently, in a reduction of the purity and/or the productivity of the collected fractions.

The solution we propose to overcome this problem is a proprietary device that predissolves the injected sample in the supercritical fluid before its introduction in the flowing eluent. The dissolution procedure is easily automated. The effect of this procedure can be seen in Fig. 2.

C. Fractions Collect

This is one of the major difficulties of the process. As can be seen in Tables 1 and 2, most authors have difficulties to trap quantitatively the purified sample after it has been purified. The main difficulty is in separating the sample from the eluent. Theoretically, a simple depressurization reduces the solvating power of the eluent and allows a simple phase separation between a liquid sample (or solid) and a gaseous eluent. But during the rapid depressurization, the sample is micronized and tends to be carried away with the large volume of high-speed gas phase. Several solutions have been reported in the literature.

High-pressure collection treats the problem of depressurization off-line [14,18, 19]. The sample is stored together with the liquid or solid eluent in cooled traps. Then the trap pressure is slowly reduced to atmospheric pressure. The collect in solid eluent is given to be quantitative while the collect in liquid high-pressure eluent gives a collect yield of 85%. Neither solution is adapted

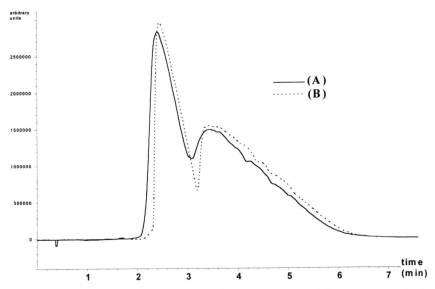

Figure 2 Comparison of peak shapes obtained with two injection modes. Eight hundred microliters of crude solution was injected on a 10-mm-id column. Injection in liquid phase (A) gives a poor resolution compared with supercritical fluid injection (B).

to production because of the necessity of storing big volumes of eluent under high pressure.

Low-pressure collection with adsorption of samples on a solid sorbent requires desorbtion of purified fraction off-line and redilution of the sample [10,15]. This solution is not adapted to the use of a supercritical eluent mixed with a modifier. The collect yields obtained are still not quantitative (90%).

Low-pressure collection with "bubbling" of the gaseous eluent in a liquid solvent is often not quantitative with 85% or less collection yields [14,16, 17]. This technique can be used only with small flow rates of eluent. In Chapter 3, this technique is described with high concentrations of modifier (30%) and recovery levels higher than 85%.

The most effective and extrapolable way to trap the samples on-line has been patented by Perrut [44]. After depressurization, the micronized droplets of sample are separated from the gaseous eluent by centrifugation in a cyclone. It is adapted to any level of flow rates, the collected samples can be collected continuously, and the gaseous eluent can be instantly recycled. The collect yield of a cyclone depends very much on its design. We have optimized cyclones to obtain collect yields always greater than 95%, often greater than 97% and up to 100%. The only limit to the efficiency of the cyclones is the vapor pressure of the sample: the vapor phase is flushed away with the gaseous eluent. This phenomenon is critical only with volatile modifiers that are quantitatively trapped only when the cyclone temperature is kept low enough (under 0°C for methanol).

D. Depressurization

Problems are often encountered in the depressurization device at column outlet. Valves or restrictors are easily plugged because of the very low temperatures created by the adiabatic depressurization. Moreover, the depressurization device must have a design that avoids the remixing of fractions previously separated on the column. Various solutions have been proposed that include heating of a needle valve [42] or use of a vibrating valve [46,47]. These solutions have been satisfactorily implemented by several users. Our solution is a simple needle valve with an internal proprietary design to avoid plugging. This solution has the advantage of being extrapolable to the high flow rates of Prod-Prep-SFC.

V. PERFORMANCE AT LABORATORY SCALE

A. Equipment

The equipment used here is a Super C12 (Fig. 3) of Prochrom [43]. The specifications are as follows: Carbon dioxide flow rate: 5–35 g/min; modifier flow rate: 0–10 mL/min; operating pressure: 100–350 bars; operating temperature: 0–150°C; column diameter: 4.6–20 mm; injection capacity: 1–200 mg/injection;

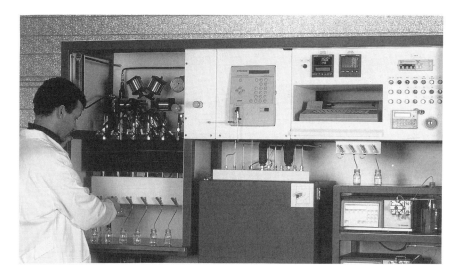

Figure 3 Lab-Prep-SFC equipment. Super C12.

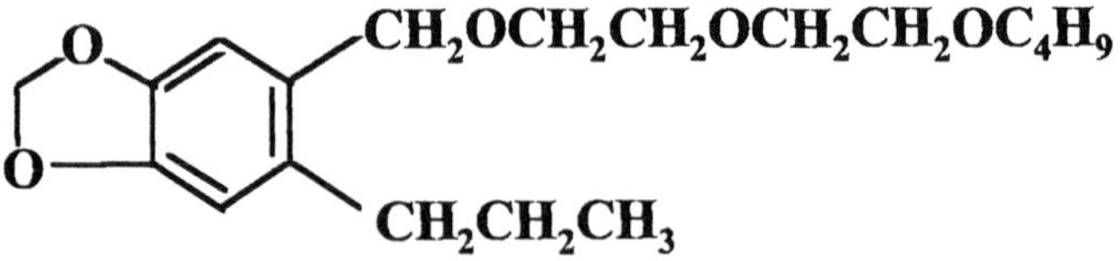

Figure 4 Piperonyl butoxide (PB) formula.

fraction collection: 6 cyclonic traps; automation allowing periodic injections for a nonstop production.

B. Application Without Modifier

Purification of the piperonyl butoxide (PB) insecticide (Fig. 4) has been carried out with Lab-Prep-SFC without modifier. Although the chosen application has only an academic significance, it demonstrates the potential of Prep-SFC without modifier. The crude contains eight main impurities analyzed by capillary gas chromatography (Table 3) and other unidentified compounds. The goal is to produce 99% + pure PB. Under optimized operating conditions, two purification steps are necessary to achieve the 99% + requirement.

1. First Production Step

Each injection has been cut into four fractions—F2, F3, F4, and F5—as described in Fig. 5. Part of the preparative chromatogram is presented in Fig. 6. The perfect regularity of peak shapes and retention times necessary for an efficient production can be seen from this chromatogram. This regularity can also be seen in Figs. 7 and 8. Quantitative results of this production step are given in Table 4. In fraction F3, 80% of the injected PB is collected at a 97.8% purity. The production rate of this fraction is 1.2 g/h. The efficiency of the

Table 3 Composition of the Piperonyl Butoxide Crude

Impurity	Mass percentage
1	0.30
2	0.64
3	0.34
4	0.36
5	3.16
6	0.70
7	0.66
8	0.78
PB	91.90
Unknowns	1.16

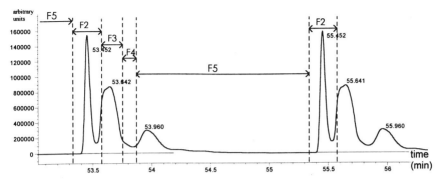

Figure 5 Preparative chromatogram. First purification step of PB. Representation of the fractions cut points. Operating conditions. Column: Amicon-Matrex C18—15 µm—100 Å—250 × 10 mm; Mobile phase: carbon dioxide; temperature: 40°C; pressure: 240 bar; flow rate; 15 g/min; injection: 50 µL/injection of undiluted crude = 52.5 mg; total number of injections: 430; cycle time: 120 s; detection: UV absorption at 312 nm.

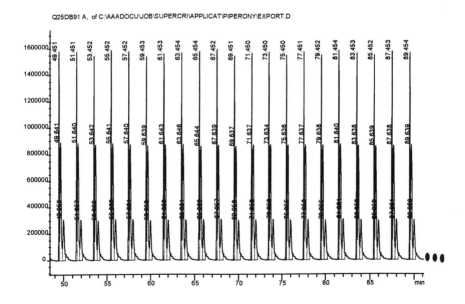

Figure 6 Preparative chromatogram of the first purification step of PB (21 successive injections among the 519 injections of the total run). Operating conditions as in Fig. 5.

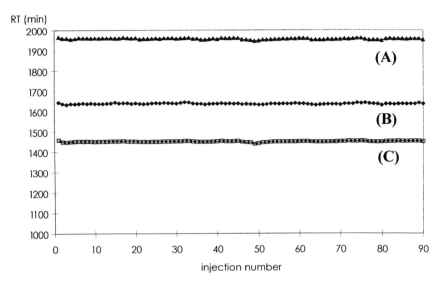

Figure 7 Retention time stability during first production step of PB. (A) Peak 3 mean retention time = 1.957 min; RSD% = 0.19%; standard deviation = 0.22 s. (B) Peak 2 mean retention time = 1.638 min; RSD% = 0.18%; standard deviation = 0.17 s. (C) Peak 1 mean retention time = 1.451 min; RSD% = 0.17%; standard deviation = 0.15 s. Operating conditions as in Fig. 5.

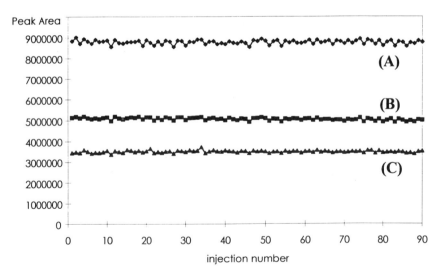

Figure 8 Peaks area stability during first production step of PB. (A) Peak 2 area RSD% = 1.17%. (B) Peak 1 area RSD% = 1.39%. (C) Peak 3 area RSD% = 1.54%. Operating conditions as in Fig. 5.

Table 4 Production Performances of the First Step of Purification of PB

	F2	F3	F4	F5	Total	Crude	Collect yield (%)
Mass (g)	0.42	16.83	1.83	3.14	22.22	22.58	**98.41**[b]
Purity PB (%)	84.71	**97.82**	93.83	53.50	—	91.90	—
Mass PB (g)	0.36	16.46	1.72	1.68	20.22	20.77	97.35[c]
$M_{PB\ frac}/M_{PB\ inj}$ (%)[a]	1.73	**79.25**	8.28	8.09	—	—	—

Operating conditions as in Fig. 5.
[a] PB in fraction/PB injected.
[b] Mass collected/mass injected = 22.22/22.58.
[c] Mass of PB collected/mass of PB injected = 20.22/20.77.

cyclones can be seen here because more than 98% of the injected material is collected in the four fractions. The F3+F4 fractions have been used as injected crude in a second purification step with different operating conditions, allowing an optimized selectivity for the impurities that were not well separated with the first set of conditions.

2. Second Production Step

Operating conditions are identical to the first step except for temperature, which has been raised to 70°C. Each injection has been cut into five fractions—F2, F3, F4, F5, and F6—as described in Fig. 9. Quantitative results of this production step are given in Table 5. In fractions F3+F4+F5, 92.4% of the injected PB is collected at a 99.0% purity. The production rate of this fraction is 1.4 g/h. Again the efficiency of the cyclones can be seen because more than 99.3% of the injected material is collected in the five fractions.

3. Total Production Performance

A crude containing 92% of PB has been purified in two steps. Of the injected PB, 81% has been collected with a purity greater than 99%. A high productivity

Table 5 Production Performances of the Second Step of Purification of PB

	F2	F3	F4	F5	F6	Total	Crude	Collect yield (%)
Mass (g)	0.030	0.05	0.356	0.403	0.042	0.884	0.89	**99.32**[b]
Purity PB (%)	69.20	**95.44**	**99.40**	**99.18**	92.72	—	97.42	—
Mass PB (g)	0.020	0.050	0.354	0.400	0.039	0.863	0.867	99.54[c]
$M_{PB\ frac}/M_{PB\ inj}$ (%)[a]	2.30	**5.75**	**40.69**	**45.98**	4.48	—	—	—

Operating conditions as in Fig. 9.
[a] PB in fraction/PB injected.
[b] Mass collected/mass injected = 0.884/0.89.
[c] Mass of PB collected/mass of PB injected = 0.863/0.867.

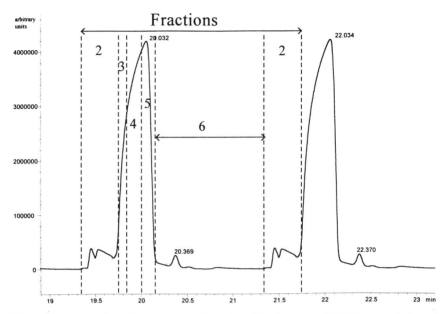

Figure 9 Preparative chromatogram. Second purification step of PB. Representation of the fractions cut points. Operating conditions as in Fig. 5 except temperature 70°C.

of 14 g/day of pure product can be obtain on a 10 mm id column. The purified PB was collected free of solvent because pure carbon dioxide was used as eluent. Unattended operation of the Super C12 with high reproducibility of the chromatograms and very high recovery of the injected crude has proven the reliability of the equipment used. Furthermore experiments with double injection quantities have demonstrated that the productivity can be increased a lot without affecting the quality of the results. These results have been extrapolated to Pilot-Prep-SFC on a 100-mm-id column (100 times larger) and experimentally confirmed as reported in Sec. VI.B.

C. Application with a Small Quantity of Modifier

The separation of the cis and trans isomers of phytol (Fig. 10) has been carried out in Lab-Prep-SFC with a low amount of modifier. The trans isomer of this fatty alcohol is used in the perfume industry. The crude contains 30.4% and 65.7%, respectively, of the cis and trans isomers. Other unidentified impurities are also present (3.9%). The aim of the study is to produce *trans*-phytol with a purity greater than 99% (respective to the phytol isomers only). A preparative chromatogram is not shown because of the detector saturation but Fig. 11 shows a chromatogram obtained under the same conditions with an 11-mg injection.

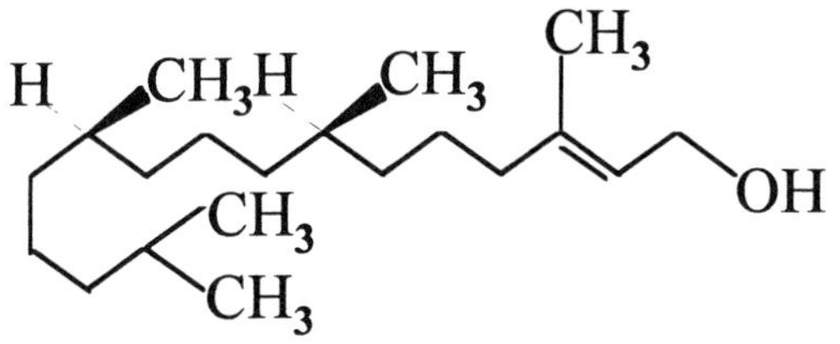

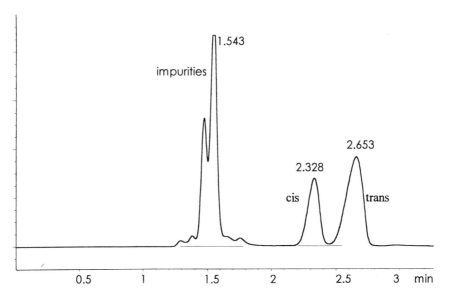

Figure 10 Phytol formula.

This example highlights one of the features of Prep-SFC compared to preparative HPLC: the cycle time is only 60 s, which allows one to make 60 injections per hour. Under these conditions, 3.14 g of the phytol isomers were injected in 1 h and 1.3 g of the trans isomer (60% of the injected trans isomer) were collected with a purity greater than 99%. The remaining 40% were recycled in the Prep-SFC process. The global performances of the process (including recycling) can be summarized as follow: 49 g of the crude are injected every day and 31.2 g of *trans*-phytol are collected with a 99%+ purity (97% of the injected isomer) on a

Figure 11 Phytol chromatogram for a 11-mg injection. Operating conditions. Column: AMICON-MATREX silica—15 μm—100 Å—250 × 10 mm; mobile phase: 13.5 g/min carbon dioxide + 1 mL/min isopropanol; temperature: 50°C; pressure: 250 bar; injection: 65 μL/injection of undiluted crude = 55 mg; cycle time: 60 s; detection: UV absorption at 223 nm.

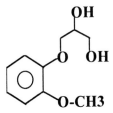

Figure 12 Guaiacol glyceryl ether (GGE) formula.

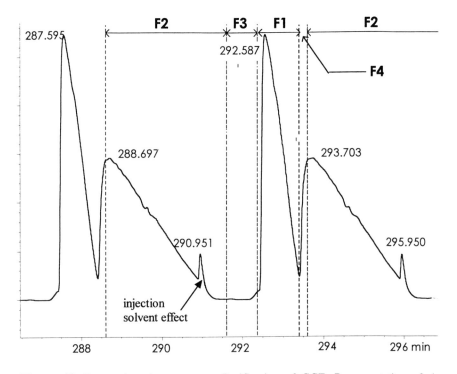

Figure 13 Preparative chromatogram. Purification of GGE. Representation of the fractions cut points. Operating conditions. Column: Cellulose 3,5-dimethylcarbamate on silica—10 µ—250 × 10 mm; mobile phase: carbon dioxide (16 g/min) + ethanol (2.32 mL/min); temperature: 10°C; pressure: 250 bar; injection: 103.9 mg of racemic diluted in 611 µL of methanol; total number of injections: 101; cycle time: 300 s; detection: UV absorption at 288 nm. A small peak due to the solvent pulse coming from the next injection can be seen on the tail of the second enantiomer; methanol should theoretically not be detected by the detector but its presence decreases the capacity factor of the second enantiomer and the methanol pulse creates a "concentration wave" of the enantiomer, which is detected but has no influence on the production process.

Table 6 Production Performances of the Purification of GGE

	F1	F2	F4	Total	Collect yield (%)[b]
Mass (g)	4.84	5.17	0.15	10.17	**97**
Purity (%)	99.9	99.3	37/63		
M_{frac}/M_{inj} (%)[a]	47.6	50.8	1.5	100	

Operating conditions as in Fig. 13.
[a] Mass of fraction/mass collected (10.17 g).
[b] Mass collected/mass injected = 10.17/10.49.

10-mm-id column. The purified isomer is collected diluted in isopropanol (about 100 mg/mL).

D. Application with a Great Quantity of Modifier

The chiral separation of the enantiomers of guaiacol glyceryl ether (GGE) (Fig. 12) has been carried out in Lab-Prep-SFC with a large amount of modifier. The work described here was initiated in the laboratory of Dr. E. Francotte (Ciba-Geigy, Bâle, Switzerland) and optimized in the laboratories of Prochrom [43] with the stationary phase provided by Dr. Francotte. Each injection has been cut into four fractions—F1, F2, F3, and F4—as described on the preparative chromatogram of Fig. 13. First enantiomer is collected in F1, second enantiomer in F2, other impurities in F3, and F4 is a fraction to be recycled. Quantitative results of this production step are given in Table 6. In fraction F1 and F2, the first and second enantiomers are collected at 99.9% and 99.3% purity, respectively. The production rate is 29 g/day. The efficiency of the collect in cyclones is 97% for the sample and 90% only for the modifier (due to the volatility of ethanol). The first and second enantiomers are collected diluted in ethanol (24 and 7 mg/mL), respectively.

VI. PERFORMANCE AT PILOT SCALE

A. Equipment

The equipment used here is a Super C100 (Fig. 14) of Prochrom [43]. The specifications are as follows: Carbon dioxide flow rate: 25–100 kg/h; modifier flow rate: 0–10 L/h; operating pressure: 100–300 bar; operating temperature: 20–90°C; column diameter: 100 mm; injection capacity: 0.1–20 g/injection; fractions collect: 3 cyclonic traps; automation allowing periodic injections for a nonstop production. The column is equipped with a dynamic axial compression system directly adapted from the HPLC technology for large-diameter columns

Figure 14 Pilot-Prep-SFC equipment. Super C100.

[43]. The use of such a technology is necessary to achieve efficiencies as high as that on a laboratory prepacked column at pilot or production scale (Fig. 15).

B. Application

The piperonyl butoxide application described on Lab-Prep-SFC in Sec. V.B has been transposed on Pilot-Prep-SFC.

1. First Production Step

Each injection has been cut into two fractions—F2 and F3—as described in Fig. 16. Quantitative results of this production step are given in Table 7. In fraction F3, 88.8% of the injected PB is collected at an 98.2% purity. The production rate of this fraction is 126 g/h (3 kg/day). The efficiency of the cyclones can be

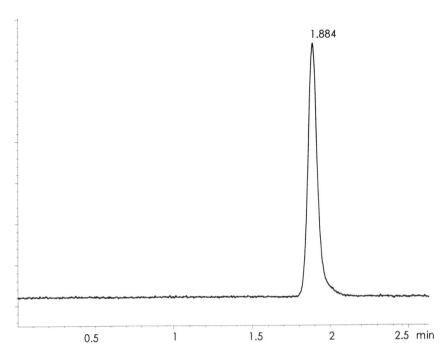

Figure 15 Efficiency test on the Super C100 column; 21,500 plates/m corresponding to a reduced plate height of 3.1. Operating conditions. Column: Amicon-Matrex C18—15 µm—100 Å—250 × 100 mm; mobile phase: carbon dioxide; temperature: 40°C; pressure: 240 bar; flow rate: 90 kg/h; injection: methyl-1 naphthalene; detection: UV absorption at 220 nm.

Table 7 Production Performances of the First Step of Purification of PB on the Super C100

	F2	F3	Total	Crude	Collect yield (%)
Mass (g)	151.2	783.8	935.0	943	**99.1**[b]
Purity PB (%)	56.6	**98.2**	—	91.9	—
Mass PB (g)	85.6	769.9	855.5	866.6	98.7[c]
$M_{PB\ frac}/M_{PB\ inj}$ (%)[a]	9.9	**88.8**	—	—	—

Operating conditions as in Fig. 16.
[a] PB in fraction/PB injected.
[b] Mass collected/mass injected = 935/943.
[c] Mass of PB collected/mass of PB injected = 855.49/866.6

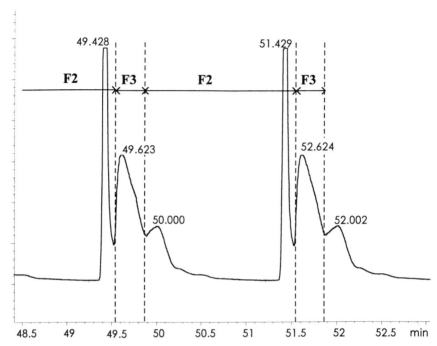

Figure 16 Preparative chromatogram. First purification step of PB on the Super C100. Representation of the fractions cut points. Operating conditions. Column: Amicon-Matrex C18—15 μm—100 Å—250 × 100 mm; mobile phase: carbon dioxide; temperature: 40°C; pressure: 240 bar; flow rate: 90 kg/h; injection: 4.8 mL/injection of undiluted crude = 5.04 g; total number of injections: 187; cycle time: 120 s; detection: UV absorption at 312 nm.

seen here because more than 99% of the injected material is collected in the two fractions. The F3 fraction has been used as injected crude in a second purification step.

2. Second Production Step

The second step of purification has been transposed with the same success as the first one.

3. Total Production Performances

A crude containing 92% of PB has been purified in two steps. Of the injected PB, 83% has been collected with a purity of 99.4%. A high productivity of 1416 g/day of pure product can be obtained on a 100-mm-id column. The purified PB was collected free of solvent since pure carbon dioxide was used as eluent.

Table 8 Comparison of the Results After a 100-fold Scale-up

	Column id (mm)	Purity (%)	Productivity (g/day)	Collect yield[a]
Injected crude	10	91.9	18.9	
	100	91.9	1824	
Collected PB	10	99.0	14	80%
	100	99.4	1416	84%

[a]Collect yield = collected pure PB/injected PB (g/g).

Unattended operation of the Super C100 with high reproducibility of the chromatograms and very high recovery of the injected crude has proven the reliability of the equipment used.

VII. SCALE-UP QUALITY

The comparison of performances obtained on the Super C12 and on the Super C100 on the piperonyl butoxide purification shows that a 100-fold extrapolation can be made directly (Table 8). The regularity of the pilot equipment is similar to the one obtained on the laboratory unit. Moreover, the superimposition of the chromatograms obtained on the laboratory and pilot equipment shows that the results obtained with Lab-Prep-SFC can be directly extrapolated to Pilot-Prep-SFC (Fig. 17).

VIII. ECONOMY OF INDUSTRIAL PRODUCTION

The above results obtained on the Super C12 and Super C100 have been extrapolated to Prod-Prep-SFC on a Super C500 (similar equipment with a 500-mm-id column) to evaluate the cost of an industrial production. Of course, the calculated cost depends very much on the application: an easy separation can cost as much as 50 times less than a difficult one. The examples given here will provide only orders of magnitude.

A. Cost of Prod-Prep-SFC Without Modifier

The Super C500 built for this application would cost about U.S.$5 million. Return on investment is chosen as 20% per year. Thanks to the complete automation of the unit (including the recycling of the carbon dioxide), a reduced number of person-hours is required (one operator). Carbon dioxide is completely recycled, so that the cost of eluent is almost negligible.

Production cost details are given in Table 9. The industrial process cost of producing 10.6 tons/year of pure PB is U.S.$240/kg. From the details of the cost

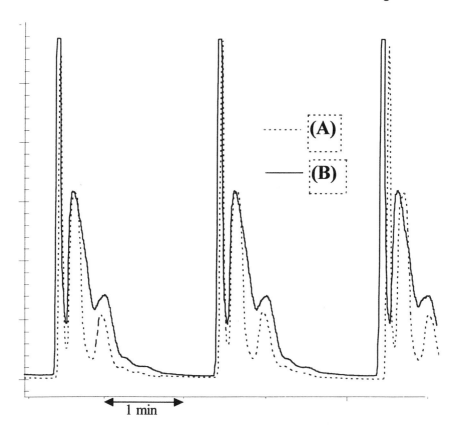

Figure 17 Superimposition of corresponding chromatograms obtained on Super C12 (A) and Super C100 (B) for purification of PB. One can see that the superimposition is not perfect. This is probably due to a default of the Super C100 detection circuit, which is on a derivation of the main circuit. The real peak profile at column outlet is not affected. This explanation is confirmed by the results, which show better performances on the pilot unit than on the laboratory unit (Table 8).

breakdown, one can see that the main share is due to investment. This is characteristic of Prep-SFC compared to HPLC. So the scale factor is very important in Prep-SFC: an increase in the production scale very much decreases the cost per kilogram. The same evaluation on a Super C200 (200-mm-id column) would give a production of 1.7 tons/year at U.S.$324/kg (35% increase). The production cost of the same quantities (10.6 tons/year) and quality by preparative HPLC has been evaluated to be about U.S.$350/kg. An important contribution to the HPLC production cost is the stripping of traces of chlorinated solvent remaining in the purified product after a simple evaporation.

Table 9 Industrial Production of PB on a Super C100

	Cost (k U.S.$/year)	Cost (U.S.$/kg of pure PB)	Cost (% of total)
Investment	1000	94.34	39.3
Operator	125	11.79	4.9
Technician (part time)	10	0.94	0.4
Engineer (part time)	10	0.94	0.4
Power	325	30.66	12.8
Stationary phase (changed twice a year)	290	27.36	11.4
Carbon dioxide losses	35	3.30	1.4
Maintenance (5% of investment)	250	23.58	9.8
Miscellaneous (taxes, insurance, administration, etc.)	500	47.17	19.6
Total	2545	**240.08**	100

Operating conditions: *Column:* Amicon-Matrex C18—15 μm—100 Å—250 × 500 mm; *mobile phase*: carbon dioxide; *temperature*: 40°C (first step) and 70°C (second step); *pressure*: 240 bar; *flow rate*: 2250 kg/h; *collect*: 2 collect lines (pure PB and waste); *automation*: completely automated industrial system; *injection*: 120 mL/injection of undiluted crude = 126 g; *cycle time*: 120 s for each purification step; *detection*: UV absorption at 312 nm; *Production period*: 300 days/year; 24 h/day; *crude injected*: 13.6 metric tons/year; 91.9% purity. Pure PB collected: 10.6 metric tons/year; 99.4% purity. Note: Production cost does not include the cost of the crude.

B. Cost of Prod-Prep-SFC with Modifier

Here we present an extrapolation of the *trans*-phytol production. For comparison purposes, the extrapolation has been done to the same scale of production as the previous example (10 tons/year). This is obtained with a 330-mm-id column. The Super C330 built for this application would cost about U.S.$5 millions; the difference of size with the Super C500 is compensated by the extra investment necessary to deal with the adjunction of the modifier.

Production cost details are given in Table 10. The industrial production cost of 10.2 tons/year of pure *trans*-phytol is U.S.$229/kg. If compared to the PB application without modifier, the total purification cost is similar. Indeed, the extra costs due to the modifier and the smallest scale factor (smaller column) are compensated by the greater productivity of this application (only one step with a cycle time of 60 s vs. two steps with cycle times of 120 s each).

IX. CONCLUSION

It has now been demonstrated that Prep-SFC on a laboratory and industrial scale is not only a technical possibility but a reliable tool that can be easily extrapo-

Table 10 Industrial Production of *trans*-Phytol on a Super C330

	Cost (k U.S.\$/year)	Cost (U.S.\$/kg of pure *trans*-phytol)	Cost (% of total)
Investment	1000	98.04	43.2
Operator	125	12.25	5.4
Technician (part time)	10	0.98	0.4
Engineer (part time)	10	0.98	0.4
Power	212	20.78	9.1
Stationary phase (changed twice a year)	78	7.65	3.4
Carbon dioxide and isopropanol losses	90	8.82	3.9
Isopropanol evaporation and recycling	42	4.12	1.8
Maintenance (5% of investment)	250	24.51	10.8
Miscellaneous (taxes, insurance, administration, etc.)	500	49.02	21.6
Total	2317	**229.13**	100

Operating conditions: *Column:* Amicon-Matrex silica—15 µm—100 Å—250 × 330 mm; *mobile phase*: 882 kg/h carbon dioxide + 65 L/h isopropanol; *temperature*: 50°C; *pressure*: 250 bar; *collect*: 3 collect lines (pure *trans*-phytol, fraction to recycle, and *cis*-phytol); *Automation*: completely automated industrial system; *injection*: 70.8 mL/injection of undiluted crude = 60 g; *cycle time*: 60 s; *detection*: UV absorption at 223 nm; *production period*: 300 days/year; 24 h/day; *Crude injected*: 16 metric tons/year; 65.7% purity. Pure *trans*-phytol collected: 10.2 metric tons/year; 99% + purity. Note: Production cost does not include the cost of the crude.

lated. In the laboratory, Prep-SFC offers the possibility of producing gram quantities of highly pure material without manipulating hundreds of liters of solvent. The high flexibility of HPLC related to the great number of solvent combinations that can be used has an equivalent in Prep-SFC, namely, the possibility of modulating the eluent properties by changing its pressure, temperature, and, to a lesser extent, composition. Prep-SFC has already been chosen by several pharmaceutical laboratories that have to routinely purify gram quantities of a large variety of enantiomers.

In the factory, Prep-SFC has the same advantages as in the laboratory plus the economical advantages demonstrated in the above examples. Of course, Prep-SFC is not a universal tool for high-efficiency separations. Its main drawbacks are related to the limited solvating power of supercritical fluids for highly polar and heavy molecules (typically proteins), and many separations that can be made by preparative HPLC are not possible by Prep-SFC. Moreover, when a separation is technically possible by both Prep-SFC and preparative HPLC, economical considerations will not always favor Prep-SFC.

The future of Prep-SFC can be evaluated by comparison with the place that preparative HPLC now holds on the market. In our laboratories, we compared the use of both techniques for the purification of tens of samples provided by customers. The figures that can be derived from our experience are the following:

More than one-third of the applications can be technically treated by Prep-SFC.

Among these, half of the applications would be treated more economically by Prep-SFC than by preparative HPLC. The purification cost by Prep-SFC varies between one-third and and three times the preparative HPLC cost.

About 80% of the applications that can be treated by Prep-SFC require the use of a modifier.

When purification costs are comparable, HPLC will be chosen because as an older technique it looks less exotic than Prep-SFC which will be retained only when it presents a substantial economical advantage.

From these considerations, we can say that the potential of Prep-SFC represents 10–20% of the preparative HPLC market. This potential has not yet been realized.

REFERENCES

1. E. Klesper, A. H. Corwin, and D. A. Turner, *J. Org. Chem.,* 27:700–701 (1962).
2. M. Perrut, Fr. Pat., 8 209 649 (1982); Eur. Pat., 0 099 765 (1984); U.S. Pat., 4 478 720 (1983).
3. P. Jusforgues, Ph. D. thesis, Institut National Polytechnique de Lorraine, Nancy, France (1988).
4. C. Berger, Ph. D. thesis, Institut National Polytechnique de Lorraine, Nancy, France (1989).
5. L. Doguet, Ph. D. thesis, Institut National Polytechnique de Lorraine, Nancy, France (1992).
6. M. SHAIMI, Ph. D. thesis, Institut National Polytechnique de Lorraine, Nancy, France (1996).
7. C. Berger and M. Perrut, *J. Chromatogr., 505(1)*:37–43 (1990).
8. G. Cretier, R. Majadalani, J. Neffati, and J. L. Rocca, *Chromatographia, 38(5–6)*:330–336 (1994).
9. J. W. Oudsema and C. F. Poole, *J. High Resolut. Chromatogr., 15(2)*:65–70 (1992).
10. Y. Hirata, Y. Kawaguchi, and K. Kitano, *Chromatographia, 40(1/2)*:42–46 (1995).
11. G. Cretier, R, Majadalani, and J. L. Rocca, *Chromatographia, 30(11–12)*:645–650 (1990).
12. S. Hogashidate, Y. Yamauchi, and M. Saito, *J. Chromatogr., 515(1)*:295–303 (1990).

13. Y. Yamauchi, M. Kuwajima, and M. Saito, *J. Chromatogr.,* *515(1)*:285–293 (1990).
14. L. Wuensche, U. Keller, and L. Flament, *J. Chromatogr.,* *552(1-2)*:539–549 (1991).
15. D. C. Messer, L. T. Taylor, W. E. Weiser, J. W. McRae, and C. C. Cook, *Pharm. Res. 11(11)*:1545–1548 (1994).
16. J. Vejrosta, J. Planeta, and M. Mikesova, *J. Chromatogr A, 685*:113–119 (1994).
17. J. Vejrosta, A. Ansorgova, and I. Planeta, *J. Chromatogr A, 683*:407–410 (1994).
18. K. D. Bartle, C. D. Bevan, A. A. Clifford, S. A. Jafar, N. Malak, and M. S. Verrall, *J. Chromatogr. 697(12)*:579–585 (1995).
19. C. D. Bevan and C. J. Lellish, *Process Scale Liquid Chromatogr* (G. Subramanian, ed.), VCH, Weinheim, pp. 163–191 (1995).
20. A. Kot, P. Sandra, and A. Venema, *J. Chromatogr., 32(10)*:439–448 (1994).
21. J. Whatley, *J. Chromatogr. A, 697, (12)*:251–255 (1995).
22. M. Hanson, *LC-GC, 14(2)*:152–158 (1996).
23. M. Hanson, *Chromatographia, 40(3/4)*:139–149 (1995).
24. K. Jinno, H. Nagashima, K. Itoh, M. Saito, and M. Buonoshita, *Fresenius J. Anal. Chem., 344(1011)*:435–441 (1992).
25. K. Ute, N. Miyatake, T. Asada, and K. Hatada, *Polym. Bull. (Berlin), 28(5)*:561–568 (1992).
26. Y. Yamauchi and M. Saito, *Fractionation Packed-Column SFC and SFE: Principles and Applications* (M. Saito, Y. Yamauchi, and T. Okuyama, ed.), VCH, New York, pp. 169–178 (1994).
27. Y. Yamauchi, M. Kawajima, and M. Saito, *J. Chromatogr., 505(1)*:237–273 (1990).
28. K. Jinno and M. Saito, *Anal. Sci., 7(3)*:361–369 (1991).
29. M. Saito and Y. Yamauchi, *J. Chromatogr., 505(1)*:257–271 (1990).
30. P. Elisabeth, M. Yoshioka, Y. Yamauchi, and M. Saito, *Anal. Sci., 7(3)*:427–431 (1991).
31. J. Neffati, Ph. D. thesis, Université Claude Bernard–Lyon I, Lyon, France (1996).
32. J. Y. Clavier, R. M. Nicoud, and M. Perrut, 3rd Italian Conference on "I fluidi Supercritici e loro Applicazioni," Grignano, Italy, Sept. 3–6 (1995)
33. K. Coleman and F. Vérillon, "Laboratory-Scale Preparative Chromatography Enhanced by Fluids Containing Carbon Dioxide under Automated Pressure Control," Proceedings of the 3rd international Symposium on Supercritical Fluids, Strasbourg, France, 1994, pp. 415–420.
34. K. D. Bartle, C. D. Bevan, A. A. Clifford, S. A. Jafar, N. Malak, and M. S. Verrall, *J. Chromatogr. A, 697*:579–585 (1995).
35. G. Cretier, J. Neffati, and J. L. Rocca, *J. Chromatogr. Sci., 32(10)*:449–454 (1994).
36. G. H. Brunner and D. Upnmoor, *NATO ASI Ser., 273*:653–668 (1994).
37. M. Perrut, *J. Chromatogr., 658(2)*:293–313 (1994).
38. P. Lembke, 7st International Symposium on Supercritical Fluid Chromatography and Extraction, 31 March–4 April 1996, Indianapolis, USA.
39. M. Perrut, *J. Chromatogr. A, 658*:293–313 (1994).
40. P. Jusforgues, *Process Scale Liquid Chromatography* (G. Subramanian, ed.), VCH, Weinheim, pp. 153–162 (1995).
41. Gilson, B. P. 45, 95400 Villiers-le-Bel, France.
42. Thar Designs, Inc., 730 William Pitt Way, Pittsburgh PA 15238, USA.

43. Prochrom S.A., 5 rue Jacques Monod, BP9, 54250 Champigneulles, France.
44. K. D-Pharma, D-66450 Bexbach, Germany.
45. M. Perrut, Fr. Pat. 8,510,468, (1985); Eur. Pat. 8,640,139, (1986).
46. Jasco Europe s.r.l., 22060 Cremella, Italy.
47. N. Periclès, Eur Pat. 0427671A1, 30 October 1990; U.S. Pat. 5,224,510, 6 July 1993.

15

Production of High Purity *n*-3 Fatty Acid– Ethyl Esters by Process Scale Supercritical Fluid Chromatography

Peter Lembke

K.D.-Pharma GmbH, Bexbach, Germany

I. INTRODUCTION

The beneficial effect of regular intake of omega-3 (*n*-3) fatty acids or their ethyl esters, which are mainly found in cold-water fish oils, has long been a matter of controversy [1–4]. In particular, eicosapentaenoic acid (EPA) and docosahexaenoic acid (DHA) are considered to be important. The physiological effects of a regular daily EPA intake have so far been proven to include reduction of the following factors: arteriosclerosis, hyperlipidemia, blood viscosity and coagulation, blood pressure (systolic and diastolic), and cholesterol content. Together these positive effects result in a significant reduction of the risk of a fatal heart disease, which is the most common disease in Western Society. DHA, on the other hand, plays an important role in biological membranes (i.e., membrane phospholipids of brain and retina). Many studies have shown that a sufficient DHA intake is essential for the proper functioning of developing brain and retina [2,4]. Thus, highly enriched DHA concentrates have an enormous potential in the infant formula, baby food, and health food markets.

To understand the above mentioned controversial discussions on the beneficial effects of the *n*-3 fatty acids EPA and DHA, one has to know that most other fatty acids found in fish oil belong to the omega-6 group. These *n*-6 fatty acids are counter active toward the positive action of the *n*-3 group. According to the ratio *n*-3 to *n*-6 in a particular fish oil (concentrate), the physiological effects of EPA and DHA are more or less strongly pronounced. Accordingly, clinical trials in this field show more or less significant results. This antagonism between the *n*-3 and *n*-6 fatty acids explains the large demand for high-purity EPA and DHA.

To isolate these interesting fatty acids in high purity (>95%) from a complex matrix such as fish oil, a chromatographic step will always be necessary to reach the required selectivity. The classical approach to solve this separation problem would be normal phase high-performance liquid chromatography (HPLC) using a polar stationary phase and a lipophilic organic solvent as a mobile phase. Unfortunately, these solvents are environmentally incompatible and are hazardous for human health if the exposure limits at production site are not respected or if residuals amounts are contaminating the final product. An alternative is reversed phase HPLC, which often uses aqueous alcoholic solvents as a mobile phase. However, every HPLC process is a *dilution* process and considerable effort has to be invested in reconcentrating the obtained EPA/DHA fraction (typically dilution factor 500–1000). This, in combination with the large amounts of organic solvents that have to be stored, reconcentrated, and eventually disposed of, pushes up the production cost price.

A further alternative for the production of concentrated EPA and DHA could be countercurrent supercritical fluid extraction (cc-SFE). This technology has the big advantages that it is a continuous process and that corresponding large-scale equipment is commercially available. However, cc-SFE on its own cannot reach the required high purities because separation only takes place according to the chain length and not according to the degree of unsaturation of the fatty acids or their esters. With cc-SFE EPA and DHA purities of up to approximately 85% are possible. Zhu et al. [5], for example, reported in 1994 the possibility of purifying EPA and DHA up to approximately 76% with this technology. However, to achieve purities above 95% the additional selectivity originating from a suitable chromatographic stationary phase is essential.

According to the unique potential economical and ecological advantages offered by supercritical fluid chromatography (SFC), it was decided to produce high-purity EPA and DHA with process scale SFC.

In this chapter we will show on behalf of our gathered experience with our SFC plants for EPA purification that process scale pSFC is worth considering as an alternative to process scale HPLC.

II. POSSIBILITIES AND CHANCES OF PROCESS SCALE SFC

When working with neat carbon dioxide as mobile phase, the collected fraction is not diluted in the mobile phase as in HPLC because the product/mobile phase separation consists only of a careful evaporation of the mobile phase. Unlike HPLC, this separation does not require a separated product recovery unit and does not generate extra costs for solvent removal, cleaning, and disposal. This is a very important fact for any process scale chromatographic separation, as it brings down the production cost significantly.

Additional factors influencing the production cost include that carbon dioxide is nontoxic, environmentally compatible, easy to recycle, and has low acquisition and disposal costs. It is also nonexplosive and nonflammable, and its use is not submitted to ex-proof conditions, which again signifies important savings in investment costs.

Due to the low viscosity of the mobile phase and the high diffusion coefficient in supercritical carbon dioxide, higher linear velocities can be used without any significant loss in separation efficiency. Hence, a higher sample throughput generates shorter production time cycles and reduces costs compared to HPLC. For example, a 1150-mm-long column (id 100 mm) packed with 30 μm stationary phase and operated at a carbon dioxide flow rate of 200 kg/h (170 bar, 60°C) generates a pressure drop of only 3–5 bar.

The low viscosity of the mobile phase, and thus the small pressure drop in the column, additionally has the advantage that much longer columns can be used than in HPLC, generating very high separation efficiencies which may be necessary for particular separations. For example, Berger and Wilson [6] showed that it is possible to generate up to 250,000 plates or 298 plates/s by combining several packed columns in series with commercial available equipment. Among others [7], Lembke and Engelhardt [8] demonstrated the possibility of creating unique new selectivities by coupling several packed columns in series differing in their length and packing material (see Chapter 6). Even with relative high pressure drops in the column, which are often said to be responsible for a severe efficiency loss [9], we had the experience that the separation still can be optimized by adjusting the temperature and pressure accordingly.

In our opinion, the main limitation of process scale pSFC with neat carbon dioxide is its limited application range to nonpolar and hydrophobic solutes. Nevertheless, fats and fat-soluble products are ideal for this technology. Once a modifier becomes necessary to elute the solutes from the column, many advantages of this technology lose in significance. An example would be the arguments that there is no need of organic solvents, no product dilution, no ex-proof conditions, and so forth.

However, the SFC technology with modifier should still be thought of as an alternative to some preparative LC separations, as shown in Chapter 14. In many cases production costs will be significantly reduced by the fact that organic solvent consumption is cut down to 90% and more. Even for the pharmaceutical industry, which traditionally uses process scale HPLC to clean up their mostly polar and water-soluble active agents, process scale pSFC with alcoholic modifier contents of up to 50% may be an alternative. Even if the mobile phase may be at a subcritical state, the advantages of pSFC are maintained, i.e., lower organic solvent consumption, lower viscosity, higher diffusion coefficient, shorter cycle times, etc., compared to the traditional HPLC separation [10].

III. SCALE-UP

First experiments to find out the best separation conditions of the fish oil fatty acid ethyl esters (FAEEs) on pSFC were carried out at analytical scale on home-made equipment using commercial 200×4 mm id columns. These experiments showed that according to the polarity of the stationary phase, either a separation according to the fatty acid carbon number or according to its double bond number took place. Only very few phases were capable of separating the FAEEs according to both parameters in such a way that a sufficient selectivity for the preparative isolation of EPA/DHA was present [11] (Fig. 1).

After optimizing the selectivity of the separation system on an analytical column (200×4 mm id), first preparative experiments were carried out on a preparative lab scale with a 300×11 mm id column. Through the very positive results we gathered during these lab scale experiments we were encouraged to build a pSFC pilot plant with increasing column dimensions (Fig. 2). The aim was to construct a representative pilot plant with a minimum of investment while maintaining an up-to-date safety standard. Additionally, all individual parts of the pilot plant had to be off-shelf products found in the high-pressure technology (up to 1000 bar and more), which were also easily available in larger dimensions for the production plant. This minimizes the technological risk for the second scale-up step to the production plant and ensured an up-to-date safety standard. These presuppositions resulted in a pilot plant that is shown in Fig. 3. It has the following features:

Optimal column internal diameter 100–200 mm (10–30 L column)
0–300 kg carbon dioxide/h, at $P_{max} = 320$ bar
Temperature range from 0°C to 90°C
Injection pump: 0–100 L oil/h, $p_{max} = 320$ bar
Material of construction stainless steel (1.4571) and PTFE
Fully automated (computer controlled)

The fact that all individual parts of the pilot plant were off-shelf products (some with slight modifications) helped to keep the investment in a very reasonable dimension. A further feature of our pSFC pilot/production plant that reduced the hardware investment significantly is the fact that we optimized the plant design in such a way that we were able to use only one separator. This enabled a fractionation under atmospheric pressure in a way similar to that long known in process scale SFE, while still recovering the majority of the mobile phase. However, this technology is only possible if a sufficient good separation selectivity is present. If not, then due to the extracolumn band broadening effects of this setup, a partial or total remixing of the previously separated elution bands coming from the column is almost unavoidable. In this case the separation of the elution bands should be done in separate separators as described in Ref. 12 and Chapter 14.

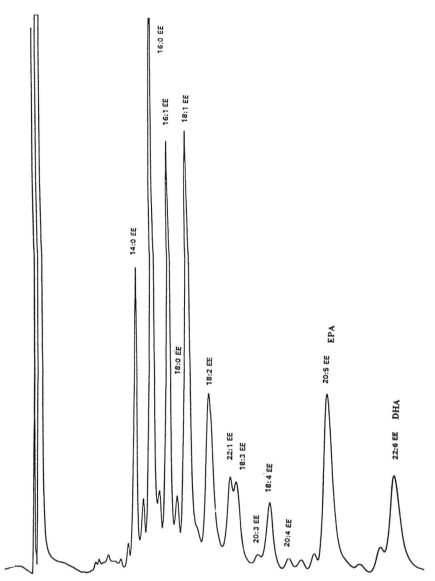

Figure 1 Analytical pSFC chromatogram of fish oil ethyl ester. Column: 200 × 4 mm id, carbon dioxide, 37°C, 170 bar, FID.

COLUMN - DIMENSION

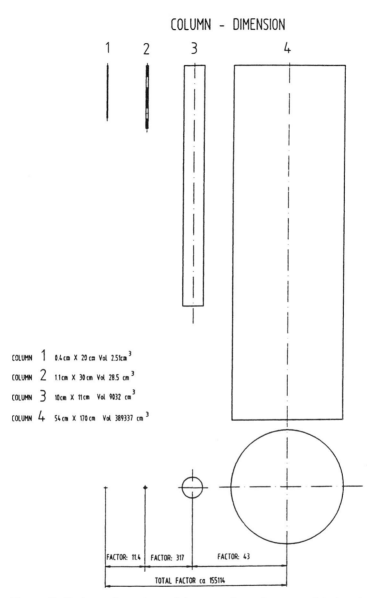

COLUMN 1   0.4 cm X 20 cm   Vol 2.51cm^3

COLUMN 2   1.1 cm X 30 cm   Vol 28.5 cm^3

COLUMN 3   10 cm X 11 cm   Vol 9032 cm^3

COLUMN 4   54 cm X 170 cm   Vol 389337 cm^3

FACTOR: 11.4   FACTOR: 317   FACTOR: 43

TOTAL FACTOR ca 155114

Figure 2 Scale-up dimensions of the separation columns used during the development of this process.

Figure 3 Photo of the pSFC pilot plant for the production of high-purity EPA.

The main impediment to cost-effectiveness was finding or constructing a suitable chromatographic column that had a large internal diameter, a compression system, and that could withstand high operating pressures (~250 bar). As can be seen from Fig. 4, the cost price per kg EPA >95% initially drops dramatically when larger column internal diameters are chosen. Considering our separation problem we would need at least two 280 mm id columns (each 110 L) to be competitive on the market, which would have an annual capacity of approximately 10 tons EPA >95%. The company decided to build a production plant with an annual EPA >95% capacity of 30–35 tons. This plant consists of two pSFC units, each equipped with an approximately 400-L column (id 540 mm). With this large column internal diameter the new plant allows us to produce high-purity EPA at only one-tenth of the cost price of our old plant equipped with a 100-mm-id column. The breakdown of the EPA cost price is shown in Fig. 5. The investment for the production plant (depreciation 10 years) adds up to only 18% of the production costs. The main part are the labor costs with 33%, which include overhead and maintenance costs for the plant. Another large portion is the stationary phase (17%), which has to be replaced every 3–4

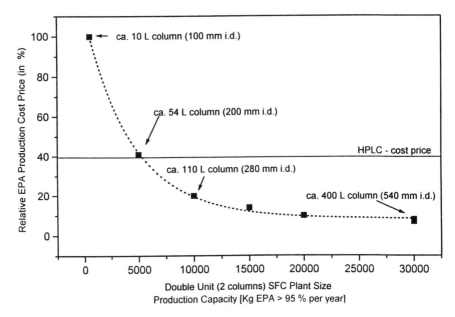

Figure 4 EPA cost price vs. plant production capacity (each plant equipped with two columns). Critical plant size for this particular application: <10 tons EPA/year.

month. Figure 5 also demonstrates that the costs for the mobile phase (carbon dioxide) are negligible and add up to only 1% of the cost price. On the other hand, energy costs (10%) are significant and indicate that this technology requires low energy acquisition costs to work profitably.

Figure 6 shows the schematic flow chart of a pSFC production plant unit.

IV. PRODUCTION PROCESS

Figure 7 shows the EPA production flow chart on the SFC pilot plant located in Bexbach, Germany. In this case the plant was equipped with an approximately 1 column (1150 × 100 mm id) packed with approximately 5 g of 40–63 μm chemically modified silica. The injection volume was 500 mL (~450 g) fish oil ethyl ester. The carbon dioxide flow rate was 200 kg/h. Each run lasted 30 min.

The EPA-flow chart shown in Fig. 7 shows that the pSFC pilot plant can isolate approximately 60 g EPA (95%), and simultaneously around 58 g DHA (60–70%) from a normal fish oil ethyl ester mixture per hour. This corresponds to a maximal annual capacity of approximately 400–500 kg of EPA (95%) and about the same amount of DHA (60–70%).

Figure 8a–h shows the corresponding gas chromatography chromatograms of the individual fractions listed in the EPA flow chart in Fig. 7. After the first run,

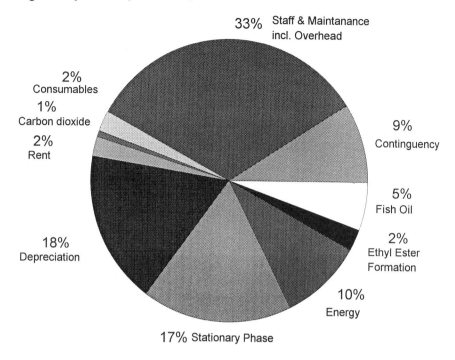

Figure 5 Breakdown of the EPA cost price generated by 30 tons per production plant.

the "saturated" fraction (Fig. 8A) is transfered to the waste and eventually used as fuel for a block power station. The second fraction contains already 63% EPA and no DHA (Fig. 8B). This fraction is used as feed for the second cleaning step. Fraction three contains 80–85% EPA and less than 3% DHA (Fig. 8C). This fraction is used as feed for the third and last clean-up step. The fourth fraction (DHA 60–70%; see Fig. 8H) collected after the first chromatographic clean-up can be sold directly on the food additive market. This is a valuable byproduct of this EPA production process.

After the second chromatographic clean-up we obtain two fractions: one that contains mainly unsaturates (<5 double bonds) and less than 10% EPA, the other with an EPA content between 50% and 60% (Fig. 8D and E). The first of these fractions is used again as fuel for the block power station, whereas the second fraction is mixed with the third fraction obtained in the first clean-up step. This new mixture is the feed for the third chromatographic clean-up and contains approximately 65% EPA and 15% DHA (Fig. 8F). From this third clean-up step we collect three fractions. The first contains a few unsaturates (<5 double bonds) and is burnt. The second is the EPA (>95%) fraction (Fig. 8G).

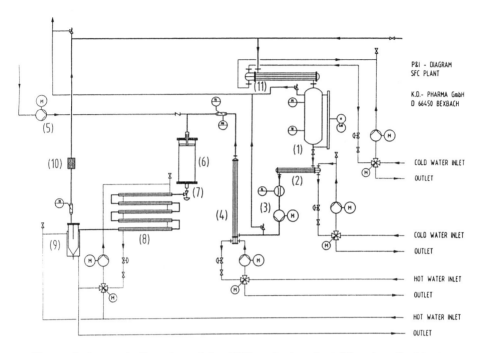

Figure 6 Schematic flow chart of the pSFC production plant. (1) carbon dioxide storage tank, (2) heat exchanger, (3) carbon dioxide high-pressure pump, (4) heat exchanger, (5) fish oil ethyl ester injection pump, (6) separation column, (7) pressure reduction valve, (8) heat exchanger, (9) separator, (10) demister, (11) condenser.

The third fraction of the third clean-up step is a DHA (60–70%) oil (Fig. 8H) which is mixed with the DHA fraction of the first clean-up step.

The absolute recovery of EPA depends on the fatty acid pattern of the original fish oil and lies in the region of 40–50%. The more concentrated in EPA and the less DHA and arachidonic acid (AA) the original feed contains, the better is the yield (>80%). However, all calculations presented in this chapter were carried out on behalf of a normal fish oil, which needs three pSFC clean-up steps to reach the desired EPA concentration of >95%.

The company can guarantee that our EPA and DHA products contains no toxic organic residues as we do not work with such solvents during the entire production process. Furthermore, the high concentrated omega-3 fractions contain no cholesterol. Polar substances like oxidized fatty acid ethyl esters and traces of free fatty acids get irreversible adsorbed onto the stationary phase during the three chromatographic clean-up steps. The same is valid for traces of polymers.

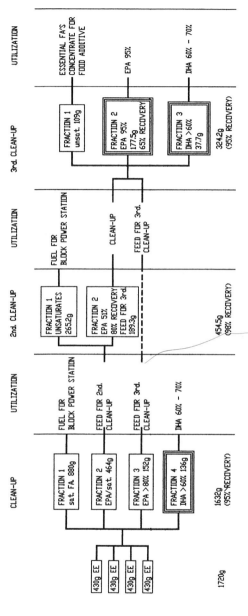

Figure 7 Flow chart of the EPA and DHA production process generated on the pSFC pilot plant equipped with a 10-L separation column (1150 × 100 mm id). Each run requires 30 min (EE: fish oil ethyl ester).

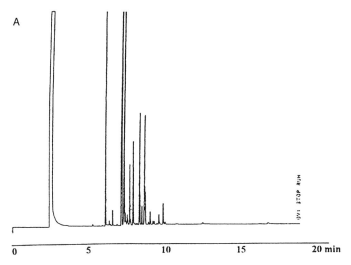

Figure 8 GC chromatograms of the individual fractions shown on the EPA production flow chart in Fig. 7. (A) first clean-up, saturates; (B) first clean-up, 2nd fraction (~63% EPA, 0% DHA); (C) first clean-up, 3rd fraction (80–85% EPA, <3% DHA), (D) 2nd clean-up step, 1st fraction (<10% EPA); (E) 2nd clean-up step, 2nd fraction (50–60% EPA), (F) feed for 3rd clean-up step (G) 3rd clean-up step, 2nd fraction (>95% EPA); (H) mixture of 1st clean-up step (fraction 4) and 3rd clean-up step, 3rd fraction (60–70% DHA).

A further advantage of this pSFC process for our EPA and DHA production is the fact that these fatty acids that are very sensitive towards oxygen are during the entire process in an inert atmosphere (carbon dioxide), which additionally acts as a preservative and prolongs the shelf life of the product.

V. CONCLUSION

Process scale pSFC has proven to be a very interesting economical technology for the production of high-purity EPA and DHA from cold-water marine oils. Compared to the classical preparative HPLC clean-up we were able to cut down the cost price for 1 kg of EPA >95% to approximately one fourth of its original price. In spite of the higher initial investment for a preparative pSFC plant, the relatively low price is explainable by the dramatically reduced acquisition, storage, recycling, and disposal costs for organic combustible solvents, which would be necessary when operating in the HPLC mode. In addition, we were able to cut down the production cost price of our process scale pSFC by in-house design and construction of most parts of the pSFC plant.

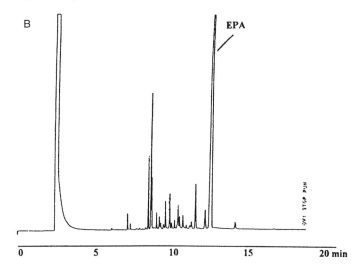

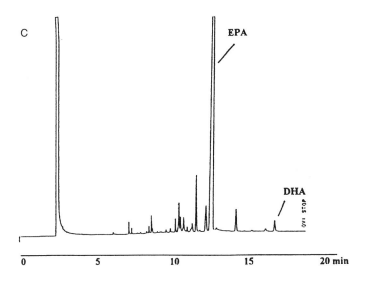

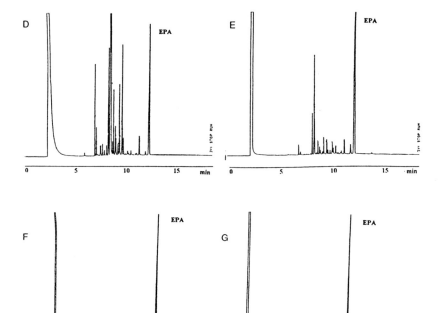

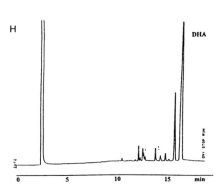

Figure 8 Continued

ACKNOWLEDGMENT

I thank Mr. Alois Hausknecht for the technical drawings. Furthermore, I thank the entire SFC project team in our company; especially Mr. A. Hausknecht and Mr. N. Schirra, for their very important technical know-how, which was the essential link between science and chemical plant engineering.

REFERENCES

1. F. M. Sacks, *J. Am. Coll. Cardiol.*, *25(7)*:1492–1498 (1995).
2. R. Uauy-Dagach and A. Valenzuela, *Prog Food Nitr. Sci.*, *16*:199–243 (1992).
3. *Deutsche Apotheker Zeitung*, *135(27)*:62–64 (1995).
4. "Proceedings of the International Conference on Highly Unsaturated Fatty Acids in Nutrition and Disease Prevention," Barcelona, Spain (1996).
5. H. Zhu, J. Yang, and Z. Shen, "Supercritical Fluid Extraction and Fractionation of Fish Oil Esters Using the Incremental Pressure Programming and a Temperature Gradient Process," Proceedings of the 3rd International Symposium on Supercritical Fluids '94, Strasbourg, France, pp. 95–99 (1994).
6. T. Berger and W. Wilson, *Anal. Chem.*, *65*, 1451–1455 (1993).
7. A. Kot, F. David, and P. Sandra, *J. High. Resolut. Chromatogr.*, *17*:272 (1994)
8. P. Lembke and H. Engelhardt, *Journal High Resolution Chromatography, Vol. 16*, December 1993, 700–702. (1993)
9 H. G. Janssen, H. Snijders, J. Rijks, C. Cramers, and P. Schoenmakers, *J. High Resolut. Chromatogr.*, *14*:438 (1991).
10. P. Lembke, H. Engelhardt, and R. Ecker, *Chromatographia*, *38(7/8)*:491–501 (1994).
11. P. Lembke, R. Krumbholz, and H. Engelhardt, KD-Pharma GmbH EP 0558, 974 B1 (1993).
12. Perrut M., Eur. Patent, 0,099,765 B1 (1984).

Compound Index

Subject Index